Rock Engineering
Applications

Rock Engineering Applications

John A. Franklin

Maurice B. Dusseault

McGraw-Hill, Inc.

New York St. Louis San Francisco Auckland Bogotá
Caracas Hamburg Lisbon London Madrid
Mexico Milan Montreal New Delhi Paris
San Juan São Paulo Singapore
Sydney Tokyo Toronto

Library of Congress Cataloging-in-Publication Data

Franklin, John A.
  Rock engineering applications.
        p.   cm.
  Includes bibliographies and index.
  ISBN 0-07-021889-7
  1. Rock excavation.   2. Rock mechanics.   I. Dusseault, Maurice B.   II. Title
  TA740.F73   1991
  622'.2—dc20   90-47250

Copyright © 1991 by McGraw-Hill, Inc. All rights reserved. Printed in the United States of America. Except as permitted under the United States Copyright Act of 1976, no part of this publication may be reproduced or distributed in any form or by any means, or stored in a data base or retrieval system, without the prior written permission of the publisher.

1 2 3 4 5 6 7 8 9 0   DOC/DOC   9 6 5 4 3 2 1 0

ISBN 0-07-021889-7

The sponsoring editor for this book was Joel Stein, the editing supervisor was Nancy Young, and the production supervisor was Suzanne W. Babeuf. This book was set in Century Schoolbook. It was composed by McGraw-Hill's Professional Publishing composition unit.

Printed and bound by R. R. Donnelley & Sons Company.

> Information contained in this work has been obtained by McGraw-Hill, Inc., from sources believed to be reliable. However, neither McGraw-Hill nor its authors guarantees the accuracy or completeness of any information published herein and neither McGraw-Hill nor its authors shall be responsible for any errors, omissions, or damages arising out of use of this information. This work is published with the understanding that McGraw-Hill and its authors are supplying information but are not attempting to render engineering or other professional services. If such services are required, the assistance of an appropriate professional should be sought.

Contents

Preface ix
Acknowledgements xi

Part 1 Rock Engineering at Surface 1

P1.1	Stone as a Natural Resource	1
P1.2	Stability of Rock Slopes and Excavations	2
P1.3	Foundations on Bedrock	3
P1.4	Water-Retaining Structures	3
P1.5	Features of Near-Surface Construction	4

Chapter 1. Quarrying and Use of Stone 5

1.1	Stone and Ore Materials	5
1.2	Evaluation of Resources	7
1.3	Rockfill	12
1.4	Aggregates for Roadstone, Concrete, and Ballast	21
1.5	Industrial Minerals	28
1.6	Building Stone	33
	References	45

Chapter 2. Landslides and Stability of Rock Excavations 49

2.1	Failures of Natural Slopes	49
2.2	Engineered Slopes	53
2.3	Slide Mechanisms	55
2.4	Site and Route Investigations	74
2.5	Slope Design	77
2.6	Excavation and Stabilization	84
2.7	Monitoring and Maintenance	101
	References	106

Chapter 3. Rock Foundations 111

3.1	Introduction	111
3.2	Types of Foundation	111
3.3	Shallow Footings	113
3.4	Deep Foundations	128
3.5	Foundation Investigations	133

3.6 Foundation Design 134
3.7 Foundation Construction 145
References 151

Chapter 4. Dams and Reservoirs 155

4.1 Overview of Dam Construction 155
4.2 Types of Dam 158
4.3 Benefits and Risks 161
4.4 Case Histories of Foundation Instability 163
4.5 Reservoir Engineering 169
4.6 Site Investigation 173
4.7 Design 177
4.8 Construction 189
4.9 Long-Term Inspection and Maintenance 195
References 199

Part 2 Rock Engineering Underground 203

P2.1 Tunnels and Urbanization 203
P2.2 Underground Space 204
P2.3 Mining and Civil Engineering—Different Objectives 204
P2.4 Reservoir Engineering 205
P2.5 Modes of Rock Behavior 206

Chapter 5. Tunnels 209

5.1 History and Applications of Tunneling 209
5.2 Site Investigation 211
5.3 Planning Considerations 214
5.4 Excavating Methods 221
5.5 Support and Stabilization 227
5.6 Control of Groundwater and Gas 248
5.7 Construction Control 254
5.8 Tunnel Maintenance 255
References 257

Chapter 6. Caverns and Underground Space 263

6.1 Natural and Artificial Caverns 263
6.2 Uses and Benefits of Underground Space 264
6.3 Underground Warehouses, Offices, and Factories 267
6.4 Power Plants and Energy Storage 271
6.5 Storage of Bulk Fluids 277
6.6 Geological Repositories for Radioactive Waste 285
6.7 Disposal of Chemical and Other Wastes 290
6.8 Cavern Design and Construction 292
References 306

Chapter 7. Underground Mining 313

- 7.1 Introduction to Mining — 313
- 7.2 Mining Methods — 317
- 7.3 Mine Planning and Design — 337
- 7.4 Mining Procedures and Equipment — 344
- 7.5 Rock Behavior and Ground Control — 347
- 7.6 Subsidence — 363
- References — 368

Chapter 8. Oil, Gas, and Geothermal Energy 375

- 8.1 Oil and Gas Reservoirs — 375
- 8.2 Geothermal Reservoirs — 382
- 8.3 Exploration, Drilling, and Evaluation — 383
- 8.4 Completion Technology — 397
- 8.5 Production Technology — 404
- 8.6 Development of Geothermal Resources — 415
- 8.7 Environmental Effects — 417
- References — 421

Index 425

Preface

This is a sequel to *Rock Engineering*, by the same publisher and authors. The earlier book described the character of rocks, their environment of water and stress, and how they can be explored, tested, modeled, excavated, and stabilized. This second volume, *Rock Engineering Applications*, demonstrates how these techniques are applied to quarrying and the use of stone; to the design of foundations, slopes, dams, and reservoirs; and to mining and civil engineering works underground.

The chapters follow the rock engineering syllabus taught to both undergraduate and graduate students at the University of Waterloo in Ontario, Canada. The same or similar materials have been presented to engineers attending short courses in Canada and Hong Kong. Both books were written not only for students, but also, and perhaps mainly, for the practicing civil, mining, or petroleum reservoir engineer.

Teachers should prepare students for a career, and in geomechanics this can follow any of a number of diverse branches from contracting and consulting, to highway engineering, mine design, radioactive waste disposal planning, or oil reservoir engineering. Few students know in advance which branch they will follow. Few will find themselves employed in research, yet professors, ourselves included, continue to dwell on research topics at the expense of the more down-to-earth technology in everyday use. A more balanced and practical approach is needed. Hence in both volumes we have sacrificed detail in favor of a broad picture. Particular attention has been given to recommending supplementary reading wherever detail has been omitted for lack of space.

John A. Franklin
Maurice B. Dusseault

Acknowledgments

We gratefully acknowledge the following for sparing the time to review chapters of the manuscript, and for their detailed, constructive, and most helpful comments: Dr. W. Bawden, Queen's University, Canada; Dr. P. Bérest, École Polytechnique, France; Professor Z. T. Bieniawski, Pennsylvania State University; Dr. B. Brady, Dowell Schlumberger, United States; Mr. R. Clarke, CANMET, Canada; Dr. B. Côme, Commission of the European Communities, Belgium; Professor E. J. Cording, University of Illinois; Dr. J. Coulson, Tennessee Valley Authority; Dr. P. Duffaut, Consultant, France; Dr. J. Emery, Consultant, Canada; Professor Charles Fairhurst, University of Minnesota; Dr. D. Fourmaintreaux, Elf Aquitaine, France; Dr. G. Herget, CANMET, Canada; Dr. O. Hungr, Thurber Consultants Ltd., Canada; Prof. K. John, University of Bochum, Germany; Mr. E. Magni, Ontario Ministry of Transportation; Dr. V. Maury, Elf Aquitaine, France; Dr. A. H. Merritt, Consultant, United States; Prof. S. Pelizza, Politechico di Torino, Italy; Dr. J. Pera, CETU, France; Dr. J.-C. Roegiers, University of Oklahoma; Mr. C. A. Rogers, Ontario Ministry of Transportation; Professor S. Sakurai, Kobe University, Japan; Dr. F. J. Santarelli, Elf Aquitaine, France; Mr. R. Schuster, U.S. Geological Survey; Dr. M. Scobel, McGill University, Canada; Dr. O. K. Steffen, Steffen, Robertson & Kirsten, South Africa; Dr. B. Stimpson, University of Manitoba, Canada.

Some of the photographs in *Rock Engineering* and *Rock Engineering Applications* are also available as part of the Educational 35-mm Slide Collection of the International Society of Rock Mechanics. Slide sets can be purchased by ISRM members by writing to the ISRM Secretariat in Lisbon, Portugal.

Rock Engineering Applications

Part 1

Rock Engineering at Surface

P1.1 Stone as a Natural Resource

Rock formations provide a wealth of construction materials including stone for use as aggregate or rock fill, shale for the manufacture of ceramics and bricks, and limestone for the production of cement and agricultural lime. Even more valuable are industrial minerals such as gypsum and sulfur; the fuels including coal, oil, and gas; and the ores from which are extracted iron and aluminum, gold, silver, copper, and many other metals.

Chapter 1 discusses the extraction of these materials by quarrying and open-pit mining, as well as quality and testing requirements. The more valuable ores and architectural stones are also worth extracting from underground, the topic of Chap. 7.

The economics of mining or quarrying depend on the available quantity of ore and thickness of overlying waste deposits, and on the quality of the deposit in terms of physical, mechanical, and mineralogical criteria, the costs of extraction, demand for the product, and proximity of a market.

Requirements vary greatly from one application to the next. Rock fill in embankments, dams, and breakwaters must be either weak enough to allow thorough breakage and compaction, or strong enough so that the blocks survive intact for many years. Riprap that protects an embankment of less durable earth must remain in place in spite of wave action and extremes of climate. Aggregates for asphalt and concrete have to pass physical, mechanical, and chemical tests, and some are intrinsically weak or react with cement, causing cracking and crazing. In cold climates, the salt used to de-ice roads penetrates cracks in the concrete and corrodes the reinforcing steel.

Stone is one of our traditional building materials, and even now there is great demand for marble and granite slabs that can be cut and polished. Building stones must be free from joints and other weaknesses. Softer rocks are easier and less expensive to cut and dress, but harder ones take and retain a high polish. Increasingly important is a stone's resistance to attack by acid rain and airborne pollutants.

Extraction costs include those of engineering, excavating, and processing of the product. Pit design must provide working faces, stable rock slopes, and haul roads that give an orderly circulation of traffic. Groundwater must be controlled to prevent flooding and slope instabilities and to maintain pit bottom access. Blast design must be optimized, in the case of aggregates, to give well-fragmented rock at minimum cost. The opposite is true in stone quarries, where smooth blasting, line drilling, or sawing with diamond studded wire rope is used to produce large undamaged blocks.

P1.2 Stability of Rock Slopes and Excavations

Chapter 2 examines the stability of slopes created naturally by geological erosion, and the design and stabilization of those excavated for open-pit mines or for civil engineering highways or canals.

Natural slopes are created by the interaction of geological uplift and erosion. As a result, vast tracts of hillside have only marginal stability and will fail when exposed to extreme conditions of pore pressure, toe erosion, or earthquake loads. Excavation of rock cuts for roads and railways through hilly or mountainous terrain often causes instability, simply as the result of gravitational forces. These cuts trigger natural slides, often by reactivation along a historic surface of sliding.

Open-pit mine excavations are higher and more extensive than those in civil works, and are mined as steeply as possible. Pit slopes can tolerate much greater movements than civil structures, and mining may even continue while slides are actively moving, sometimes by 1 or 2 m/day.

To analyze or stabilize an ongoing slide one needs information on the mechanisms and rates of movement. Slides can be classified according to geometry as wedge, block, or slab types or according to velocity, for example, creep or avalanche slides. To design and construct a stable slope requires data on the joints and on the water pressures acting in the jointing system.

The engineer is fortunate in having available a wide range of methods of stabilization, including cut and fill to reduce the slope angle, drainage of the slope, erosion protection, and reinforcement and retaining systems. Other systems that catch, deflect, or bypass falling rock can be employed to avoid the consequences of instability when prevention of rock falls and slides is impractical or uneconomic.

P1.3 Foundations on Bedrock

Rock is usually considered a reliable foundation material, but this is true only if the rock is in sound condition and not undermined by caverns, decomposed by weathering, broken by jointing, or shattered by blasting. This topic is elaborated on in Chap. 3. For example, shallow foundations can fail by shearing and punching, sliding, differential settlement, collapse of hidden sinkholes or mine stopes, fault displacement, frost heave, or swelling of the foundation rocks. Deep basements can, in addition, suffer damage from squeeze of walls, heave and buckling of excavation floors, and buoyancy uplift of the building from its foundation. Piled foundations can suffer from pile defects or failure to reach a satisfactory bearing stratum.

If they can be found, cavities and weak zones can be plugged or bridged by concrete. Structures can be founded on deep caissons or piles reaching sound rock below the level of the suspected problem. Otherwise, some accommodation must be made in the structure or foundation. Flexible structures are designed for areas where there is a high risk of settlement. Jacking facilities may be provided for releveling.

P1.4 Water-Retaining Structures

Dams offer the substantial benefits of irrigation for crops, water, and hydroelectric power for developing urban areas, and regulation of river flows to prevent flooding. However, in spite of greatly improved technology, they remain the most accident-prone of all engineered structures, and when they fail, the consequences are severe.

Most dam failures have been caused by inadequate spillway capacity or foundation seepage. Challenges of a different sort center on the reservoir. They include flooding of valley lands, excessive leakage or siltation, landslides around the perimeter slopes, earthquakes induced by the weight of water, and modifications to the groundwater regime that can have a profound and not always favorable effect on agriculture.

Most of the problems are by now well understood and easily avoided, although at a price. In recognition of the importance of a thorough knowledge of ground conditions, the budget for investigating a dam and reservoir project can amount to 10 percent of the cost of construction. Nearly all medium and large dams include extensive instrumentation to monitor their behavior.

Investigations are phased, starting with regional feasibility studies and becoming more detailed later. The complexities of the site progressively unfold, and only during construction can the full extent of unusual features be appreciated.

P1.5 Features of Near-Surface Construction

There are substantial differences between conditions at shallow depth and those encountered in deep underground excavations. Near-surface works have to contend with soil and with hidden features of the soil-rock interface. Costs of construction depend on whether rock is present and how deep. Although shallow rock often provides a sound foundation, costs escalate because of the need for blasting.

The near-surface rock is often weathered, weakened, and loosened by stress relief and by chemical, physical, and biological agencies. The rock may be anisotropic with open bedding or sheeting joints or heterogeneous with soft and hard zones. Rock quality greatly improves with depth.

Groundwater plays an even more important role at shallower than at greater depths in determining stability and the cost of construction. Near the surface, the rock is more permeable and the inflows more severe. Near-surface works run the risk of releasing large volumes of water, such as when a dam fails or a tunnel breaks through to a zone of high groundwater pressure or a body of open water.

On the positive side, shallow site investigations are easier and cheaper because the strata of interest can be reached by shorter drillholes. Conditions tend to be more thoroughly explored and better understood. Construction costs are nearly always lower because of easier access.

Chapter

1

Quarrying and Use of Stone

1.1 Stone and Ore Materials

On many projects, rock is just an inconvenience to be removed by blasting, but on other occasions it is a valuable resource to be recovered from the ground by quarrying or mining. It is an inexpensive and durable material for construction, and depending on the composition of the rock, it can sometimes yield chemicals or minerals. It may be excavated as a primary resource, or recovered as a byproduct from tunneling or mining.

Rock resources can be classified into the following seven categories in a sequence that reflects increasing market value:

- Rockfill
- Roadstone, railway ballast, and aggregates for concrete
- Riprap and armor stone
- Industrial minerals such as talc, potash, salt, gypsum, shale, limestone and clay for production of cement, brick, etc.
- Architectural stone including dimension stone, paving blocks, and stone cladding
- Fossil fuels including coal and oil
- Ores that yield aluminum, iron, copper, and gold

At the inexpensive, high-volume end of the market spectrum are rockfill materials used for embankment construction and mine backfilling. They are the least demanding in terms of required quality, and many rock types are suitable if available in sufficient quantity with appropriate strengths, sizes, and shapes. Low market value places a limit on the cost of production. These materials are quarried at locations of little or no overburden, and mined underground only as a byproduct of other mining activity or cavern development (Fig. 1.1). To limit haulage costs, a "wayside" pit or quarry is often opened near the construction site, then abandoned after construction is complete.

Central in the market value spectrum are the rock resources used for making brick or cement, and the aggregates for riprap, railway ballast, road bases, pavements, and concrete. These must conform to stringent requirements of strength, durability, and chemical and mineral content. They have intermediate value and may be extracted economically from beneath thick overburden, or even by underground mining. They can be hauled large distances while still remaining profitable, if materials of appropriate quality cannot be found locally. For example, crushed granite from a quarry in Scotland finds markets in the Caribbean, North America, and Europe. Efficient quarrying and water transportation permit use of Finnish rock products for reinforcement of dikes in the Netherlands. Several centuries ago, a 40-km canal from quarries in the Vosges Mountains to Strasbourg was built expressly to transport building stone to the Rhine, from where it was barged, to building sites hundreds of kilometers away.

At the low volume–high-cost end of the spectrum are metal ores and industrial minerals. These include metallic ores such as pure metals or metal silicates and sulphides; evaporites such as rock salt, potash, nahcolite, and dawsonite; industrial minerals such as talc, vermicu-

Figure 1.1 Nineteenth-century shallow underground limestone mine north of Paris, France. (*Photo courtesy V. Maury*)

Quarrying and Use of Stone 7

Figure 1.2 Open-pit mining of 110-m-thick brown coal seam at the Morwell pit near Melbourne, Australia. The coal is extracted by bucket-wheel excavator and burned in the nearby power station. (See also Franklin & Dussealt, 1989, Sec. 14.2.1)

lite, quartz crystals, fluorite, and gypsum; energy sources such as coal (Fig. 1.2), lignite, and peat; and decorative stone for architectural use. Materials with the right combinations of properties are in short supply and therefore valuable, and can be exploited even though development costs are high. They can be transported over great distances: Australian and Canadian coal are shipped to Japan; Italian marble and Swedish granite are used for construction work in North America. Marble from Boulonnais, France, was used in the railway station in Tokyo, Japan. Phosphates from Florida strip mines are shipped worldwide.

The rock engineer often is the person charged with locating and evaluating materials and planning the quarrying operation. This includes providing for access roads, assessing stability and drainage of the pit, and selecting the mechanical plant such as drilling, loading, crushing, and screening equipment.

1.2 Evaluation of Resources

Four major factors influence the economics of quarrying:

- Available quantities of salable materials
- Quality of the deposit in terms of physical, mechanical, and mineralogical criteria

- Costs of establishing and running the quarry
- Demand for the product, and proximity of a market (transportation economics)

1.2.1 Quantity and quality of resources

Estimation of available resources in a deposit of uniform quality requires only mapping of its extent by outcrop observations, drilling, and sampling. The costs of mining increase as the pit becomes deeper. This increase is most pronounced for a quarry developed below the water table—the greater the water head, the greater the inflow and pumping requirements. Even in a dry pit where the rock joints are tight and widely spaced or the water table deep, costs increase as the pit deepens because of increased in-pit haulage distances, reduced stability of the higher pit walls, and more constricted and difficult conditions in the pit bottom.

Lateral and in-depth uniformity of the deposit is important. Local variations in quality can make quarrying uneconomical even though much of the deposit may be valuable product. For example, uniform and attractive colors are valuable in building stone or fired bricks, and inconsistencies in the color of stone or fired clay products greatly reduce their value. Interbeds and lenses of deleterious material in an otherwise excellent source of concrete aggregate call for expensive selective mining, further screening and washing, and more product quality control.

Requirements of quality vary from product to product. Physical and mechanical criteria such as block size and durability are important for rockfill and armor stone, whereas chemical criteria control the value of an industrial mineral deposit, and block size, appearance, and ease of dressing are important factors when selecting an architectural stone.

1.2.2 Quarry planning

1.2.2.1 Requirements for site exploration. An investigation must assess not just the quantity and quality of the resource, but also the best mining methods and equipment for ripping, excavating, drilling, blasting, and processing; the requirements for land acquisition, access roads, and servicing; the optimum pit and stockpile layout; and the likely environmental effects of extraction. Investigation costs form a part of the overall start-up costs, and allowance must be made for ongoing exploration during the life of the quarry or mine.

Investigation of a potential quarry site includes evaluation of rock, overburden, and groundwater conditions (*Rock Engineering,* Chap. 6).

It should be phased, starting with broad-based, inexpensive methods such as a review of geological data and an outcrop reconnaissance. Completing the picture requires core drilling to obtain samples, and superficial test pits and diamond coring to map the top of bedrock and depth of weathering. Geophysical methods can be useful to determine overburden thicknesses, to correlate to material quality indices from core, and to interpolate between the coreholes.

1.2.2.2 Land acquisition and protection of the environment. Ownership of the land and mineral rights must be established to determine royalties, rents, and land acquisition costs. Access permission and impact on adjacent properties should be considered before any on-site activity begins. Further costs are often incurred in upgrading access routes, electrical power, and water supplies.

In those jurisdictions requiring a quarrying permit, the owner must submit to the planning authorities and at public environmental hearings a development program that will minimize environmental problems during quarrying, and assure rehabilitation of the site after quarrying has ended. Site rehabilitation only after the quarry is mined out is rarely permitted today; ongoing reclamation is necessary (Bauer, 1970).

Environmental hazards include:

- Damage to natural vegetation and wildlife
- Effects on the water table and water supplies
- Increased traffic on neighboring roads
- Visual impact, blast vibrations, dust, and noise

Information is gathered on the site, the environment, the deposit, and the operation itself to determine:

- Suitable locations for the processing plant
- Sites for inconspicuous disposal (or sale) of waste rock and soil
- Sites for storing landscaping material
- Stripping and stockpiling patterns and locations
- Schedules for simultaneous mining and rehabilitation

Whereas gravel pits and excavations into soils and soft rocks can be landscaped (contoured) to disappear into the scenery when the pit is closed, a hard rock quarry will often remain as a permanent hole in the ground. Depending on whether the water table is high or low, mined-out quarries can be used as artificial lakes or as waste disposal sites. The latter use is of particular value near large urban centers, where there is often an acute shortage of waste disposal facilities as well as a demand for quarried rock. An increasing concern to prevent pollution means that use of a quarry for waste disposal must be studied carefully. Landfills in quarries often require the construction of engineered barriers of clay, with gravel filters to control infiltration, and with an intensive system of groundwater control and monitoring. Only certain types of wastes may be acceptable.

1.2.2.3 Stripping, waste storage, and disposal. Overburden is defined as the waste soil or rock that must be removed to gain access to the underlying resource. Often expressed as a "stripping ratio" (volume ratio of overburden to mined resource), it is important to profitability. The following equation, for example, gives the volume V of overburden with thickness t for a conical pit of depth Z and average slope angle A:

$$V = \frac{[\pi t]}{\tan^2 A} \left(Z^2 - tZ + \frac{t^2}{3} \right)$$

For large, shallow pits, the increased costs of stripping are a weaker function of depth, usually normalized depth raised to a power of 1.2 to 1.5. For complex geometries, computer-aided graphics or manual block summing will give stripping ratios for various assumptions of pit development.

The stripping ratio increases rapidly as the overburden becomes thicker, and decreases with an increase of pit slope angle. Profitable mining demands the steepest possible slopes consistent with safety. The value of materials extracted from greater depths is offset by the cost of additional stripping and by the need for additional land, which also increases in proportion to the square of pit depth. At some depth, depending on the value of the commodity being mined, it may become economical to convert from open pit to underground mining.

Stripping is often split into phases to distribute capital costs over a longer time, rather than doing all stripping initially. An evaluation must consider capital costs, interest on loans, and the cost of reclamation. The quarry operator must locate waste dumps and tailings ponds carefully. Lack of foresight can lead to a later need to relocate stockpiles, dumps, ponds, pit offices, and processing plant.

Sources of dead weight and water in particular must not be placed where they may induce slope failure, or where they will cause reclamation difficulties. Careless stocking of wastes can lead to rock mass contamination and loss of valuable resources. Long-term topsoil stockpiling may result in alteration of physical properties through bacterial or chemical degradation.

Disposal may be costly if suitable sites are not available nearby. On the other hand, byproducts of mining can often be sold to offset mining costs. Topsoil always finds a ready market near towns and cities, but some may need to be stockpiled for use in reclamation. Waste rock or granular soils may be suitable as general purpose fill or for more specialized applications in roads and embankments. These "wastes" may be cheaper to process and sell than to rehabilitate. Even old waste dumps may become, with time and increasing demand, economically exploitable.

1.2.2.4 Quarry layout. Quarry design must consider the need for at least two working faces and for haul roads at grades and alignments that suit the equipment and give an orderly circulation of traffic in the pit. Conveyors may be considered as an alternative to "unitized" (truck or rail) transportation. A hillside quarry will give more favorable conditions than one excavated downward into horizontal terrain because gravity assists bench blasting, loading and haulage, and drainage conditions.

Pit wall design methods are described in Chap. 2. The walls of shallow quarries in hard rock are often vertical. In poorer ground or deeper pits, walls should be benched to reduce the overall slope angle and to catch falling rock. The upper slopes in soil and weathered rock are excavated at reduced angles. A wide bench is left on the competent rock surface to catch debris as it erodes or ravels from the less competent overburden deposits. Haul roads in particular have to be stable and protected from bouncing rock debris.

1.2.2.5 Groundwater control. Groundwater must be controlled to prevent flooding and slope instabilities and to maintain pit bottom access. Small inflows through tightly jointed rock in shallow pits can be removed easily by pumping from sumps within the pit. Pumping requirements need to be estimated in relation to the hydraulic conductivity of the rock and overburden, and the preexisting water table. Groundwater may need to be controlled before it reaches the pit, by curtain grouting or by installing a perimeter dewatering system (*Rock Engineering,* Chap. 4, and 18).

High groundwater pressures can affect pit wall stability, and, in combination with high lateral stresses, can induce heave and buckling in the pit floor. Pressures are relieved by drilling drain holes, or in extreme cases, by lowering the water table or excavating perimeter drainage galleries. Drainage measures can often be justified in terms of steeper slopes and reduced stripping requirements.

The erosive effects of surface runoff water can lead to raveling, loss of benches and haul roads, and costly pit bottom cleanup. Perimeter runoff channeling and control are often needed, even to the extent of diverting streams and damming up lakes to gain access to a valuable resource. A hydrological study is then required to consider high water levels and the likelihood of flooding during intense storms.

1.2.3 Costs of development and operation

1.2.3.1 Development costs. Contributing to the development costs are those of land acquisition, engineering and exploration, clearing, stripping, excavation, stockpiling, processing, transportation, and maintenance or rehabilitation of the quarry or pit.

1.2.3.2 Costs of quarrying. Direct quarrying costs include those of engineering, stripping, blasting, loading, conveying, crushing, and screening. Drilling and blasting patterns need to be optimized (*Rock Engineering,* Chap. 13). The aim is to minimize secondary blasting and to produce materials of satisfactory quality, block shape, and gradation. When quarried rock is to be processed further, size and shape limitations on blasted rock are often imposed by capabilities of crushing equipment.

1.2.3.3 Costs of haulage. Quarry evaluation must consider market proximity; the less valuable the rock, the more important are haulage considerations. Rockfill must be quarried within a few kilometers of the construction site; aggregates can be transported tens of kilometers; industrial minerals, large intact stone blocks, and ornamental stone can be exported to international markets. Trucking costs for aggregates can approach the combined costs of quarrying and processing, typically from 30 to 70 percent of the delivered price, depending on distance. Costs can be reduced if materials are extracted in quantities sufficiently large to justify rail, canal, or marine transportation. In the Great Lakes region of North America, crushed rock from nearshore quarries, transported hundreds of kilometers by barge, can be competitive with granular aggregate that must be surface hauled only 20 to 30 km.

1.3 Rockfill

Quarried rockfill is used mainly in the construction of highway embankments, dams, and breakwaters (Bruun, 1985); as armor stone for protection against erosion along coasts, river banks, and earth dams; and as backfill for support of underground mine openings.

1.3.1 Rockfill embankments

1.3.1.1 Types of embankment. Embankments for roads, dams, and breakwaters can be either homogeneous, consisting of a well-graded mixture of all sizes, or zoned to make better use of the available materials. Zonation proceeds with fine-grained soil or crushed rock near the embankment core, and larger-sized, more resistant rock in the outer shell, or as special toe drains and buttresses (Brandl, 1980). Embankment dams require an impermeable core or blanket, usually of compacted clay or shale, with more freely draining and erosion-resistant rock closer to the upstream and downstream faces. Highway and other nonmarine embankments can be protected by sodding or seeding, and seldom require rockfill, except in the form of buttresses or counterfort drains to control sliding (*Rock Engineering,* Secs. 17.1.2 and 18.1.5.2).

Rockfill embankments are constructed by either *end dumping* or *compaction*. The end-dumping method is used for strong, durable rock, whereas compaction methods are used for shale or soft, weathered rocks that are expected to break down to form a dense, earth-like fill..

1.3.1.2 End-dumped embankments. The only compaction given to rockfill composed of hard, durable fragments is that imposed by self-weight and by the bulldozers and dump trucks. Lifts are sometimes as high as 60 m, both for economy and to achieve enhanced compaction by impact. The rock is brought from the quarry in trucks that drive out over completed sections of the embankment. It is dumped down the sloping face of the construction lift, and often is sluiced with high pressure water jets (*monitors*).

The main purpose of sluicing is to wash fines from the surface of the lift and into the void space so that two consecutive lifts bond together. A secondary reason is to reduce the strength of block-to-block contacts so that degradation occurs during construction rather than later in service when the embankment becomes thoroughly wetted for the first time. Sherard et al. (1963) suggest that a volume of water equal to 30 to 50 percent of the volume of rock should be sufficient in most cases to achieve the necessary water-weakening.

Even if the quarried rock is itself dense, rockfill embankments have a substantial porosity by virtue of the voids between blocks. Macroporosity, the ratio of the volume between blocks to the total fill volume, may vary between 18 and 35 percent; lower macroporosities and higher densities will be achieved for embankments constructed by compaction using the weaker types of rock. Volumes of rockfill required are often estimated from experience; however, test compaction of typical materials may be undertaken. Once the range of macroporosities achievable is known, the bank volume V_{quarry} of rock to be quarried can be calculated:

$$V_{quarry} = V_{fill} \frac{(100 - n_m)}{100}$$

where n_m is the macroporosity determined as a percentage.

1.3.1.3 Compacted rockfill embankments. Weaker rock types, especially shales, are first ripped, then compacted to achieve maximum size degradation during placement and therefore to limit degradation and associated settlement thereafter. The rock is placed in lifts of thickness of about 1 m, no thicker than 1 to 2 times the diameter of the largest permissible rock block (Fig. 1.3).

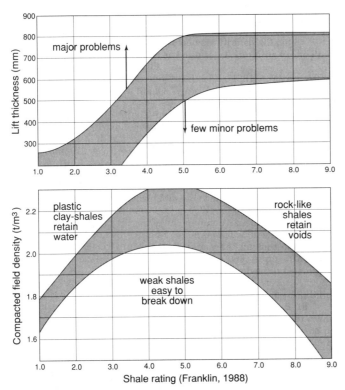

Figure 1.3 Lift thickness and compacted density of shale embankments as a function of the type of shale (Franklin, 1983).

Soft rocks may be broken into appropriate sizes by blasting or ripping in the borrow pit, or they may be left to weather for a season, taking advantage of natural breakdown. They are usually wetted after placement to assist compaction. Sheep's foot, spiketooth, or chiseltooth tamping rollers and heavy smooth-drum vibrating rollers will further break down and compact the blocks. Thicker lifts can be permitted using *dynamic compaction* methods by repeated dropping of a large weight from a crane. The impacts break down the block contacts, greatly increasing the density of placement.

1.3.1.4 Specifications. Construction specifications may stipulate either the *end result* (the required quality of the completed embankment), or the *methods* to be used for compaction. The end result (quality) type of specification calls for determination of moisture-density relationships using soil mechanics compaction tests, followed by measurement of in-place densities and moisture contents to check that sufficient compaction has been obtained (Lovell, 1983). These measurements are more difficult and less reliable in rockfill than in earth fill, and often a preferred alternative is to give a procedural specification. These are standardized in terms of specific equipment energy (Q) and surface area (S) for particular materials in lifts of particular thickness (e.g., Brauns et al., 1980). Highway departments and other construction authorities provide these specifications to bidding contractors.

1.3.1.5 Embankment performance. Settlement of a rockfill embankment during its construction is seldom of concern; long-term settlements are what create problems. Settlement typically amounts to 1 percent of embankment height after about 10 years and is unavoidable, although it can be minimized by good construction practices (Fig. 1.4). It is the result of degradation (weakening and disintegration) of blocks caused by high stresses at the block contact points, aided by slaking and sometimes swelling. Excessive settlements are most often associated with rock types that are difficult to compact well, yet are too weak to serve in an end-dumped embankment. Rockfill with more than about 3 percent of smectite clay minerals can give long-term settlements much larger than 1 percent (Dusseault et al., 1985).

1.3.2 Riprap and armor stone

1.3.2.1 Riprap. Riprap is placed on the shoulders of an embankment of earth, less durable rock, or small-sized rockfill to protect against erosion and weathering. Protection is provided by loose rock, mortared or grouted rock, concrete in bags, or concrete slabs and rubble. The

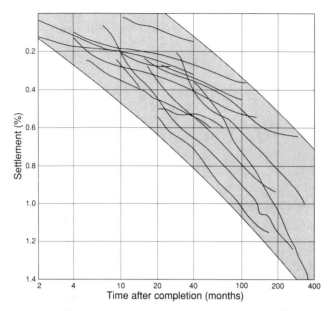

Figure 1.4 Typical postconstruction settlements of rockfill dams (from Franklin, 1983, after Bragg and Zeigler, 1975).

low cost and high durability of many types of rock, when available, make them ideal for this purpose.

Riprap can be enddumped and graded with a bulldozer working from the toe of the embankment either during or after construction. On flat embankments, it may be placed by enddumping directly onto the face. Hand work with bars may be needed to adjust the positions of blocks and to ensure that the voids are filled (Campbell, 1966). Riprap can also be hand-placed, block by block. However, a survey by the U.S. Corps of Engineers found hand-placed riprap to have failed 6 times more frequently than dumped riprap of equivalent thickness.

In river and canal protection work, the toe of the riprap should extend beneath the stream bed into a toe trench or over the surface of the stream bed as a blanket to protect against turbulent erosion. Conditions of erosion are especially severe where the riprap meets natural unprotected earth. Particular care is needed to design and construct smooth transitions at these points.

A filter layer must be placed beneath riprap unless the bank materials themselves meet the filter requirements. Filters are designed to ensure that particles will not wash through the riprap, causing undermining, channeling, and loss of integrity of the protective layer. The filter can consist of a geotextile fabric or a granular blanket of gravel or crushed rock with an appropriate gradation. Often a granular filter and a geotextile are used together. Granular filter blankets are usually 150 to 375 mm thick, with a carefully selected grain size. The *filter ratio* is defined as the 15 percent particle size (D_{15}) for the coarser layer divided by the 85 percent particle size (D_{85}) for the finer layer. The filter ratio should be less than 5 for dumped riprap. As an additional requirement, the ratio of D_{15} sizes for riprap and bank should be greater than 5 but less than 40. For multilayered filters, gradation curves of adjacent layers should be parallel to each other to achieve a minimum total filter thickness (U.S. Army Corps of Engineers, 1987).

1.3.2.2 Breakwaters and armor stone. Key aspects of *breakwater* design are to ensure that the rock remains in place in spite of wave action, and intact in spite of extremes of weathering (Bruun, 1985). A densely packed, zoned embankment is protected by large blocks of durable armor stone.

When large blocks are not available, smaller sizes can be confined in gabion wire baskets or placed loose and cemented together by grouting. Gabion baskets filled with stone can be tied together to form a mattress, or may be freestanding or anchored to the underlying rock or soil (*Rock Engineering,* Chap. 17). Local rock can be used as aggregate to make large concrete "tetrapods," which, because of their interlocking nature, are effective in breaking the energy of waves.

1.3.3 Mine backfill

Rockfill is also used in backfilling and stabilizing underground mine openings. It can be used as an alternative to other wastes such as tailings, or to supplement these materials in special circumstances. Mine backfilling is discussed in Chap. 7.

1.3.4 Quality and testing requirements

1.3.4.1 Overview of requirements. Rockfill for embankment construction should be either weak enough to allow thorough breakage and compaction, or strong and durable enough to allow end-dumping without compaction. Riprap and armor stone have to be both strong and durable. All materials must conform to grading requirements; usually they are required to have a well-graded distribution of block sizes (linear on a logarithmic scale) to permit high-density packing. They must be available in sufficient quantity close to the construction site.

1.3.4.2 Strength and durability. Local rock materials are often less than ideal, but embankments can be constructed with any kind of rock, provided the limitations of the materials are recognized and the embankment design and construction methods are chosen accordingly.

Shales, siltstones, chalks, and severely weathered rocks can often be readily compacted, but the deposits should be of uniform rock quality. High porosity chalk may present problems if it has been allowed to dry before compaction: large subsidences may occur upon wetting.

Lovell (1983) and Franklin (1983) discuss in greater detail the procedures for testing and classifying shales for embankment construction. Compaction characteristics measured for design include the moisture-density and compaction-degradation relationships, one dimensional compression behavior in an odometer, and consolidated-undrained triaxial test data on saturated samples. The odometer and triaxial test results allow prediction of settlement and assessment of stability of the embankment slopes. Hale et al. (1981) describe methods for measuring shale degradation in terms of the percentage change in mean particle size during compaction. With alterations, these methods can be used to evaluate siltstones, chalks, and other transitional materials, but special precautions are required where large quantities of swelling shales are present.

Rock to be used as armor stone in marine works needs to be particularly durable because of the aggressive action of repeated wetting and drying, often combined with freezing and thawing, and salt crystallization when the rock is placed in a saltwater environment. The most suitable rocks are massive igneous types such as granite or massive sedimentary beds of limestone or sandstone, free from fissuring or weakness planes, and with a high intergranular cohesion and low porosity and permeability. Quartzite and dense crystalline carbonates are often used.

Tests to evaluate the suitability of the harder types of rockfill, riprap, and armor stone can, for example, include point-load strength evaluation both parallel and perpendicular to weakness planes, porosity and density measurements, and evaluation of cementing materials and rock texture by examination of hand specimens and thin sections under the microscope. Tests on small pieces of rock, however, give information of limited value when evaluating large blocks for construction purposes.

Large blocks remain large only if they are free from defects such as microfissures and shaley or micaceous layers. The evaluation should be completed by inspecting the weathering behavior of rock blocks exposed to the elements for a number of years in old embankments, quarries, or natural rock outcrops. Weathering accentuates natural weaknesses, and provides excellent qualitative information on long-term rock behavior.

Various abrasion and accelerated weathering tests are suggested for evaluating the durability of stone to be used for riprap. Resistance to salts from road salting and other sources can be evaluated by the sulfate soundness test (AASHTO Test T104 for ledge rock using sodium sulfate). Stone should have a loss not exceeding 10 percent after five cycles. In the Los Angeles abrasion test (AASHTO Test T96), the stone should have a percentage loss of not more than 40 percent after 500 revolutions. In the freezing and thawing test (AASHTO Test T103 for ledge rock, Procedure A), the stone should have a loss not exceeding 10 percent after 12 cycles of freezing and thawing. Rock deterioration caused by salt and freeze-thaw effects may be evaluated with sonic velocity measurements.

Numerical modeling to predict the behavior of a rockfill embankment (usually a dam) calls for large-scale triaxial testing to measure such properties as modulus of deformability and strength. The maximum particle size that can be tested is in the range 50 to 200 mm, depending on the available test cell. The ratio of maximum particle size in the prototype to that in the test should not exceed 6.0 to obtain meaningful results (Lo, 1982). Large-scale odometer testing can be performed to determine consolidation characteristics, with caution exercised because of scale effects. Small-scale trial compaction, followed by wetting, may be a useful indicator of behavior.

1.3.4.3 Gradation. The mechanical behavior of fill depends on the size distribution and absolute size of blocks, the shape of blocks, and the state of packing as defined by porosity or void ratio (Lo, 1982). Rockfill should be well-graded; that is, its size data should follow a straight and flat gradation line on a log scale. The small blocks occupy spaces between larger ones, giving a dense fill, and reducing the long-term settlement potential.

Grading specifications generally exclude oversized or slabby rock fragments, although in some embankments, blocks can be as large as a meter in diameter. Admixtures of large blocks, particularly if slabby, lead to compaction difficulties and voids in the completed embankment. Large blocks and slabs are sometimes removed during placement of the fill, and used on the shoulders or toe of the embankment where they present no problem and can even contribute to protection against erosion.

Riprap stone also should be well-graded. The U.S. Waterways Experiment Station recommends that for any required 50 percent size k, 100 percent of the riprap should be smaller than $3\ k$; 80 percent smaller than $2\ k$; 50 percent smaller than $1\ k$; and not more than 10 percent smaller than $0.1\ k$ (Campbell, 1966).

Careful blasting is needed to produce the required gradation, and to avoid production of oversized, undersized, or slabby blocks. This is achieved if natural joint spacings give approximately the correct block shapes and gradations without secondary breaking. Oversized blocks may be left in the quarry, or they may be broken by secondary blasting, pneumatic rock breakers, or a drop ball.

1.3.4.4 Block size to resist water flow. Large-sized material is essential for marine works since all but the largest-sized blocks can be carried away by wave action. Blocks as large as 3 mm have been noted to move many centimeters in storm conditions on the Atlantic coast of France.

The resistance of riprap to displacement by moving water in a river or canal depends on the weight, size, shape, and gradation of the stone; the depth and velocity of the water; the steepness of the protected slope; the effectiveness of the underlying filter; and the protection of the toe and upstream and downstream edges. The required size of stone as a function of water velocity and embankment side slope is shown in Fig. 1.5. When the depth of flow exceeds about 3 m, a value of 0.4 times the total depth is used in determining the flow velocity against the stone (Campbell, 1966).

1.3.4.5 Block shape. Stone for use in rockfill, riprap, and armor stone applications, except soft rock to be broken up during compaction, should have a block shape close to equidimensional. Slabby or prismatic blocks are difficult to place and may cause voids in an otherwise interlocking fill. This excludes the use of thinly bedded sedimentary rock and metamorphic rocks with mica along planes of fissility. The

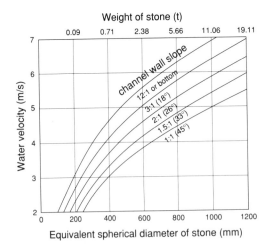

Figure 1.5 Size of stone riprap, density 2.64 t/m^3, as a function of water velocity and side slope of a channel (from Campbell, 1966).

stone should consist of angular quarried rock, and rounded stones or boulders are seldom acceptable.

1.4 Aggregates for Roadstone, Concrete, and Ballast

1.4.1 Types and uses

Crushed rock aggregates have been used since ancient times. As far back as 600 B.C. the Babylonians constructed roads using a base of tiles set in asphalt mortar and covered by flags with the gaps filled by broken stone (Hosking, 1970). Some Roman roads were paved with layers of aggregate bonded with mortar, the particle size diminishing toward the surface. The more heavily traveled roads were provided with a wearing course of flags set in mortar. Bituminous binders were used by the Incas in South America; thousands of kilometers of asphalt-based roads were built in the Andes.

Today, binders are better and more varied. Flexible pavements incorporate thermoplastic binders, usually asphalt, and rigid pavements have a binder of portland cement. A *subbase*, usually 150 to 450 mm thick, is constructed of aggregate that distributes the load of passing vehicles to the underlying *subgrade* of soil or rock. The overlying road *base* is constructed of higher grade materials, usually between 150 and 250 mm thick. These can be lean concrete, coated crushed rock, rolled asphaltic concrete, cement-stabilized granular soils, or unbonded aggregate (gravel). Often there are two surfacing courses, a base course and a wearing course. A layer of chipped stone is often pressed into a hot asphalt layer on traffic lanes to give a skid-resistant surface.

Quarried and crushed rock competes with natural aggregates (sands and gravels), which are cheaper to excavate and require less processing (Shergold, 1960). In Ontario, for example, about two-thirds of requirements are satisfied by sand and gravel, and one-third by crushed rock. Overall demand is increasing by about 100 percent per decade. Crushed rock is often stronger and more durable with better grain shape and surface roughness characteristics; the friction angles for crushed rock may be as much as 13° greater than those of a rounded gravel (Eerola and Ylosjoki, 1970). This converts to a 57 percent increase in shearing resistance in fill, bituminous surfacings, and concrete. Crushed rock is less likely to contain clay and silt particles and other harmful contaminants.

1.4.2 Quality and testing

1.4.2.1 Overview of requirements.
Aggregate quality must conform with various specifications relating to physical, mechanical, and

chemical properties. Each agency has its own specifications and tests (e.g., ASTM, 1978; Arquie, 1980; Tourenq and Denis, 1982; Aitcin et al., 1983; Rogers and Magni, 1987). Techniques for the evaluation and testing of highway aggregates and railway ballasts are similar to those for concrete aggregate, although compatibility with portland cement is no longer an issue, and the acceptance criteria for particular tests are often different.

1.4.2.2 Particle strength and contaminants. Crushed rock aggregate for concrete usually must be at least as strong as the cement mortar matrix. High-porosity rocks, weathered rocks, and most shales are avoided because of low strength and a tendency to degrade with use. *Deleterious materials* include clay and shale lumps and weathered rock materials that weaken the aggregate, and salt grains and coatings that reduce bond strength and alter the setting time of a concrete mix. Sulphides cause rust-staining, and salt contaminants give rise to severe corrosion of reinforcement steel in concrete. These materials should be identified as part of the petrographic examination (Mielenz, 1963).

Natural gravels may contain several rock types, whereas crushed rock is usually from a single source. With various rock types, a measure of overall quality is required. One way of assessing this is to measure the *petrographic number* (PN), which reflects the percentages of different rock types and qualities. The method of assessment is visual and makes use of a stereoscopic microscope, pocket knife, and weak hydrochloric acid. The petrographic category is based largely on rock type, weathering, and strength. It is a useful and quick test, although operator experience is needed to obtain reproducible results.

A sample of about 200 particles of each size is subdivided into rock-type categories and then into good aggregate (weighting factor 1), fair aggregate (weighting factor 3), poor aggregate (weighting factor 6), and deleterious aggregate (weighting factor 10). PN is obtained by multiplying the percentages of each quality group by the corresponding weighting factor and then adding the products. The lower the PN, the better the aggregate; a sample entirely of good aggregate would have a PN of 100. For concrete paving in Ontario, only aggregates with a PN of 125 or less are used, and for structural concrete the maximum allowable PN is 140.

The mechanical competence of rock as an aggregate material can be assessed using standard rock engineering index test procedures, of which the point load strength test (*Rock Engineering,* Chap. 2) is one of the more convenient. It can be used on either core or irregular outcrop samples from a potential quarry site, or on the individual crushed aggregate particles themselves.

As an alternative to measuring the properties of specimens of rock from the quarry, bulk samples of the crushed aggregate can be tested. For example, the *Los Angeles abrasion test* subjects a 5-kg sample of aggregate to impact in a rotating drum. The sample, mixed with a charge of steel balls, is rotated for 500 revolutions at 33 rpm. Breakdown is measured by removing the sample and screening it on a 1.7-mm mesh sieve; the amount passing the sieve is expressed as a percentage of the original sample weight. Typical permitted losses are 50 percent when the stone is for use in concrete, and 60 percent when for use as a granular base.

1.4.2.3 Absorption, durability, and soundness. Water absorption tests estimate the porosity of a rock. Porous rocks are weaker and require more water for a given workability of concrete mix; therefore they give lower concrete strengths. Absorptive stone also is more expensive to dry for use with hot mix asphalt. Offsetting these disadvantages, porous and absorptive rocks give a somewhat better bond with cement and asphalt, because of their greater surface roughness.

In a typical *absorption test,* a 3-kg sample of aggregate is placed in water for 24 h, after which the particles are removed and dried with a towel. The saturated-surface-dry sample is weighed, and then is oven-dried and reweighed. Absorption is expressed as the percentage ratio of water absorbed to weight of dry rock. The Ontario Ministry of Transportation specifies an absorption of less than 2 percent for concrete aggregates, although this requirement can be relaxed if field performance has been shown to be satisfactory.

The slake-durability test (*Rock Engineering,* Chap. 2) is insufficiently aggressive to assess either frost susceptibility or resistance to the salts of coastal environments and to those used to melt ice on highway pavements. Accelerated weathering is simulated by freezing and thawing or "soundness" tests in saturated solutions of calcium or magnesium sulphate salts. Substantial pressures are generated in the pore space by salt crystallization, hydration, water absorption, and ice growth, and these simulations may therefore not always be realistic.

ASTM C-666 measures *freeze-thaw durability* by subjecting concrete (or rock) specimens to rapid freezing in air and thawing in water. A test continues for 300 cycles or until the sonic modulus of elasticity has decreased by 40 percent. A *durability factor* is calculated in terms of the reduction in sound velocity that results from internal cracking.

In the *magnesium sulfate soundness test* (ASTM C-88), a sample of aggregate is submerged in a saturated solution of magnesium sulfate for 16 h, then dried for 6 h at 110°C. After five such cycles, the sample is washed for 36 h to remove the salt, dried, and regraded on the original sieves. The amount of aggregate passing the original sieve size is expressed as a percentage of the original sample weight. Ontario specifications require the loss to be less than 12 percent. If they have satisfied field performance criteria in concrete, aggregates with losses of up to 20 percent are sometimes accepted.

The simplest and most reliable assessment of aggregate durability is obtained, if time is available, by recording the performance of rock types in service. Dense, strong aggregates and those composed of proven rock types are seldom vulnerable to frost damage or other forms of degradation. This information must be collected in a databank, and upgraded as needed to reflect changes in road use and treatment.

1.4.2.4 Alkali-aggregate reactivity. A reaction occurs slowly between certain undesirable types of aggregate rocks and the alkalis (sodium and potassium compounds) in cement paste (Diamond, 1975, 1976). It produces a gel that expands, generating cracks that weaken the concrete (Fig. 1.6).

In cold climates, the cracks give access to water which freezes. Salt used to de-ice roads also gains access through cracks, which accelerates disruption of the concrete, and also corrodes the reinforcing steel. Often the resulting secondary freeze-thaw deterioration in northern latitudes is more important than the primary cracking.

Structural and aesthetic damage can be severe, as shown by a survey of more than 300 concrete structures and engineering works in and around Quebec City (Fournier et al., 1987; Bérubé and Fournier, 1987; Fournier and Bérubé, 1989).

Alkali-aggregate reactions can develop over several decades and are difficult to recognize and predict. Exterior signs are spalling and pop-outs, polygonal cracking and crazing, milky aureoles around cracks, aggregate debonding, joint material extrusion, rust from deteriorating steel, and pillar bulging.

The reactions are classified into *alkali-carbonate,* and *alkali-silica* (or -silicate) types. The most important process in the alkali-carbonate category is expansive dedolomitization of a specific type of carbonate rock containing equal amounts of calcite and dolomite, and substantial amounts of illite (Chap. 41 in ASTM, 1978). Reactive rock types in the alkali-silica category include glassy ones such as opal and chalcedony, some types of chert, and microcrystalline quartz, particularly when strained (Chap. 40 in ASTM, 1978).

The problem can be avoided either by using low-alkali cement, or by excluding the reactive aggregates. The alkali content of cements has been increasing in recent years because of changes in methods of manufacture, and can be as high as 1.35 percent (Na_2O equivalent). A content of less than 0.6 percent is considered low-alkali and generally

Quarrying and Use of Stone 25

Figure 1.6 Alkali-aggregate reactions. (a) Crazed concrete in an overpass girder (limestone aggregate); (b) desiccated siliceous gel in a fissure (rhyolitic tuff aggregate). (*Courtesy M.-A. Bérubé, Laval University, Québec, Canada*)

produces negligible reaction. Reactive rocks can be identified by microscopic and chemical techniques, or more reliably by measuring expansion of rock cylinders or concrete prisms containing the suspect rock immersed in an alkaline solution. Testing usually takes at least 1 year, although accelerated tests are available. Provincial and state transportation agencies keep records of aggregate sources that are known to have caused deterioration, and this is a valuable means of determining reactivity.

1.4.2.5 Crushing, gradation, and particle shape. Crushability of a rock for aggregate can be related to its properties in use. Rock for processing into aggregate material should break readily into equidimensional fragments without an excess of fines (silt and clay sizes). Brittle, dense, isotropic, and crystalline materials are better from this point of view than porous, friable, or laminated rocks.

Size gradation is an important characteristic of crushed rock and natural gravel aggregates in most applications; the product usually needs to be well graded. Well-graded aggregate gives a denser particle packing, leading to greater interlock, strength, and cyclic load resistance. Bituminous asphalt strength, rigidity, and trafficability are functions of the density achieved during hot or cold compaction (Huschek and Angst, 1980), hence gradation is a critical parameter for high-quality road surfacing. Single-sized or *gap-graded* aggregates are used in limited applications, such as free-draining pavement layers and for insulation.

In aggregates, a limit is placed on the acceptable percentage of *fines,* defined as particles of silt or clay size passing a 75-μm sieve. Fines are harmful in concrete because they require extra water in the mix to achieve sufficient workability, thereby reducing the strength of the concrete. Weak and deleterious minerals (such as clay or mica) are concentrated in fine-grained fractions because they are the most readily crushed. Excess fines may also result from inadequate crushing equipment and methods. Washing and screening may be needed as part of the aggregate manufacturing process to reduce fines and contaminants to acceptable levels.

A faceted, equidimensional *particle shape* is the ideal for aggregates. Flaky or elongated particles impart poor workability to a concrete mix, and are weaker and more difficult to compact in other applications. Metamorphic rocks such as schists and thinly bedded sedimentary siltstones are the most likely to give flaky aggregates when crushed. The flat surface of flaky particles is often smooth, and gives a poorer bond with the cement paste.

Because of requirements for angularity and roughness, railway ballast is exclusively a crushed product. It is a uniformly sized (poorly graded) stone with a minimum of fines. Because it is placed without a binder, it must not display excessive vibratory compaction with time and traffic. Only strong and durable rocks are employed as ballast, and field performance must be evaluated (Selig et al., 1980).

1.4.2.6 Surface roughness and frictional characteristics. In concretes, surface roughness is required for a satisfactory bond between an aggregate and cement or asphalt, except that carbonate rock particles bond well with cement paste even when smooth. In road surfacing applications, aggregates make up about 95 percent of the wearing surface of a pavement, so their properties control frictional resistance to skidding. A stable, high stone content mix is required to give a macrotexture of protruding particles that break the water film and provide drainage to reduce the chance of hydroplaning (tires losing traction because of gliding on a film of water).

Crushed quarried rock consists of rough and angular particles, whereas natural gravels are rounded and have to be processed through a crusher to provide surfaces that are broken and rough. The mineralogy of the rock also affects bond, but probably to a lesser extent than roughness. Porosity improves bond; porous but strong limestones give excellent bonding characteristics.

The best polishing resistance is afforded by rocks with minerals of contrasting hardness or those with grains that are plucked rather than worn smooth by traffic. Those most prone to polishing are monomineralic quartzite and limestone, which wear uniformly over their entire surface. In contrast, rock types like granites, which contain quartz and feldspar of different degrees of hardness, tend to remain rough when abraded by wheeled vehicles. Although higher-porosity rocks wear more rapidly, they tend to develop a skid-resistant surface texture (Fourmaintreaux, 1970).

Special laboratory tests measure susceptibility to surface abrasion and polishing (Rogers, 1983). In the *aggregate abrasion value test* (BS 812, 1975), 24 or more cubical particles of aggregate, size 9.5 to 13.2 mm, are held in an epoxy binder. A 2-kg weight presses the specimen of aggregate against a 600-mm-diameter steel lap, rotated for 500 revolutions at 30 rpm. Standard sand is fed at 800 g/min in front of each specimen. Aggregate abrasion values, expressed as loss of mass, typically range from 2 to 5 percent for dense quartzitic rocks; from 5 to 15 percent for more porous sedimentary rocks, denser limestones, and dolomites; and up to 30 percent for blast furnace slags and the more porous, poorly cemented sedimentary rocks.

The amount of polishing produced is measured by a special frictional resistance tester. The *polished stone value test* (BS 812, 1984) uses a sliding pendulum to measure the frictional properties of the aggregate after 3 h of abrasion with coarse grinding powder and 3 h of polishing with fine grinding powder. In situ methods for measuring frictional properties of pavement surfaces include measurements of the distance skidded by an automobile with locked wheels (ASTM E445-76), side force friction (ASTM E670-79), and the brake force skid trailer (ASTM E274-79). Indirect methods of estimation involve the measurement of surface texture, for example, using stereo photography (ASTM E770-80).

1.4.3 Quarrying and processing requirements

Quarry blasting aims to produce large volumes of well-fragmented rock at minimum cost, and with minimum environmental disturbance (*Rock Engineering,* Chap. 13). Breakage must be sufficient to produce feed rock for the primary crusher, with a small percentage of fines, and with few slabby or oversized fragments requiring secondary breakage. The trend is toward longer and larger blastholes, and the use of AN/FO and slurry explosives is now widespread.

Factors affecting the design of the quarry treatment facilities include the output requirements, capacity and type of crushers and screens, stockpiling facilities, and storage capacity. Quarried rock is blasted, then passed through a "grizzly" to scalp off oversized blocks before being fed to the primary crusher (Fig. 1.7). Oversized blocks can be broken by secondary blasting, or with a mobile hydraulic impactor which is also useful in clearing blocked crusher jaws.

Crusher types include impact breakers, hammer mills, jaw crushers, and gyratory crushers. A jaw, gyratory, or cone crusher reduces sizes to those that can be handled by an impact crusher. Equipment is selected to give optimum particle shapes and gradations.

After crushing, rock is screened and stockpiled. Screens should be selected to operate at the maximum anticipated output from the crushers to avoid overloading. Friable and weak particles are removed as fines from the screening process. During crushing and screening, the quality of the product can be improved by blending with imported materials. Crushed rock may be used to improve poor-quality natural sands and gravels, which frequently are deficient in coarse sizes and rough-faced particles.

1.5 Industrial Minerals

The mineralogical and chemical characteristics of certain rock types make them valuable resources for the manufacture of various products. Fluorite, graphite, agricultural lime, borax, talc, and silica sand

Quarrying and Use of Stone 29

(a)

(b)

Figure 1.7 Aggregate quarrying. (a) Bench blast; (b) Primary cone crusher at Dufferin Aggregates Quarry, Milton, Ontario.

for manufacture of glass are among the many lesser-known products. More familiar are bricks, ceramics, and cement.

1.5.1 Bricks and ceramics

Bricks and ceramics such as clay pipe and roofing tiles are manufactured from clay soils and shales. These consist mainly of clay minerals, although some quartz in the form of sand and silt grains is desirable to limit shrinkage and improve workability and ease of molding. Typically, for bricks and ceramics, softer shales are used; their slake-durability index should be low enough (typically less than 80 percent) so that they can easily be broken down into a workable clay mixture. Brick-making shales have a mineralogical composition that minimizes shrinkage and distortion during firing and gives dense, durable products. Smectitic shales are the worst for bricks and ceramics, whereas kaolinitic clays and shales are the best for firing and are used for china clays and porcelain.

Color is also important; red, green, and buff colorations are imparted by impurities such as iron and calcium carbonate. Color consistency in a product calls for a consistent impurity content throughout the deposit, or careful blending. Because the color can change substantially during firing, tests are carried out on samples taken from representative locations throughout the brick-shale deposit and fired over a range of temperatures. Shrinkage, strength, and absorption characteristics are measured, and the color of the fired briquettes is registered by comparison with standard color charts.

Durability and frost resistance depend mainly on the firing process, but also to some extent on the character of the shale raw material. Durability can be measured using tests similar to those for building stones and aggregates, and many highly specialized tests are available for ceramic applications.

1.5.2 Cements

The earliest cements produced by the Romans were manufactured from a mixture of lime, water, and a volcanic ash, *pozzuolana,* from Naples. *Portland cement* is made from a kiln-fired mixture of lime (calcium carbonate, $CaCO_3$) and clay or shale with some silica sand. The resulting cement *clinker* is crushed and ground to a fine powder, ready for mixing with water. The process was invented in 1824 by Joseph Aspdin, and was improved in 1845 by Isaac Johnson, who introduced firing at higher temperatures. The name derives from the Portland limestone in the south of England, used extensively as a building stone and also, in ground form, to make the early cement.

In the early days of portland cement manufacture, an ideal clay-bearing (argillaceous) limestone was found in the Portland stone quarries; today, blending is used to achieve the appropriate mixture. Cement works are often located along the geological boundaries between limestone and clay formations, and raw materials are quarried from adjacent pits. Limestones with a substantial clay content and with a moderate to low strength for easy crushing are the most suitable, although an excess of clay will contribute undesirable alkalis. Chalks and porous limestones are ideal as sources of $CaCO_3$. Chlorides must be avoided, because they cause blockages in the kiln, and give poor quality clinker.

High-alumina cement (Ciment Fondue), noted for its rapid gain of strength, is made by sintering a mixture of limestone, coke, and bauxite (which is a clayey weathering product with a high content of aluminum oxide).

1.5.3 Industrial and agricultural lime

Lime is a calcined or burned form of limestone, commonly known as quicklime or calcium oxide. When water is added, it turns to calcium hydroxide or slaked lime. The term is commonly used to denote almost any form of crushed limestone or dolomite (Lefond, 1983). Nearly 90 percent of lime is today used in chemical and metallurgical industries, for example, as a flux or acid neutralizer.

Most agricultural limes are manufactured by crushing limestone and dolomite, although quicklime and hydrated lime are also used. The quality of these materials for use as fertilizer depends on the Ca:Mg ratio, which is measured by chemical testing.

1.5.4 Talc

Talc ($Mg_3Si_4O_{10}(OH)_2$) is a widely used mineral with industrial applications arising from its softness and platy mineral structure. In powdered form it gives a slippery, dry surface of low friction. Talc is inert, and can be combined with materials such as oils and other powders. In addition to use as a high-quality filler in papers and paints, talc is used to dilute ammonium nitrates, to coat fertilizers and other products to avoid sticking, as an additive to plastics, as a carrier for fungicides and insecticides, and in cosmetic products where it is mixed with oils, colors, and perfumes.

The major sources of talc are in ultrabasic and schistose metamorphic dolomitic limestones that have experienced strong hydrothermal activity. In the largest European talc quarry (Fig. 1.8), production of more than 300,000 t/year is achieved by bench blasting techniques (Py and Grange, 1979). Three grades of talc are produced:

(a)

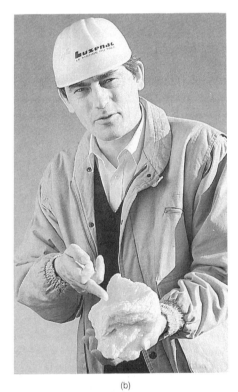

(b)

Figure 1.8 Talc mining at Trimouns, in the Pyrénée front ranges of Southern France. (a) the open pit; (b) the talc. (*Photos courtesy Talc de Luzenac S.A.*)

white, intermediate, and gray, depending on chlorite content. Slope stability problems are common in talc quarries because the geological structure is usually steeply dipping, and because the friction angle of the talc-containing rocks is low.

1.6 Building Stone

1.6.1 Ancient use of stone

Stone and wood are traditional building materials (Fig 1.9).The Pyramid of Cheops in Egypt was constructed 4700 years ago, of 2.3 million blocks (nearly a million cubic meters) of limestone on a 224-m^2 foundation, with a finished height of 147 m. The exposed blocks weighed 2.5 t on average, but some of them weighed as much as 15 t. The outer blocks matched so neatly that, in the words of the archaeologist Petrie, "neither needle nor hair" can even today be inserted at the joints.

The pyramids were built of layers that were inclined inward at 74°, but they were benched to give an average exterior inclination of 52° (Kerisel, 1987). The exterior casing was made of high-quality dressed blocks, but behind this exterior, the builders used stone of much lower quality that was roughly quarried and often ill-fitting. Today, the pyramids stand as monuments to the durability of stone, but they are also showing signs of age. Kerisel points to foundation problems, to stress-induced fractures, to differential slip, and to the use of poor-quality friable rock.

An exceptional skill in joining masonry blocks was developed by the Incas, and displayed in their cities and temples high in the Andes. A rough-dressed stone was placed in its course and moved slightly, marking the high points. The stone was removed, the high points were dressed manually, and the stone repeatedly refitted until the match was extremely close.

Whole cities have acquired their character from the stone from which they are made: Petra, Jordan, the "rose red city, half as old as time", and Aberdeen, Scotland, the "city of granite." The light-colored porous limestone used in Paris buildings comes from both underground and surface quarries; old underground workings of doubtful location and stability remain as a present-day hazard.

Certain rock formations with exceptional qualities of color, massive bedding, and easy cutting have earned an international reputation. Portland stone, a cream-colored oolitic limestone first used by the Romans, has been used in the construction of buildings, dock walls, and monuments in London for centuries. The Carrara marble of Italy has also been a favorite since the early days of the Roman Empire (Fig. 1.10). To this day in many countries it is a highly valued ornamental stone (Piga and Pinzari, 1984).

(a)

(b)

Figure 1.9 Monumental stones. (a) Taj Mahal, Agra, India, of marble inlaid with semi-precious stones. (b) monument in Leningrad, USSR, constructed from a single block of granite.

Figure 1.10 Quarrying of dimension stone, Italy. (a) Underground room and pillar mine in Carrara marble, excavated by diamond wire sawing; (b) quarry in "Peperino di Viterbo," a pyroclastic rock with a strength of about 30 MPa, excavated in the upper part by helicoidal wire sawing, and in the lower part by splitting. (*Courtesy Prof. Renato Ribacchi, University of Rome*)

36   Rock Engineering at Surface

1.6.2 Stone production in modern times

Although buildings are nowadays seldom constructed entirely of stone, the modern ornamental stone industry is large, and is expanding rapidly (Fig. 1.11). World production of ornamental stone increased tenfold between 1926 and 1982, when the annual production reached 15.3 million metric tons. Key producers included Italy (35

(a)

Figure 1.11 Quarrying of dolomitic limestone. (*a*) Line drilling of blocks for construction of (*b*) the Canadian Embassy in Washington, D.C. (*Photos, Arriscraft Corp.*)

(b)

Figure 1.11 (*Continued*)

percent), West Germany (20 percent), France (11 percent), and the United States and Japan (6 percent each) (Piga and Pinzari, 1984).

Architectural demand is greatest for marble and granite *facing slabs* (*cladding*) that can be cut and polished. Sandstones, siltstones, limestones, and granites are used for ornamental paving in the form of *flagstones,* and slates for paving and roofing.

Many countries still use *ashlar* blocks for walls in buildings of several stories, reinforced with wood or steel tension members. Smaller quantities of large *dimension stone* are employed for monuments. Specialized products include microgranites that are turned and polished to make curling stones, and quartzitic sandstones, which are still used for grindstones.

1.6.3 Quality and testing requirements

1.6.3.1 Block size, shape, and jointing. The most important characteristic of rocks suitable for quarrying as building stone is the presence of well-defined planar and persistent joints that define the faces of the quarried blocks (Godfrey, 1979). Usually between one and three major joint sets are present at right angles, so the stone splits naturally into cubes or rectangular prisms. The jointing should be relatively flat, planar, and widely spaced. Sets that are wavy or at odd angles tend to form irregular blocks and make the source of stone unsuitable except as a crushed product or for ornamental use. Fig. 1.9b shows a monument in Leningrad machined from a single very large block of granite. In less favored areas where the stone is irregular and slabby, it has still found use in the construction of foundations, dry stone walls, and flag pavements (Figs. 1.12 and 1.13).

When there are two widely spaced joint sets and a third at a closer spacing, the stone naturally splits into flags suitable for paving or facing stone. Thinly bedded sandstones, limestones, and slates with a fine cleavage are the most common types of flagstone. Slabs can also

Figure 1.12 Dry stone retaining wall composed of rectangular blocks of limestone and polygonal blocks of columnar basalt. (*Photo, MBD*)

Figure 1.13 Floor of quarried stone tiles, Norway. (*Photo courtesy E. Broch*)

be cut from the larger blocks in a rock deposit with widely spaced jointing. Closely jointed materials and those containing microfissures, significant clay mineral content, or shaley bedding partings are unsuitable in most applications.

Joint orientations and spacings can be measured in the field using various methods described in *Rock Engineering,* Chap. 3. These include the direct use of a tape measure and geological compass, and the alternative methods using digitized video tapes or photographs.

Open, persistent joints are sometimes accompanied by invisible planes of weakness that affect quarrying operations and the durability of the stone. These are much more difficult to detect. Techniques used to identify weaknesses in the rock fabric include dye penetrants, ultrasonic velocity measurements, thin section microscopic examination, and point load testing of core loaded in various directions.

1.6.3.2 Machinability. The ease with which a stone can be cut and dressed largely determines the cost of production. Softer rocks are easier and less expensive to cut and dress, but harder rocks take and retain a high polish. Their attractive appearance, durability, and ease of maintenance may result in a high market value, which can outweigh any difficulties and costs of preparation.

Maximum use is made of natural jointing and cleavage surfaces along which blocks can easily be split. Stone blocks and slabs are separated from the mass either by splitting or by sawing.

1.6.3.3 Color, appearance, and homogeneity. Uniformity of color is essential when large quantities of rock are to be quarried over many years, and should be confirmed by exploratory drilling. Rocks of variegated color can be used in special ornamental applications but care is required in selecting quarry locations and in stockpiling the different varieties.

The most famous ornamental stone is marble, which geologically is limestone or dolomite that has been recrystallized by thermal metamorphism. It has a crystalline texture and often is uniformly white, although impurities lead to pink and green coloration and banding.

Also commonly called marble in the building trade are the fossiliferous limestones, even though these have been subjected to little or no metamorphism. When cut and polished, they are buff and cream in color with fossils or reef textures. An example is Tyndall stone, a devonian limestone quarried north of Winnipeg, Manitoba, and used in buildings from Vancouver to Quebec in Canada. It contains large porous fossils and spiral shells that give the slabs paeleontological interest as well as architectural beauty. Travertine, another famous facing stone, is an open-textured limestone formed by the evaporation of lime-bearing waters.

Granite makes an attractive building stone because of its appearance and durability. Coarse-grained varieties are the most attractive, and colors may be pink, red, or gray depending on the predominant feldspar type. Porphyritic igneous rocks, those containing large crystals in a fine-grained crystalline matrix, often are more attractive than those of uniformly fine or medium grain. Stone with large, light pink feldspar crystals in a darker, fine-grained granite gneiss matrix is a common and appreciated facing material. Dark colored feldspathic rock (anorthosite) containing the mineral labradorite is well known for the iridescence displayed by the large crystals as the angle of incident light changes.

In rocks with attractive textures, such as gneisses and marbles, mirror images are made by polishing the two adjacent faces of slabs made with a single saw cut. When mounted side by side on a wall or floor, symmetrical shapes are created.

1.6.3.4 Durability and strength. An essential characteristic of a building stone is its durability. Except for facing stones which are thin and must have an appreciable tensile resistance, strength is not a primary criterion, and many widely used building stones are quite weak. The porous fragmental limestone on the island of Bermuda is friable and so soft that it is cut by chain saws into blocks and even thin slabs for use as roofing tiles. Portland stone and Bath stone in southwest England are also limestones of high porosity and low strength. Their weakness and the absence of hard, abrasive minerals make them easy to cut and dress. Although weak, these rocks are durable and have lasted in buildings for many centuries, but they are now threatened by acid rain and urban pollution (Winkler, 1978).

The Gothic cathedrals of the Rhine valley, such as the Munster in Freiburg and the cathedrals of Strasbourg and Thann in Alsace, are made with local sandstones. Notre Dame cathedral in Strasbourg is made from the grès rouges des Vosges, a red sandstone that is easy to cut and shape. Because it weathers quite rapidly, continual replacement of blocks is required, and a permanent rock dressing facility exists in Strasbourg.

Several aspects of durability require testing using methods described in *Rock Engineering*, Sec. 2.3.3, and discussed further below. In northern climates, frost action is probably the most important cause of damage. In coastal regions, where stone is used extensively for the construction of sea walls, expansions caused by salt crystallization are equally disruptive and have a similar mode of action.

Very porous stones are usually resistant to frost and salt because of their large pores which permit drainage and preclude high internal stresses when the water within the stone freezes. Low-porosity rocks, 1 percent or less, are similarly durable because they do not become saturated, and because the freezing point of moisture in the pores is depressed as a result of small pore size and capillary tension. Also, dense rocks are stronger and more resistant to any expansive forces that develop. The rocks most susceptible to frost and salt crystallization damage are those of intermediate porosity and pore size.

An increasingly important aspect of durability is a stone's resistance to attack by acid rain and airborne pollutants. This problem is most acute in the case of limestones which react with the weak carbonic and sulphuric acids in rainwater. Resistance can readily be tested by the stone's reaction to acids in the laboratory.

Potential new sources are evaluated by examining their microstructure and comparing them to widely used stones, and by accelerated weathering tests. Cubes or cores are partially immersed in a salt solution and exposed to drying or freezing and thawing. The mechanisms of accelerated weathering are similar to those that develop in the magnesium sulphate soundness test for aggregates, and this method offers an alternative and more standardized method for measuring durability.

Large-scale features not present in small laboratory test specimens must be considered. Weathering of partings and development of iron staining by oxidation are important in decorative stone evaluation. These can cause splitting of larger blocks upon weathering, particularly when they are shaley or contain expansive minerals. The durability of a building stone becomes known over long periods in use. Another way to assess these subtle features is to observe the way the rock has weathered in old outcrops and quarry walls.

Resistance to wear is a requirement of paving stones, which are manufactured usually from abrasion-resistant igneous rocks or quartzitic sandstones. Various specialized testing methods are available to measure wear resistance. Quartz content, however, can often be taken as a reliable indicator of resistance to wear, and its measurement requires only a thin section and a polarizing microscope (*Rock Engineering*, Chap. 2).

Special applications call for unusual types of stone; curling stones, for example, have to be particularly tough to withstand impact. Most are made from microgranite quarried from the island of Ailsa Craig in the Firth of Clyde, Scotland. This rock has a texture of interlocking feldspar laths that makes it unusually tough. To determine whether the granite is tough enough, a weight is dropped from increasing heights until a cylindrical specimen fractures. Even in the well-known Ailsa Craig rock, microfissures have led to manufacturing problems. Ultrasonic and radar methods, dye penetrants, ultraviolet light, and magnetic particle procedures can be used to detect microfractures so that flawed stone can be rejected before the start of expensive machining.

1.6.4 Quarrying methods

1.6.4.1 Extraction. Building stone blocks must be separated from the mass without fracturing. Primary cuts separate panels of 1000 to 3000 m^3, which are divided by secondary cuts into vertical blocks of 100 to 300 m^3, which are then overturned and cut down into commercial blocks. Quarry benches may be as high as 10 to 15 m.

Blasting is widely used in dimension stone quarries of strong silicate rocks. Traditionally black powder, not dynamite, was employed, splitting the rock by gas expansion rather than by shock wave propagation. The black powder method is giving way to smooth blasting with high explosives (small diameter decoupled cartridges or detonating cord) and with a 150- to 300-mm blasthole spacing (Piga and Pinzari, 1984).

In many quarries, particularly those in softer rocks, the blocks are removed by *wedging* or *sawing* (*Rock Engineering*, Sec. 14.3). Stone masons with long experience in working with a particular source of rock develop great skill in recognizing the "grain" of the rock, and make use of its anisotropy and the almost invisible planes of weakness. In *plug-and-feather* splitting, shallow holes are drilled in a line, and "feathers" of steel are inserted between which "plugs" are driven.

The wedge action generates a fracture that separates the block from the rock mass. The operation in large quarries is often mechanized using hydraulic splitters.

In the Swedish system, a line of holes is drilled, thus avoiding the risk of an uneven break. A bar is inserted in an adjacent hole to ensure parallel drilling. Either the holes can touch, or they can be closely spaced and the rock bridge broken using a broaching tool (Shadmon, 1989).

Wire sawing can be used to separate large masses of rock, and more marble is produced by this method than by any other (Fig. 1.10). In a typical operation, parallel adits or boreholes are driven into the base of a rock cut and pulley wheels are installed. A braided (helicoidal) wire cable looped around the pulleys and against the base of the rock face is fed with an abrasive silica sand and water to give a continuous undercutting action (Piga and Pinzari, 1984).

This type of wire saw is being superseded by diamond beads mounted on a wire, which reduces wastage and gives straighter smoother cuts (Pinzari, 1983; Berry et al., 1988). Diamond wire sawing was introduced in the Carrara marble quarries in 1978. The beads are strung 30 to 40/m of wire, separated by springs or plastic spacers. The cable is tensioned and driven at 20 to 40 m/s. The method can undercut a block at a rate of 5 to 10 m^2/h. A water stream cools the beads and removes dust (Fig. 1.14).

1.6.4.2 Dressing of stone. Further splitting, sawing, chiseling, grinding, and polishing are required for many architectural applications. Dressing of stone is commonly done at facing stone plants, which are modern, automatic facilities at central locations.

Depending on requirements, the stone may have rough-split, sawn, ground, or polished faces. *Ashlar* blocks are dressed on five faces and rough on the sixth. Types of finish include rock or pitch faced (a naturally broken face with no tool marks), fine pointed (projections removed leaving chisel or pick marks), drafted margin (edges chiseled around a rock-faced block), sawn (with saw grooves), bush hammered (with a hand-tooled pitted surface), sand blasted (bringing out the natural textures of a sedimentary rock), honed (velvety smooth with no gloss), or polished (which reveals minerals and fossils). Further information on surface dressing is given in Shadmon (1989).

For sedimentary rock, frame and gang saws fed with abrasives in suspension are used in the traditional method of dividing a block into slabs. Modern reciprocating or rotary saws are more often equipped with tungsten carbide, silicon carbide, or diamond-impregnated cutting edges. Diamond-impregnated blades, disks, or wires are used mainly for the

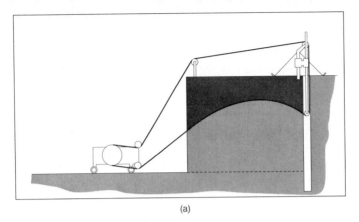

(a)

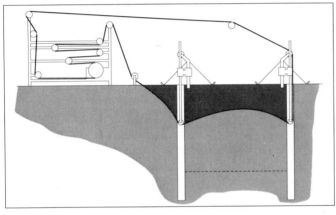

(b)

(c)

Figure 1.14 Diamond wire cutting techniques. (*a*) Traditional cut into the rock face; (*b*) cut into the floor (from Pinzari, 1983); (*c*) face cut. (*Photo, R. Ribacchi*)

stronger rock types. Multiple wire saws can be used to cut many slabs simultaneously to the correct thicknesses for grinding and polishing.

The better quality slates have well-developed, uniform planar *cleavage,* and can be split (cleaved) into plates just a few millimeters thick. These *roofing slates* are typically fine-grained, having been formed by low-grade metamorphism of shales with a high clay and low quartz content. The sandier varieties of slate cannot be cleaved in this manner although they can often be split into thicker slabs, usually with an irregular, wavy surface that can be attractive in architectural applications.

Grinding and *polishing* expose mineral grains and microtexture. Smoothing or *gritting* of the stone surface is achieved with three to six successively finer grades of silicon carbide or carborundum abrasive powder applied in a water slurry to the surface of large lapping equipment. In one type of radial arm polisher, a rotating polishing disk on a universal head can be swung by an operator over the stone surface. Water is fed through the polisher spindle. Automatic versions have interchangeable facings for the polishing head, and track automatically across the surface.

Facing stones are equipped with drilled epoxied hooks to hoist and hang the slabs on walls. Flexible silicone rubber sealants prevent the development of excessive stresses caused by heating or cooling, and the exclusion of moisture from the back of the slabs preserves the stone's appearance, and reduces the chances of splitting.

References

Aitcin, P. C., Jolicoeur, G., and Mercier, M. 1983. *Technologie des Granulats.* Les Editions du Griffon d'Argile, Paris, 372 pp.
Arquie, G., 1980. *Granulats.* Editions Anciens, Ecole Nationale des Ponts et Chaussées, Paris.
ASTM, 1978. *Significance of Tests and Properties of Concrete and Concrete Making Materials.* American Society for Testing and Materials, ASTM STP 169-B, Chap. 33 (pp. 540–572) and Chaps. 35–42 (pp. 584–761).
Barton, W. R., 1968. *Dimension Stone.* U.S. Bur. Mines Information Circular 8391, 147 pp.
Bauer, A. M., 1970. *A Guide to Site Development and Rehabilitation of Pits and Quarries.* Ontario Dept. Mines Industrial Mineral Report No. 33, 62 pp.
Berry, P., Bortolussi, A., Ciccu, R., Manca, P. P., Massacci, G., and Pinzari, M., 1988. "Optimum Use of Diamond Wire Equipment in Stone Quarrying." *Proc. Conf. on Applications of Computers in Mining,* Las Vegas, Nev., 15 pp.
Bérubé, M.-A., and Fournier, B. 1987. "Le Barrage Sartigan dans la Beauce (Québec), Canada: Un Cas-Type de Détérioration du Béton par des Réactions Alcalis-Granulats." *Can. J. Civ. Eng.,* vol. 14, no. 3, pp. 372–380.
Bragg, G. H., and Zeigler, T. W., 1975. "Design and Construction of Compacted Shale Embankments," vol. 2, *Evaluation and Remedial Treatment of Compacted Shale Embankments,* Rept. FHWA-RD-75-62, U.S. Federal Highways Administration, Washington, D.C.
Brandl, H., 1980. "Construction and Compaction of 100–120-m-High Highway Embankments." *Proc. Int. Conf. on Compaction,* LCPC, Paris, vol. 1, pp. 221–226.

Brauns, J., Kast, K., and Blinds, A. 1980. "Compaction Effects on the Mechanical and Saturation Behavior of Disintegrated Rockfill." *Proc. Int. Conf. on Compaction,* LCPC, Paris, vol. 1, pp. 107–112.

Bruun, P., (ed.), 1985. "Design and Construction of Mounds for Breakwaters and Coastal Protection." *Developments in Geotechnical Engineering,* no. 37. Elsevier, 938 pp.

Campbell, F. B., 1966. "Hydraulic Design of Rock Riprap." Department of the Army Corps of Engineers, Waterways Experiment Station, Misc. Paper no. 2-777, Vicksburg, Miss.

Diamond, S., 1975, 1976. "A Review of Alkali-Silica Reactions and Expansion Mechanisms." *Cement and Concrete Res.,* no. 5, pp. 329–346; no. 6, pp. 549–560.

Dusseault, M. B., Scott, J. D., and Moran, S., 1985. "Smectitic Clays and Post-Reclamation Subsidence of Strip-Mined Areas." *Clay Minerals Science,* vol. 1, pp. 163–172.

Eerola, M., and Ylosjoki, M., 1970. "The Effect of Particle Shape on the Friction Angle of Coarse-Grained Aggregates." *Proc. 1st Int. Conf. Int. Assoc. Eng. Geol.,* Paris, vol. 1, pp. 445–456.

Fournier, B., and Bérubé, M.-A., 1989. "Alkali-Reactivity of Carbonate Rocks from the St. Lawrence Lowlands (Québec, Canada)." *Proc. 8th Int. Conf. on the Alkali-Aggregate Reaction.* Kyoto, 6 pp.

———, and Vezina, D., 1987. "Condition Survey of Concrete Structures Built with Potentially Alkali-Reactive Limestone Aggregates from the Québec City Area." *Proc. Mather Int. Conf. on Concrete Durability,* Atlanta, Ga., ACI-SP-100, pp. 1343–1364.

Fourmaintreaux, D., 1970. "Interprétation Minéralogique de la Résistance au Polissage des Roches." *1st Int. Conf. Int. Assoc. Eng. Geol.,* Paris, vol. 1, pp. 344–355.

Franklin, J. A., 1983. "Evaluation of Shales for Construction Projects—An Ontario Shale Rating System." Rept. RR229, Ontario Ministry of Transportation and Communications, Research & Development Branch, 99 pp.

Godfrey, J. D., 1979. "Chipewyan Granite—A Building-Stone Prospect in Alberta." *CIM Bulletin,* May, pp. 105–109.

Hale, B. C., Lovell, C. W., and Wood, L. E., 1981. "Factors Affecting Degradation and Density of Compacted Shales." *Proc. Int. Symp. on Weak Rocks,* Tokyo, Japan, vol. 1, pp. 321–326.

Hosking, J. R., 1970. "Road Aggregates and Their Testing." *Proc. Symp. on Quarrying,* Bristol Univ., England, 13 pp.

Huschek, S., and Angst, C., 1980. "L'influence du Compactage sur les Propriétés Mécaniques des Enrobés." *Proc. Int. Conf. on Compaction,* LCPC, Paris, vol. 1, pp. 405–410.

Kerisel, J., 1987. *Down to Earth. Foundations Past and Present: The Invisible Art of the Builder.* A. A. Balkema, Rotterdam, 149 pp.

Lefond, S. J., 1983. *Industrial Minerals and Rocks.* (5th ed., 2 vols.), Soc. Mining Engrs., Am. Inst. Min. Met. Petrol. Engrs., New York, vol. 1, 722 pp., vol. 2, 1446 pp.

Lo, K. Y., 1982. *Rockfill in the Foundation Design of Highway Structures.* Ontario Ministry of Transportation and Communications, Res. Rept. RR227, 50 pp.

Lovell, C. W., 1983. "Standard Laboratory Testing for Compacted Shales." *Proc. 5th Int. Cong. Rock Mech.,* Melbourne, Australia, vol. A, pp. 197–202.

Mercer, J. K., 1968. "Some Considerations Involved in Opening up a Quarry." *Quarry Manager's J.,* April, pp. 143–153.

Mielenz, R. C., 1963. "Reactions of Aggregates Involving Solubility, Oxidation, Sulfates and Sulfides." *Highway Research Record,* no. 43, pp. 8–18.

Moebs, N. N., Sames, G. P., and Marshall, T. E., 1986. *Geotechnology in Slate Quarry Operations.* U.S. Bur. Mines Rept. Investigations RI9009, 38 pp.

Penman, A. D. M., 1971. *Rockfill.* Building Research Station (U.K.), Current Paper CP 15/71, 10 pp.

Piga, P., and Pinzari, M., 1984. *The Quarrying of Ornamental Stones in Italy* (in English), Bollettino della Associazione Mineraria Subalpina, Anno XXI, no. 1–2, pp. 48–72.

Pinzari, M., 1983. "Quarrying Stone by Diamond Wire in Italy." *Industrial Diamond Review,* vol. 5/83, pp. 231–236.

Py, L., and Grange, J. P., 1979. "Talcs de Luzenac." *Industrie Minérale,* 61, 7, pp. 371–383.

Rasolofosaon, P., Lagabrielle, R., Rat, M., and du Mouza, J., 1983. "Reconnaissance de Formations par Étude Fréquentielle de Signaux Sismiques." *Bul. Int. Assoc. Eng. Geol.*, pp. 26–27, pp. 285–293.

Rogers, C. A., 1979. "Alkali Aggregate Reactions, Concrete Aggregate Testing and Problem Aggregates in Ontario—a Review." *Ontario Ministry of Transportation and Communications Rept.* EM-31, 27 pp.

———, 1983. "Search for Skid Resistant Aggregates in Ontario." *Proc. 19th Forum on Geology of Industrial Minerals*, Ontario Geological Survey, Miscellaneous Paper 114, pp. 185–205.

———, 1986. "Evaluation of the Potential for Expansion and Cracking of Concrete Caused by the Alkali-Carbonate Reaction." *Cement, Concrete and Aggregates*, CCAGDP vol. 8, no. 1, pp. 13–23.

———, and Magni, E. R., 1987. *Assessment of Construction Aggregate Suitability of Rocks Underlying the Toronto Area*. Ontario Ministry of Transportation & Communications, Eng. Matls. Rept. EM 83.

Selig, E. T., and Yoo, T. S., 1980. "Compaction of Railroad Ballast." *Proc. Int. Conf. on Compaction*, LCPC, Paris, vol. 2, pp. 463–468.

Shadmon, A., 1989. *Stone, an Introduction*. Intermediate Technology Publications, London, England, 140 pp.

Sherard, J. L., Woodward, R. J., Gizienski, S. F., and Clevenger, W. A., 1963. *Earth and Earth-Rock Dams*. John Wiley, New York.

Shergold, F. A., 1960. "The Classification, Production and Testing of Roadmaking Aggregates." *Quarry Manager's J.*, vol. 44, no. 2.

Tourenq, C., and Denis, A. 1982. "Les Essais de Granulats." *Rapport de Recherche* LPC no. 114, LCPC, Paris, 87 pp.

U.S. Army Corps of Engineers, 1987. *Hydraulic Design of Navigation Dams*. Engineer Manual EM 1110-2-1605.

U.S. Department of Transportation, 1967. *Use of Riprap for Bank Protection*. Hydraulic Engineering Circular no. 11, U.S. Department of Transportation, Hydraulics Branch, Bridge Division, Washington, D.C., 43 pp.

Winkler, E. M., 1975. *Stone: Properties, Durability in Man's Environment*. 2d ed., Springer-Verlag, Vienna, New York, 230 pp.

———(ed.), 1978. *Decay and Preservation of Stone*. Eng. Geol. Div., Geol. Soc. Amer., Boulder, Colo., 104 pp.

Chapter

2

Landslides, Open-Pit Mines, and Rock Cuts

This chapter addresses the stability of slopes created by geological erosion and also the design and stabilization of excavated slopes. Excavations are created either for open-pit mining or for civil engineering projects such as construction of highways or canals.

2.1 Failures of Natural Slopes

2.1.1 Extent of the hazard

A landslide hazard can take one of two forms (Fig. 2.1) depending on the relative positions of the slope and the threatened structure or community:

- *Hazard from above,* where down-slope structures are in danger of being destroyed by rockfall, slide, or avalanche material (Fig. 2.2).
- *Hazard from below,* where structures sited on a slope or near its crest are threatened by progressive undermining of their foundations as a result of landslides or creep. This then becomes a foundation problem (Chap. 3).

A large landslide can be a disaster as severe as an earthquake or volcanic eruption. A Norwegian study group estimated that between 1876 and 1976 more than 1400 people in that country were killed by sliding rock, soil, snow, or ice. Half of the deaths were in snow avalanches, one-quarter in rock falls, and one-quarter in earth slides. The greatest damage was caused by slides into fjords or lakes, which set up

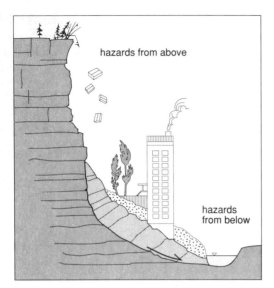

Figure 2.1 Landslide hazards from above and below.

flood waves as high as 75 m. In France in the year 1248, the Granier mountain avalanche in the Alps completely destroyed several villages (Panet and Rotheval, 1976). Similar disasters can be cited for mountainous regions throughout the world.

With the spread of urbanization, flat land is used up first; then, cities grow and swallow up surrounding unstable hillsides, and advance into steep-sided valleys in the path of rockslides and mudflows. In Los Angeles some of the most attractive coastal hillside sites are susceptible to sliding, and for California as a whole, the annual cost of landslide damage amounts to about 4 times the world average.

Geomorphology is the study of the changing shape of the earth's surface. Natural slopes are created by the interaction of geological uplift and erosion. Hills and mountains are formed and broken by sun and frost, sculpted by water, wind, and glaciation, and the fragments are washed down rivers to be deposited in lakes and seas. Processes are most rapid where the rates and amounts of uplift and erosion are greatest, along recently formed mountain chains, active fault scarps, and sea coasts. In tectonically active areas, the land is being raised faster than it can be eroded. Consequently, vast tracts of hillsides have marginal stability, and the landscape evolves mainly as a result of landslides at various scales.

Accumulating evidence from aerial photographs and surveys show the true importance of landsliding in mountainous areas; in a study of just 880 km^2 of the Canadian Rockies in Alberta, 228 landslides were

Figure 2.2 High-rise apartments in Hong Kong, in the shadow of steep slopes in weathered rock.

identified, with 8 km² of slide debris and 96 km² of talus, most of which had probably accumulated from small block falls and toppling (Cruden, 1985; Cruden and Eaton, 1987). All this happened since deglaciation, about 10,000 years ago.

2.1.2 Landslide dams

Masses of earth and rock deposited by landslides often block the valleys in which they come to rest. The resulting *landslide dams* obstruct rivers to form reservoirs, but differ from constructed dams in several important aspects. They are typically much wider and encompass larger volumes of material. The 1959 Madison Canyon rockslide that dammed the Madison River in Montana forming Earthquake Lake had a base width 5 to 8 times as great as would have been used in building a rockfill dam of the same height. An extreme example is the debris avalanche blockage of Spirit Lake at Mount St. Helen's, which extends 24 km downstream from the lake. An excellent review of landslide dams and their consequences is provided by R. L. Schuster (1986).

An impoundment may last for several minutes or several thousand years. Sooner or later, nearly all landslide dams are overtopped by their impounded lakes, and usually this leads to destruction of the dam and to downstream flooding of disastrous proportions. Most landslide dams are remarkably short-lived. In 63 cases documented by Schuster, 22 percent failed in less than 1 day after formation, and half failed within 10 days.

One of the first recorded flooding catastrophes occurred in central Java in 1006. The southwestern part of the cone of Merapi volcano sheared off in a large rockslide that created a dam behind which a flourishing countryside, famed for its Hindu temples and monuments, was submerged by a deep and extensive lake. Casualties of some of these floods have reached into thousands. For example, at least 2400 people died in the 1933 flood impounded by the Deixi landslide dam on the Min River in central China.

These hazards continue. Professor Sun Guangzhong of Beijing reports the case of Liangziya cliff on the south bank of the Yangtze River. Large-scale avalanches occurred in 100, 377, 1030, and 1542. The slide in 1030, precipitated by an earthquake and torrential rain, blocked the Yangtze River's navigation for 20 years, and the 1542 slide blocked the river for 82 years. The cliff is 700 m long and now contains more than 30 long and deep fissures, obvious signs of movement. The State Council has set up a study group to address the problem.

Some landslide dams are naturally stable, whereas others are protected against failure by the installation of engineered control measures. The simplest and most commonly used method to prevent overtopping is the construction of spillway channels either across adjacent bedrock abutments or over the landslide dam itself. An example is the spillway through adjacent bedrock constructed by the U.S. Army Corps of Engineers in 1981 to control the level of Coldwater

Lake, one of the impoundments of the 1980 Mount St. Helen's debris avalanche.

2.2 Engineered Slopes

2.2.1 Civil engineering excavations

Typical engineered slopes are rock cuts along highways and railways through hilly or mountainous terrain. Transportation corridors must maintain acceptable gradients through rugged topography, requiring embankments of fill (compacted earth or broken rock) and slopes cut into the natural earth and rock materials. A balance is maintained between the quantities of cut and fill by excavating only the amount needed for constructing embankments. The ideal route minimizes haulage distances and the costs of separate quarrying and disposal operations, while reducing risks of rockfalls and slides to an acceptable level. Some stability is achieved within the initial construction budget, and often the remainder is developed over the years as part of the long-term costs of highway maintenance.

The engineer can control the side slope and quality of fill in an embankment, and thus its stability. However, the stability of cut slopes depends not only on slope angle, but also on natural conditions that vary from place to place. Hillside excavations are troublesome because they undercut weathered and weakened natural slopes that are often marginally stable. Ancient sliding surfaces, perhaps hundreds or thousands of years old, now buried and hidden, can be reactivated by excavation and by changes to the surface or underground drainage pattern. The discovery of old slides can disrupt planning and increase construction costs.

Many other types of construction require the design of stable slopes in rock; these include spillway channel walls, dam abutments, and tunnel portals. Basements and underground parking for buildings often have to be excavated vertically next to existing structures and services. Little or no movement can be tolerated, much less full-scale sliding.

2.2.2 Open-pit mines

Open-pit mine excavations (Fig. 2.3) are higher and more extensive than those in civil works; they are mined as steeply as possible to avoid excavating waste rock. Also, trends continue toward deeper pits and steeper slopes in the search for more economical extraction of minerals and rock products. Pit slopes can tolerate much greater slope movements than civil structures since they have to stand stable for

Figure 2.3 Kidd Creek open-pit mine, Timmins, Ontario. Open pit operations have now ceased and extraction continues underground. (*Photo, Kidd Creek Mines Ltd./Texasgulf*)

just a few months or years. Mining may even continue while slides are actively moving, sometimes by 1 or 2 m per day.

A survey found that more than half of the open-pit mine slopes in the United States had large wall slides, often more than one active at any particular time (Ross-Brown, 1973). Of 104 operating pits surveyed, 35 had minor slope problems and 27 had serious ones. Five out of the 40 inactive pits had been closed prematurely because of slope problems, at an estimated cost of between $7 million and $20 million for the closure of a single mine.

Unusually difficult conditions were encountered when mining for copper on the island of Bougainville in Papua-New Guinea. The open pit was excavated into intensely fractured and weathered volcanic rocks. Problems included an average rainfall of 5 m/year, and up to 10 percent of the world's earthquake occurrences (Baumer et al., 1973). Limiting equilibrium analysis led to a curved slope profile for a pit wall up to 1000 m high, with overall slopes of 45° in competent rock, but 30° to 35° in weathered rock and at the foot of the highest slopes where the highest stresses prevailed. Pit wall safety depended on an efficient drainage system, a recurring theme throughout rock slope engineering. The mine access road was built to normal highway standards through extremely rugged terrain, with design and construction running concurrently. Slides developed in the surface layer of soil, volcanic ash and weathered rock in such large quantities that the design had to be revised. The volume eventually excavated amounted to 10 million cubic meters of weathered superficial materials.

In addition to landslide hazards that may develop in intact or weathered rock in mines, instability in mine wastes constitutes a serious problem. These wastes may range from coarse-grained rock rubble dumped down a mountain slope to cast-back smectitic spoil from coal strip mines to mine tailings behind dikes (Stead and Singh, 1989; Coulthard, 1979).

2.3 Slide Mechanisms

2.3.1 Classification of slides

Slides can be classified according to the geometry of the sliding mass, for example, wedge, block, or slab slides, deep-seated or shallow rupture surfaces (Fig. 2.4). They can also be classified according to movement, distinguished in terms of velocity, as in creep, slip, or avalanche, or in terms of type of motion, as in planar, translational, rotational, or toppling failures (Varnes, 1978; Brunsden and Prior, 1984).

A distinction can also be made between the mechanisms that occur in hard and in softer rock types, such as the mud and debris flows characteristic of soils but also common in closely jointed rocks, the erosional failures found mainly in shale and weathered rock slopes, and the raveling that occurs in closely jointed and steep hard rock faces. Careful classification gives the rock engineer insight into mechanisms and possible corrective measures.

In stratified rocks, a classification of the slope can also be based on the orientation of the bedding with respect to the geometry of the slope; categories of dip and overdip slopes, reverse-dip slopes, oblique and strike-dip slopes, and underdip slopes have been used for this purpose (Selby, 1982; Cruden and Eaton, 1987).

2.3.2 Importance of jointing and rock strength

Slides in hard rock are governed almost entirely by movement along preexisting joints. Whether or not the slope will remain stable depends very largely on the joints and their orientations with respect to each other and to the slope face, their roughnesses, persistence and strengths, and the water pressures acting within them. Rock free from joints, in theory, could be cut vertically to heights of many hundreds of meters while remaining stable (*Rock Engineering*, Sec. 5.1.2), whereas

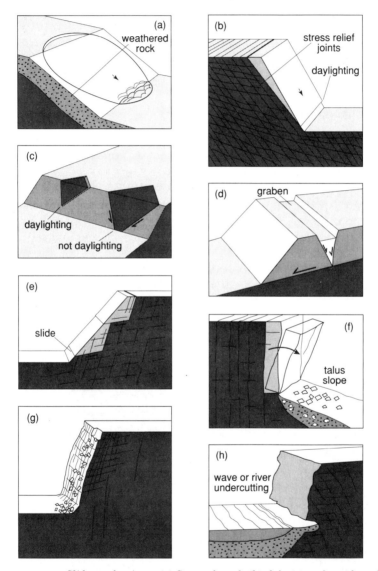

Figure 2.4 Slide mechanisms. (*a*) Spoon-shaped; (*b*) slab; (*c*) wedge; (*d*) graben; (*e*) stepped path; (*f*) toppling; (*g*) raveling; (*h*) erosion and undercutting mechanisms.

rock that is intensely fractured will in the limit behave much like a soil, adopting an angle of repose of 32° to 38°.

Slides in soft rocks can occur by the generation of new sliding surfaces through intact material, although even in the weakest rocks, the majority of the failure plane is usually along preexisting joints. The sliding surface may occasionally resemble the classical slip circle of soil mechanics; more often a shallower but nevertheless curved surface is seen, such as in shale slopes along the banks of the Panama Canal during its construction. In near-horizontal bedded strata, block slides with crest grabens are common.

2.3.3 Translational rockslides

2.3.3.1 Spoon-shaped and stepped-path slides. Deep-seated slides in jointed rock are often spoon-shaped like their counterparts in soil (Fig. 2.4a and 2.5). Often the sliding surface is formed by two or more joint sets in the form of "risers" and "steps" in a staircase pattern, called *stepped path failure* (Fig. 2.4e). Barton (1972) suggests that these slides often are brought about by the development of an overstressed zone leading to progressive failure as discussed in Sec. 2.3.3.4.

Slopes in soft rocks such as the smectite-rich clay shales beneath the eastern foothills of the Rocky Mountains present a particular challenge (Morgenstern, 1977). The river valleys, having been eroded in geologically recent times, are only marginally stable and are cut

Figure 2.5 Deep-seated spoon-shaped slide in clay-shales, Alberta, Canada.

many times by highways and railways that steepen the slopes. Many residential districts are built at desirable valley crest locations, which are often among the most susceptible to landsliding.

Aerial photographs of the Missouri, Peace, Athabasca, and the North and South Saskatchewan river valleys show individual landslides that include the entire valley wall, which is perhaps 200 to 400 m high, and "single" complex slides that extend for 5 to 10 km along the river (Mollard, 1977). In some areas, the entire valley wall is made up of disturbed terrain, so it is difficult to distinguish individual slides. Geomorphological studies show valley slopes to average 33° in the foothills, progressively reducing to 16° to 20° at a distance of 250 km, and 7° to 10° at distances of 700 km from the mountains. Associated with this is a change of clay mineralogy from smectite in the plains to illite and chlorite in the high foothills. Cementation is greater and the density is higher in the more competent rocks nearer the mountain front.

2.3.3.2 Slab and wedge slides. Slab and wedge slides are movements of rigid blocks on planar joint surfaces. Slab slides in particular are often small and shallow, containing only a few rock blocks resting on a single basal plane. Slabs are formed when joints strike more or less (within about 20°) parallel to the slope face (Fig. 2.4b). The key features controlling slab sliding are often either bedding joints in a sedimentary rock, or stress-relief (*sheeting*) joints that develop parallel to free surfaces for example in granite terrain (Fig. 2.6).

A wedge slide is three-dimensional resting on two joint surfaces, and can encompass a larger volume of rock. Wedges are formed when the line of intersection of two joint sets dips at a lesser angle than the slope; that is, it "daylights" in the slope face (Figs. 2.4c, 2.7).

2.3.3.3 Graben slides. A slab slide may develop with a crest graben in which downward movement provides an outward thrust on the horizontally slipping block (Fig. 2.4d). The mechanism allows sliding on weak flat-lying bedding planes, and the affected mass can be substantially greater than in a simple slab slide. In the coal strip mines of Forestburg southeast of Edmonton, slopes 25 to 30 m in height with a 60° to 70° angle become unstable, developing a series of horizontally outward moving blocks with grabens that may pond water and thus aggravate the problem. Backslope tension cracks in these materials have been found at a crest distance of 5 times the slope height (Stimpson et al., 1987).

2.3.3.4 Progressive mechanisms. Many of the slides in softer rocks occur even when their shear strength is much greater than the calculated average shear stress along the sliding surface (Bjerrum, 1967).

Figure 2.6 Sheeting joints parallel to valley wall.

This is caused by a mechanism known as *progressive failure* in which the sliding surface, instead of shearing all at once, propagates from toe to crest. Shear stresses are initially highest near the slope toe, and remain concentrated around the propagating "crack tip" at a level much higher than the "average" usually assumed for purposes of design. The localized stress concentrations exceed the peak strength of the shale, which then becomes weakened, often to a residual strength condition (*Rock Engineering*, Sec. 11.2.5).

Banks (1972) reports that for slides in clay shales along the upper Missouri River and Panama Canal, back-analysis indicated a large decrease in shear strength with time following initial sliding, perhaps caused in part by progressive failure, and in part by weathering and softening of shales along the rupture plane.

Morgenstern (1977) describes a similar progressive mechanism. Erosion and the creation of valleys is accompanied by unloading and

Figure 2.7 Wedge slide in road cut near Trondheim, Norway. (*Photo courtesy E. Broch*)

uplift. This causes bedding plane slip along the weakest layers, bringing them to a condition where all cohesion has been lost, and the frictional resistance reduced to a minimum. A slide is triggered when seepage water enters the crestal tension cracks, or when river erosion undermines the toe. Often much of the slide surface lies within a single horizontal stratum 10 to 20 mm thick, so that identification is very difficult, requiring great care to preserve and inspect core samples.

In the Australian Bowen Basin, weak, *slickensided* clay shale surfaces are common just above the coal seams; they are perhaps of geological origin but are enhanced by highwall differential straining during excavation. Mining causes inward propagation of a plane of strain-weakening, a "shear band." Reduced angles of shearing resistance of 7° to 13° are common, and cohesion has been destroyed along these surfaces (Brawner, 1974).

2.3.4 Rock avalanches and debris flows

2.3.4.1 Mechanisms. Avalanches are extremely rapidly moving slides. Although popularly associated with ice or snow, they often consist partly or completely of rock fragments that behave more or less like a liquid. The mechanism can be likened to that of a submarine turbidity current (also a landslide) in that the fragments are carried in a fluid that may be water, air, or a mixture of the two. Fluid trapped within the fast-moving debris keeps the particles in suspension, with near-zero effective stresses as in a liquified soil.

Ice avalanches occur where a glacier terminates on a smooth rock slope inclined steeper than about 30°. Ice-rock avalanches occur where an ice avalanche can entrain closely jointed rock or talus during its descent. *Rock avalanches* are often generated by the sudden detachment of a large, relatively intact rock mass along a well-defined rupture surface. Seismic shaking is a common trigger mechanism in mountainous terrain.

Where avalanches meet and mingle with water courses, they form *debris torrents (debris flows)*, streams of rock, mud, and water, often confined in channels and flowing at 10 to 40 km/h (Fig. 2.8; Hungr et al., 1984). In Japan, a seismically active country, debris flows occur every year and are the most harmful type of natural disaster (Ikeya, 1989).

A debris torrent in 1845 on the mountain Nevada del Ruiz in Colombia killed at least 1000 people. A similar event on the same mountain in 1985 triggered by a volcanic eruption killed at least 23,000 people and seriously affected the lives of another 200,000. It was a virtually liquid mass of rock, ice, and water that descended and entrained soil, vegetation, and more water in its deadly course to lower

Figure 2.8 Debris flow near Squamish highway, British Columbia (Hungr et al., 1984). *(Photo courtesy O. Hungr)*

elevations. These volcanically triggered liquid slides are called *lahars*. Contributing factors to these huge events are varied, the only consistent ones being mountainous terrain and a relatively high rainfall or snowmelt (VanDine, 1985; Voight et al., 1987).

2.3.4.2 Reach and volume. The "mobility," or "reach," of a dry rock avalanche appears to be much greater for large slides than for small. Davies (1982) plotted deposit length against volume for a large number of avalanche slides and found a linear relationship on a log-log scale:

$$L = 9.98 V^{0.33}$$

where L and V are length and volume of the debris in metric units.

When the volume of the slide exceeds perhaps between 100,000 and 1 million cubic meters, the free tumbling of individual blocks changes into a coherent streaming of disintegrating rubble. A large quantity can be carried many kilometers in this manner, and can climb a ridge several hundred meters high. The Frank slide (Fig. 2.9) and seven others in the Canadian Rockies are described by Cruden (1976). Each had a volume of more than a million cubic meters. The Frank slide debris dropped 1 km vertically and traveled at least 1.5 km horizontally, destroying half of a village.

Figure 2.9 Frank slide. (*Photo courtesy O. Hungr*)

2.3.4.3 Velocity. Whereas most normal rockslides travel at less than 4 km/h, catastrophic speeds greater than 10 and up to 400 km/h can be attained in a debris torrent or an avalanche. Avalanches at Elm and Goldau in Switzerland were seen to move at 140 to 250 km/h (Heim, 1932). The slide that precipitated the Vaiont disaster probably reached 90 to 110 km/h (Müller, 1968).

Large rock avalanches tend to be preceded by periods of accelerating creep. At Vaiont in Italy, episodes of movement corresponded to three attempts at raising the water level in the reservoir. On the first two occasions, the movement stopped as soon as the water level was lowered. The third time, after a displacement of nearly 4 m, the slide accelerated and surged through the reservoir and up the opposite mountain slope. The resulting wave overtopped the dam, which remained stable, but hundreds were killed by the catastrophic flow of water in the valley (Chap. 4).

Debris flow activity at Cathedral Mountain in the Rocky Mountains of British Columbia, Canada, began in 1925 and has since increased in frequency. A typical debris flow involves about 100,000 m^3 of material, moving at 5.5 m/s and discharging 210 m^3/s at the foot of the slope (Jackson et al., 1989).

When accelerating movements are preceded by creep, monitoring can be used to give warning. However, earthquakes can trigger an avalanche in seconds; in such cases timely warnings may be impossible.

2.3.5 Shallow failures in hard rocks

2.3.5.1 Toppling. Near-vertical slabs and columns of rock are often formed by steeply inclined joints that strike parallel to the slope crest

Figure 2.10 Toppling failure. Roadside cut (*Photo courtesy E. Magni*); Inset: Toppling diagram (discussed in Sec. 2.5.5).

(Figs. 2.10 and 2.4f). Toppling occurs when these are undermined and the center of gravity moves outside the area of basal support (De Freitas and Watters, 1973; Goodman and Bray, 1976). The slabs can rotate either independently, or under the action of forces transmitted by blocks higher up the slope. Toppling may be aggravated by water or ice pressure acting in the joints.

2.3.5.2 **Raveling and rockfalls.** Raveling is the continuous separation and falling of rock blocks from centimeters to a meter or so in size. Hazards can develop both from the cumulative undermining of the slope and from falling and rolling blocks which can reach substantial speeds and heights of bounce. Debris accumulates as a talus on the slope face and at the toe.

Raveling occurs in the outermost few meters of a rock face, where the rock is often closely jointed and weathered. Daily and seasonal heating and cooling, long-term relief of stresses, and the large forces generated by water pressures and ice formation cause joints to extend and the blocks to separate. Most rockfalls occur in the spring and fall at times of freezing, thawing, and high rainfall (Peckover, 1975, and Fig. 2.11). Fractures allow easier access for water, so that once started, the process of disintegration is an accelerating one. It is only stopped when the slope has regressed to a stable angle by the accumulation of a deposit of talus. If the talus is continuously removed by erosion and further sliding, the process continues.

Small block falls are more frequent than large ones. Frequency is governed by many factors, and can vary from zero to continuous raveling depending on local conditions.

Raveling is an important factor in oil sand slope stability

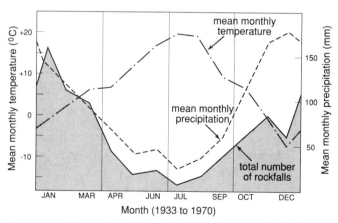

Figure 2.11 Seasonal variations of rockfall frequency in the Fraser Canyon, British Columbia, Canada (after Piteau and Peckover, 1978).

(Dusseault and Morgenstern, 1978). Closely spaced, stress-relief fractures develop parallel to the slope face and the slabs ravel and deteriorate. Talus removal by river erosion keeps slopes steep.

2.3.5.3 Bouncing and rolling. Rock can arrive at the base of a slope by free fall, bouncing, rolling, or sliding. Trajectories are difficult to predict, requiring estimates of coefficient of restitution (the energy absorbed per bounce) and of the breakage of the blocks. The shape of the block has a great effect on movement (Falcetta, 1985); subrounded blocks build up rotational momentum, whereas slabs do not.

Bouncing and rolling behavior was studied in field tests at Mont San Martino, Italy, where a limestone debris slide in 1969 cut through Route Nationale 36 and the nearby railway, reaching parts of the town of Lecco (Broili, 1977). A model was constructed at scale 1:160, but realistic scaling was difficult, and field confirmation was needed. Therefore, seven movie cameras were arranged to follow about 400 m^3 of blasted rock as it fell from a cliff 190 m high and rolled and bounced down a talus slope inclined at 24° to 36°. Three stages were identified: impact and disintegration, bouncing in parabolic paths more than 200 m long and up to 12 m high, and rolling and deceleration over the last 400 m. Average bounce height was between 4 and 8 m. Stones larger than 1 m^3 started rolling within the first 100 m after impact, but smaller ones continued to bounce. Stones of 0.3 to 0.5 m^3 reached velocities of between 40 and 120 km/h.

In a similar problem near the small town of Springdale, Newfoundland, houses were damaged by boulders which, after becoming detached from rock bluffs, bounced and rolled down a talus slope (Fig. 2.12). Some fragments actually penetrated the wall of a house. A child was killed by a loose block while trying to climb the slope. A first step in solving the problem was to diagnose it as one of raveling, aggravated by freeze-thaw loosening of microfissured rock (a fault zone is visible in Fig. 2.12a). Because of the microfissuring, the rock could not be drilled to install rockbolts. Instead, a catch system was designed consisting of a gabion wall topped by wire mesh fencing. At some locations, draped mesh and cable slings were employed. Design was aided by a computer model that simulated the heights and velocities of bouncing fragments.

2.3.6 Shallow failures in soft rocks

2.3.6.1 Creep. Creep is slow movement, usually evident only from instrument readings or damage to buildings and services. Usually it is confined to the upper few meters of weathered rock and soil where surfaces of shearing develop, often at the interface between weathered and unweathered material. Most creep is seasonal, with rapid move-

(a)

(b)

Figure 2.12 Raveling rock slope at Springdale, Newfoundland. (*a*) View from above showing the fault zone and the downslope community; (*b*) view from below, showing the rock bluff and talus slope. (*Photos, JAF, courtesy Nolan Davis & Associates Ltd.*)

ments at times of high water table and when provoked by freezing and thawing.

Creep velocities are typically a maximum of a few millimeters per hour or day. Contrast this with movements of 60 m/h measured in one of the Panama Canal slides, and with the "catastrophic limit" of 10 km/h for avalanches in the more brittle rocks.

2.3.6.2 Ice and clay damming and slab sliding. "Ice damming" caused by freezing of water in joints and pores in cold regions promotes a rise in the water table that can trigger instability. Water that is normally free to flow during the summer becomes trapped behind a frozen rock face during the winter. The accumulation of high water pressures can lead to rupture of the face slab or to a more deep-seated instability.

Similar "clay damming" is caused by the formation of a skin of clayey weathering products; the skin of weathered, clayey rock forms a dam, a natural barrier to the escape of groundwater. Volcanic rocks weather rapidly in a tropical climate, and can form a clayey skin of permeability lower than the open-jointed rock at greater depth. When the skin is sufficiently thick, and after a prolonged period of heavy rainfall, groundwater rises until the weathered skin explodes from the rock face. Escaping water mixes with rock and soil debris, forming debris torrents or mud flows that travel down the slope and into adjacent river valleys, often with catastrophic effect. Mud avalanches along valley slopes in many parts of South America are spectacular examples.

Ice and clay damming combine to cause small-scale seasonal slab slides along river banks in the shales of southern Ontario. Within a few decades of exposure, the slopes develop a skin of softened shale. Movement is encouraged by undercutting where the river meets the toe, and also seasonally by frost and a fluctuating groundwater table. The weathered slab is less permeable than underlying fresh rock, and creep movements further interrupt natural drainage. The mechanism follows a cyclic pattern; sliding leaves fresh rock exposed to weathering, and is followed by formation of a new skin.

2.3.6.3 Erosion. *External erosion* is caused by the action of water (more rarely wind or ice) on exposed rock surfaces. The two most common mechanisms are *undercutting* by a river or waves acting on the slope toe, and *gulleying* when surface runoff cuts into an unprotected slope face. Erosion is usually confined to soils, weathered rocks, and shales (Figs. 2.13, 2.14). However, jointed hard rocks can also be affected when scoured by rapid flow, such as in spillway channels where even hard rocks need reinforcement or protection (Chap. 4). When the rock formations contain soluble beds such as gypsum, *solution* can erode and undermine the slope.

Figure 2.13 Erosion gully in weathered rock and residual soil, Swaziland. (*Photo, JAF*)

Internal erosion generated by flow along water conduits within the rock mass seldom occurs except under the considerable hydraulic gradients that pertain at dam sites and adjacent to pressure tunnels. In such cases, groundwater can emerge from a slope face at a velocity sufficient to carry large pieces of strong and erosion-resistant rock material.

At the foot of a river bank or sea cliff, toe erosion steepens a slope and reduces its factor of safety (Fig. 2.4g). Slide debris is removed, which, if left to accumulate, would allow the slope to evolve toward a flatter, more stable condition. Thus, banks that are being eroded actively are steeper than their older counterparts, abandoned by river currents or sea waves. Taking measurements of actively eroding and abandoned cliffs cut into the London Clay Formation, Skempton (1964) established a correlation between slope angle and the decades or centuries since the last active undercutting took place. A similar historical pattern no doubt exists for slopes in shale and other weak rocks.

In a gullied slope face, down-slope flow, once concentrated in a slight crack or depression, removes the weathered products from its flow path. Slight depressions are thus deepened, and carry more water at higher velocity. This fast-flowing water in turn becomes more capable of removing weathered rock fragments. Rocks containing swelling clay are particularly sensitive to gullying; the Badlands of the northern United States, Alberta, and Saskatchewan are formed in these smectitic shales and sandstones.

(a)

(b)

Figure 2.14 Windrows of cast-back spoil from coal strip mines. (a) The topsoil is removed, then the overburden is stripped by dragline and cast into windrows, exposing the coal seam; (b) the windrows must be leveled and resoiled for reclamation. After 15 years, an unreclaimed strip mine traps ponds that remain murky from suspended clay

2.3.7 Triggering mechanisms

2.3.7.1 Water pressures. Evidently, in most of the above mechanisms, groundwater is one of the most important slope destabilizing agencies, and it follows that drainage is often the most effective remedy for instability (*Rock Engineering*, Secs. 4.2 and 18.1). Pressure of water causes uplift, thus reducing the effective normal stress and the resistance of joint surfaces to sliding. Also, tension cracks that develop parallel to the crest and penetrate to depths of many meters can fill rapidly with surface water, and hydrostatic forces that develop in them during periods of rainfall or snow melt can be considerable.

Figure 2.15 provides a unique demonstration of the sometimes very harmful effects of fluid in joints. Taken from the Canadian side of the Niagara Gorge looking toward New York State, it shows a large slab of rock falling onto a hydropower station. It was reported that just before the collapse, attempts were being made to grout the vertical jointing that strikes parallel to the crest of the gorge. The hydrostatic pressure of grouting, even without any additional pressure of injection, can generate very large forces that are often sufficient to trigger a collapse.

Horizontally bedded strata containing coal, sandstone, siltstone, and shale, often with *clay mylonites,* are exploited by *open-pit and strip mines* in Queensland, Australia; the Rhur valley in Germany; in Wyoming, the Dakotas, and Colorado in the United States; and in Alberta and Saskatchewan in Canada. If these mines are close to river valley walls, where the coal seams are well drained, slope stability problems are minimal. In regions where the water table is well above the pit floor, however, slides are common, particularly in mines where

Figure 2.15 Rock slab collapse onto power station, Niagara Gorge.

draglines excavate 70° slopes to expose the coal (Brawner, 1974; Christiansen and Whitaker, 1976; Richards et al., 1981).

A sudden increase in water pressure is often the final "trigger" that precipitates a landslide. Such is typically the case in Hong Kong, where densely populated urban communities sit on and below steep slopes in weathered volcanics and igneous rocks. Making matters worse, rainfall averages about 2.2 m annually, and often reaches intensities of 50 mm/h, occasionally 400 mm/day. A single rainstorm during May 1982 was followed by widespread landsliding and flooding with a loss of 28 lives. Air photographs after the event revealed more than 1000 separate slope failures (Brand et al., 1983).

Conditions of instability in urban areas can be aggravated by leaky pipework for sewage and water supply; as much as 10 percent of a city's water supply leaks from its underground distribution network through pipe joints and fractured mains. After a lengthy case in Edmonton, Canada, a court ruled that because the city could not account for all the water in the system, there was reason to accept the claim of a valley-crest homeowner that slope instability in shales had been triggered in part by water main leakage.

Rain and sewage infiltrations have been a key cause of failures of cliffs surrounding the ancient town of Orvieto, 100 km north of Rome (Cestelli-Guidi et al., 1983; Lembo-Fazio et al., 1984). The community is built on a volcanic tuff caprock 700 m wide by 1500 m long, whose vertical columnar joints form cliffs up to 60 m high. The caprock rests on weak Pliocene clays, which in turn are covered by a debris of rock talus on the lower slopes (Fig. 2.16). Records of instability dating back to the year 1280 show deep-seated landslides in the clays, undercut-

(a)

Figure 2.16 Cliff instability at Orvieto, Italy (from Lembo-Fazio et al., 1984). (*a*) View of Orvieto.

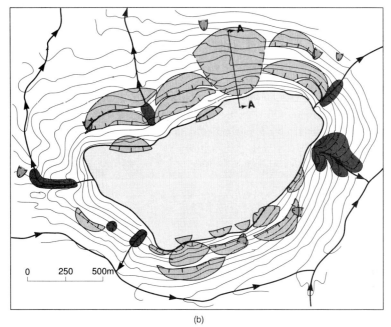

(b)

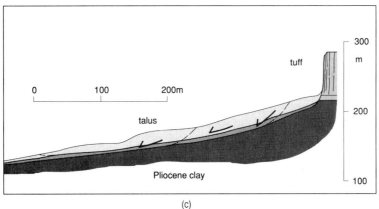

(c)

Figure 2.16 (*Continued*) Cliff instability at Orvieto, Italy (from Lembo-Fazio et al., 1984). (*b*) plan view of Orvieto Hill showing landslide scarps; (*c*) cross-section A-A of the Porta Cassia slide located to the northeast of the town.

ting the caprock and triggering toppling failures. Mudflows, creep, and erosion affect the talus deposits. Investigations have revealed a fourfold increase in domestic water consumption and losses into the ground in the preceding 100 years. A program of remedial work begun in 1980 has included rebuilding of the sewage and stormwater collection systems, drainage of the slopes and damming to control erosion in surface channels, doweling and anchoring of unstable rock in the cliff face, and revegetation of the lower slopes.

2.3.7.2 Earthquakes. In seismically active areas, a major earthquake usually will trigger large-scale landslides and rock falls. Not only do vertical and horizontal accelerations directly induce sliding, but also they give rise to surges in water pressure and contact stress normal to the joints, with a sometimes catastrophic reduction in the frictional forces that resist sliding. Also, earthquakes can modify external or internal slope drainage, causing, for example, blocking of drains by a sudden inflow of fine-grained material, permitting pore pressures to increase. Earthquakes may trigger temporary wave surges in reservoirs or lakes, adding to the effects of toe erosion.

2.4 Site and Route Investigations

Site investigation methods relating to rock cuts, natural slopes, and landslides are discussed here, following the general principles introduced in *Rock Engineering,* Chap. 6.

2.4.1 Investigation of active and ancient landslides

2.4.1.1 Mapping of hazards and risks. Landslide investigations are aimed at diagnosing the cause and extent of the problem, assigning appropriate land uses according to the degree of risk, and designing and estimating the costs of stabilization.

Hazards include the possibility of slides undermining the route and the risk of rockfalls or avalanches from above. Hazard maps assist in planning; they show the extent of threatening processes such as landslides, floods, and subsidence. Risk maps can be prepared to quantify the vulnerability of an area, often using a qualitative scale such as "high," "medium," and "low" risk.

Many countries now publish hazard maps for hilly urban areas. Perri (1983) describes the preparation of a landslide hazard map for a residential community in the hills southeast of Caracas, Venezuela, a region of calcareous schists containing graphite and mica. There 1000 km^2 were zoned to show different degrees of stability, using three-dimensional stereographic projection to evaluate joint set orientations in relation to variations in topography. Terrain classification techniques are used by the Geotechnical Control Office in Hong Kong to assist in planning, land management, and assessment of engineering feasibility (Styles et al., 1984). In many hazard-prone regions (California and Japan, for example) terrain in hilly areas is classified with respect to landslide hazards associated with earthquakes. These data sources are invaluable aids to design.

2.4.1.2 Evidence of sliding. Evidence of landslide movement may be visible on the ground, but is usually more apparent from textural variations in air photographs (Norman, 1970; Mollard, 1977). Oversteepened *crown scarps* and *tension cracks* are obvious features. They are often arcuate (curved), but may follow preexisting joints in a stepped pattern. Tension cracking is more likely to be visible on smooth pavements or grassed areas.

Slumped blocks and *hummocky or jumbled topography* are diagnostic of large landslips. Diverted streams and ponded water result from the interruption of preexisting drainage. The *patterns of vegetation* can be different on and off the landslide; species of trees are sometimes replaced by different species. Individual trees often slope uphill or downhill, or have curving trunks that can be used as evidence of movement as well as of the age of the landslide. In addition, the thickness of tree growth rings can change as a result of landslide disturbance.

2.4.1.3 Assessing activity and depth of movement. To analyze or stabilize a slide one needs information on the *mechanisms and rates of slope movement* and on the *locations of the sliding surfaces*. When a landslide appears dormant, the engineer must decide whether movement is likely to resume, particularly if conditions are to be modified by construction. A first step is to estimate how long ago the most recent movement occurred, for example, by observing the rounding of terraces, the filling of crevices with secondary deposits, and disturbances to the pattern of runoff and surficial drainage.

A common method to determine the depth of movement is to drill and sample the boundary between stationary and moving material, usually marked by a shear zone or by several subparallel slickensided surfaces. Finding this may require luck and careful drilling; moreover, exploratory drilling through broken and displaced rock or talus is particularly slow, difficult, and expensive. An alternative, when time is available, is to install inclinometers (*Rock Engineering,* Sec. 12.2.5) and to monitor movements for a few weeks or months. If active, the slide planes soon become apparent, and the movements can be measured.

In a study for a proposed cable car ropeway in the Gros Morne National Park, Newfoundland, careful examination of stereoscopic air photographs revealed a 100-million-ton landslip at the summit of the mountain. The quartzite block had become dislodged probably many thousands of years ago, and after moving several meters, had come to rest against a "keystone" of questionable long-term stability. Could the block start moving once more, and would the movement be gradual or rapid? Drilling to probe the depth and strength of the slide plane would have been very expensive at this remote location. Stability calculations were therefore initially based on data from surface geological mapping. In a parametric study, shear strength and groundwater conditions were varied within estimated limits. Because stability could not be guaranteed within the available budget for investigation, a cable car route was selected to avoid the path of any possible future movement.

2.4.2 Investigations for transportation routes

Rock excavations for highways, railways, canals, power lines, and pipe lines extend for considerable distances through a variety of terrain. Engineers must select the route and grade to avoid the worst of problems, including those of slope instability; the site investigation is therefore mainly aimed at identifying and avoiding hazards, and to a lesser extent at providing information for the design of individual cuts.

By adjusting the route and grade (elevation), often the choice is available to place a highway, pipeline, or canal below ground (in tunnel), above ground (on a bridge or viaduct), or at the surface. In this manner, the worst of naturally unstable slopes can be avoided. The choice is largely a matter of economics, balancing the extra costs of tunnels or bridges against the costs of landslide stabilization, protection, and maintenance.

Because of their extent, route investigations rely heavily on existing geological maps, and use inexpensive and rapid air photo and geophysical methods. Ground confirmation is obtained from a field reconnaissance, relying heavily on outcrops and test pits, supplemented by widely spaced drillholes at key locations.

The *Panama Canal* provides a classic example of poor to nonexistent site investigation being largely to blame for delays and greatly inflated costs. Construction called for excavation of the infamous Culebra Cut, 109 m deep, through the backbone of the Isthmus of Panama. By the time the Canal was opened in 1914, the French and American contractors had excavated around 170 million cubic meters of material, instead of the 46 million announced by de Lesseps (Kerisel, 1987). In the Culebra cut, the French withdrew after excavating only 14.5 million cubic meters, little realizing that after the Americans had finished, the total would amount to 75.5 million cubic meters. The almost vertical sides envisaged by the French Academy of Science became increasingly unstable as the cut deepened, and quickly gave way in landslides until they achieved a gradient of about 1:4. The rocks consisted of weak, faulted, and weathered shales with beds of lignite and swelling volcanic rock. "The more one dug, the more slides there were, and the more material was left to be excavated." In 1913, the canal was partially filled with water so that digging could proceed more economically with powerful dredgers. Soon after it was opened to traffic, small islands emerged here and there and in 1915 the canal was completely blocked by landslides from both banks. Occasional sliding continued until 1933.

2.4.3 Investigations for open-pit mines

Open-pit mine cuts and deep basement excavations are at predetermined locations, so the engineer has to make the best of existing conditions. The investigation is site specific, not regional. The focus is on predicting stable slope angles and on the requirements, costs, and benefits of stabilizing measures such as anchoring or drainage.

Slope design requires data on the orientations and strengths of joints, and on the water pressures acting in the jointing system (*Rock Engineering,* Chaps. 3, 4, and 7 through 11). For slopes yet to be excavated, postexcavation groundwater pressures must be predicted by extrapolating from current piezometric measurements using hydrogeological modeling and data on the hydraulic conductivity of the rock mass.

An optimum exploration program is designed to satisfy the needs both of ore evaluation and of geomechanics. Since the ore body is removed by mining, the final pit walls are largely or entirely in waste rock. Usually the ore and waste rocks have different strength and jointing characteristics. Thus, drillholes in the center of the pit intended mainly for assay purposes can provide data that is misleading if applied to pit wall design. Additional holes, drilled in and around the final pit perimeter, provide the missing information. All available holes can be logged, instrumented, and tested for geomechanics purposes using methods described in *Rock Engineering,* Chap. 6.

However, exploratory drilling often is not the best way to provide data for pit slope design. The best and cheapest data are obtained by direct observation when full-scale exposures become available during the early stages of mining, which proceeds in stages with initially flat and stable slopes. Observations made when overburden has been stripped and initial benches excavated are more reliable than properties inferred with difficulty from reconstructed drill core. The main application of the initial drilling is to establish a preliminary design for purposes of mine feasibility assessment, and to allow installation of monitoring and groundwater instruments.

2.5 Slope Design

2.5.1 Benches

Rock cuts of significant height are excavated in a series of stepped benches. These are a natural outcome of blasting, in that only 10 to 15 m of rock can be drilled and blasted with sufficient accuracy in any one round. Also, benches provide access for the equipment needed to excavate each slice or "lift," and allow one lift to be thoroughly scaled and bolted before the next lower one is removed. Another important function is to catch, or at least slow down, falling rock. This is a requirement during mining and for long-term public safety in the case of highway and similar cuts. Benches should be kept clear of broken rock: otherwise they become useless in protecting against falling material, and even act as a ramp for bouncing rock fragments.

In highway cuts, benches add to the cost of construction, but significantly reduce subsequent maintenance costs. The Colorado Department of Highways, for example, found that cleanup and maintenance work after rockfalls cost about 10 times the unit cost of the original excavation.

Stable bench heights depend on rock quality and the choice of blasting methods. The width is usually at least 7 m, depending on the size of equipment working on the bench. The "batter face angle" (bench angle as opposed to overall slope angle) is governed by the orientation of unfavorable joints to avoid excessive rockfalls onto the berms. Berm surfaces should be graded to assist in collection of water, and drainage ditches divert surface water away from problem areas. Ditches may need lining to limit infiltration.

In high cuts in open-pit mines, it may be necessary to design lower benches wider than upper ones to account for higher stresses. These benches may be mined toward the end of the useful pit life when the consequences of major instability are lessened.

2.5.2 Overall slope geometry

"Overall slope" refers to the steepness of a rock cut on a large scale, ignoring benches by drawing an imaginary line through their centerpoints. It is this slope that governs overall stability of an open-pit mine or deep highway cut.

The sequence of development of an open-pit mine often will permit the overall slopes to be gentle during initial mining and progressively steepened to achieve the lowest safety factor as the *final perimeter* is approached. Thus the mine engineer can gather both data and experience over time, and achieve final slopes much steeper than would otherwise be selected.

Gloryhole mining is an extreme case in which slopes are excavated almost vertically, and may be only marginally stable. Its purpose is to extract rich ore from the center of a pit with the least possible amount of low-grade ore and waste rock. Gloryhole mining is commonly a last phase, carried out in the knowledge that any instability that might develop is no longer of consequence. It should be employed with caution or not at all when the plan is to proceed from open pit to underground mining. In South African underground mines developed beneath a gloryhole, rim failures of wedge, toppling, or combined mechanisms have been known to adversely affect the underground operations (Bartlett and Raubenheimer, 1988).

2.5.3 Limiting equilibrium analysis

2.5.3.1 Sequence of design. Whereas individual benches are often dimensioned empirically, more rigorous methods are applied when designing the overall slope. Design usually employs the simple and powerful *limiting equilibrium* method (*Rock Engineering*, Sec. 7.2.6, and Hoek and Bray, 1973).

The two-dimensional methods common in soil mechanics may be inappropriate when applied to rock problems; they assume that a rupture plane can find its own circular or curved path through the ground. Rockslides are almost always bounded by noncircular surfaces and often by three-dimensional patterns of jointing. Analyses of planar sliding with optimally located tension cracks are provided by Kullmann and Barron (1989). Three-dimensional extensions of limiting equilibrium models are available for design where the conditions of the potential surfaces of sliding are known (Hungr, 1987). They are often limited to a linear Mohr-Coulomb strength criterion, which is not appropriate for rocks, but methods using curvilinear strength envelopes are now available (Charles and Soares, 1984).

The first step in a limiting equilibrium analysis is to assess which mechanisms of slope failure are most likely (slab or wedge sliding, etc.), and to identify the beds, joints, or faults that control movement. Persistent discontinuities such as faults, bedding planes, or old failure surfaces are more likely to form slide planes than impersistent or rough features (Priest, 1980). Schematic drawings are prepared to show the potential sliding masses, and their geometry is defined in the form of one or more simple models. Analyses of more complex mechanisms are also available, for example, the graben sliding described in Sec. 2.3.3.3 (Stimpson et al., 1987).

Rough calculations at this stage help to assess whether a problem exists, and whether a more detailed analysis is justified. A preliminary "kinematic" screening eliminates from further consideration those blocks that cannot slide because of interlocking, particularly those where the sliding surfaces and their intersections do not "daylight" in the slope face. Graphical methods are most often used, but computer methods are being developed to aid this screening process.

Kinematically possible slides are analyzed in a *parametric study* where worst and best estimates for each input parameter give upper and lower bounds for the computed "safety factor" of the slope (Salcedo and Tinoco, 1983). These calculations can be made using a pocket calculator or design charts and with approximate data for jointing, shear strengths, and water pressures determined from an initial reconnaissance of the site.

More rigorous calculations call for improved data. The parametric study serves to indicate which properties are the most critical and therefore most in need of careful measurement. New estimates are obtained by detailed joint mapping, by shear testing or back analysis of existing slides to determine shear strength, and by piezometric measurements and permeability testing to allow modeling of water pressures within the slope.

Analysis is complicated by the presence of discontinuous joints containing rock bridges alternating with joint segments. One approach is to replace joints and rock bridges by an equivalent continuous surface whose shear strength is a combination of joint and intact material properties. This can lead to an overestimate of the factor of safety, because rock bridges can rupture in tension rather than shear (Stimpson, 1978) or can be progressively destroyed by straining. In a more detailed analysis it is possible to incorporate statistical distributions of cohesion, friction, and tensile strength, leading to predictions of a probabilistic nature (CANMET, 1976).

2.5.3.2 Earthquake-resistant design. Assumed earthquake accelerations, typically 0.15 to 0.25 g, may simply be incorporated in static limiting equilibrium analyses as force vectors in the appropriate directions. In a more realistic simulation described by Ghosh and Haupt (1989), the seismically induced force vector acts in the direction of maximum ground movement, and is calculated at various times from a graph of acceleration versus time for the tremor. The decelerating forces are the frictional resistances developed on the surfaces of sliding. Velocity of the block is determined by step-wise integration of the acceleration-time curve and the resisting forces. Effects of rock anchors are included as a resisting force that may be overcome at the instant that accelerating forces exceed total resisting forces, which is the same as introducing, in vectorial form, a type of instantaneous strain-weakening.

This method permits calculation of the factor of safety for the support conditions provided. It is therefore amenable to risk analysis, where the probability of a seismic event of a certain size in a specified return period can be weighed against the cost of additional support or the design life of the slope.

Earthquake acceleration estimates are typically based on experience and on ground vibration spectra for a typical earthquake. A "characteristic" earthquake ground motion response is derived from many historical seismological records.

2.5.3.3 Factors of safety. Factor of safety is defined in the limiting equilibrium method as the ratio of forces tending to resist and to promote failure. The only really meaningful factor of safety value is 1.0, the point at which these two opposing forces are equal and sliding starts; values above and below 1.0 are difficult to relate to real degrees of safety. Actual safety, as opposed to a factor of safety, is influenced by the reliability of the analytical assumptions, observations, and measurements, and requires statistical assessment as discussed below.

Acceptable values vary depending on the application and particularly on the consequences of sliding. Mine slopes (often in the range $F = 1.0$ to 1.5) are less conservatively designed than dam abutments (often $F = 3$ or more) because fewer people are endangered, and because gradually developing movements combined with active surveillance give time for evacuation of the pit. Even in a mine, a range of safety factors may apply, with higher values for important slopes beneath the haul road, mill, and tailings dam, and lower values for other slopes that are less critical.

The Geotechnical Control Office of Hong Kong requires slopes to have a factor of safety of 1.1 to 1.4 in areas where there is a high risk to life, but permits 1.0 to 1.2 in areas of low risk to life. These values are not as low as they might appear, because conservatively selected values of shear strength are used in the calculations. The low values also reflect the need for construction in steep terrain. Hong Kong is an unusual case, combining high population densities with steep slopes on deeply weathered rock.

2.5.3.4 Probabilistic methods. An alternative to safety factor methods is to quantify safety in a *probabilistic analysis* by determining that a slope has, for example, one chance in 10,000 of failing in a given year. Probabilistic methods of slope design generally use a version of the limiting equilibrium method (McMahon 1971, 1975; Morris and Stoter, 1983; Young, 1985). The method is similar to standard limiting equilibrium design, but with statistical distributions assigned for joint orientations (whose dispersion is determined from a polar stereonet), for incremental roughness angles i, and for friction angles ϕ'. Monte Carlo simulation techniques are then used to sample possible distributions of slope geometry and slide plane characteristics to obtain graphs of probability of failure and expected volume of failure.

This approach is also imperfect, again because of uncertainties in data reliability, and limitations of the limiting equilibrium method itself. Just a single weak or continuous joint can be enough to trigger a slide—how is this plane to be located and tested?

A further problem is that owners of slopes may not be in a position to admit a remote chance of failure, even if they recognize the validity of a probabilistic approach. They prefer the less realistic but seemingly more conclusive factor of safety concept. This is also the reason that quantitative risk analyses are seldom carried out; analyzing risk is equated to admitting that a risk exists.

2.5.3.5 Slope design charts.
For the simple case of no cohesion along joints, the stability of a wedge bounded by two slide planes and a tension crack is a function of just six variables, the dips, dip directions, and the angles of friction of the slide planes (Hoek, 1973). Design charts can be prepared by combining these variables into groups to give a rapid assessment of safety factor (Hoek and Bray, 1974).

Only for the worst cohesionless case is stability independent of the height of slope and the angle of the slope face, and then only when the rock joints are dry. The assumption of zero cohesion introduces a degree of conservatism that is often justified for high slopes containing prominent joints or faults, but rarely for small benches or for discontinuous or rough joint surfaces. Stability back calculations of natural slopes usually show some amount of cohesion acting. With this limitation in mind, charts are an excellent and practical tool for design.

2.5.4 Stress analysis

Limiting equilibrium methods, although the most useful for slope design, do not predict stresses, displacements, or rates of movement. For this purpose, continuum mechanics methods such as finite element analysis are typically employed. They can predict stresses and displacements and even velocities if creep models are used, and can predict blasting or earthquake effects. Until recently, the models have seldom been employed for slope design because behavioral parameters remain poorly defined and difficult to measure, particularly if progressive failure is to be modeled.

Stress analysis and limiting equilibrium calculations can work well in combination (Popescu et al., 1985). Computed stress distributions can be used in place of averaged stress values as input to limiting equilibrium analysis. Displacements predicted using this hybrid approach give a useful comparison with the results of slope movement monitoring. Limiting equilibrium methods cannot easily account for weakening of a part of the failure surface, something that can be done with elastoplastic models of surfaces, accounting for progressive failure development (Lee, 1985). The same analysis can be performed using viscoplastic criteria, incorporating time effects, although little is yet known about rheological behavior of rock joints.

The most powerful use of finite element analyses in rock slope engineering is as a preliminary to limiting equilibrium calculations, to identify regions of high strain and potential weakening, changes of stress distributions during mining, and potential failure modes (Dolezalová et al., 1985). Then, using rock mass fabric information and strength parameters, limiting equilibrium methods are applied to the critical modes. This information also guides the selection and placement of monitoring equipment.

2.5.5 Analysis of toppling, bouncing, and rolling

The method introduced by Goodman and Bray (1976) for analyzing the toppling mode of slope failure has since been further developed by various authors (Choquet and Tanon, 1985). An individual block reaches a condition of limiting equilibrium when its weight acts exactly through the block's lower corner (Fig. 2.10b). This can be expressed as follows:

$$\frac{y}{x} = \cot A$$

where y and x are respectively the length and width of block, and A is the angle of dip of the base plane.

For flexural slip to occur, the slope angle must exceed the sum of the angle of friction between adjacent blocks and the angle of their bases. The kinematic condition for flexural slip, which precedes toppling, can be shown on a stereoplot that provides a "kinematic test," which is helpful in predicting potentially troublesome cases.

Other researchers have used finite and distinct element methods to investigate toppling failure (Kalkani and Piteau, 1976). Hocking (1978) describes the distinct-element method for simulation of toppling-sliding mechanisms. Developed initially by Cundall (Cundall et al., 1975), and now achieving general use, particularly in toppling analyses (Ishida et al., 1987), this method uses newtonian mechanics to examine translational and rotational velocities and accelerations of rock blocks, which are assumed to be rigid or elastic bodies. All such models are limited by the quality of input data and the need for accurate estimates of the energy absorbed at each bounce (i.e., coefficients of restitution and crushing, both poorly understood at present). The computer simulations are difficult to calibrate without in situ measurements using high-speed cameras (Sec. 2.3.5.2).

Dynamic analysis of rolling and bouncing blocks is a particularly difficult task. Piteau and Peckover (1978) describe a computer simulation in which falls were introduced at different heights above the slope to determine whether large blocks at the crest would reach facilities at the foot of the mountain some 1.5 to 2 km away. Models like this can be used to evaluate the effectiveness of catch walls, ditches, benches, and berms at various positions and heights.

2.5.6 Distinct-element method

Distinct-element approaches, such as described in a dynamic modeling application in the preceding section and also in *Rock Engineering,* Sec. 7.2.5, can be applied to static as well as to dynamic analyses. This versatile tool is perhaps the most promising for rock slope design. Its strengths, both in the two- and in the evolving three-dimensional versions, are its use of highly realistic rock structural geometries, automatic kinematic evaluation and generation of failure modes, use of fundamentally sound principles of mechanics, ability to consider rock stabilization and reinforcement, and adaptability to interactive graphics.

2.6 Excavation and Stabilization

2.6.1 Slope mining. Soft rocks can be removed by heavy mechanical excavators, if necessary with the aid of percussive breakers and rippers (Golosinski, 1989). Cuts in harder rock types require blasting (*Rock Engineering,* Chaps. 13 and 14). Careful smooth-wall blasting costs a little more because of the need for more drilling and explosives; however, it reduces overall costs by leaving the slope in a sounder condition (Fig. 2.17), and gives a broken rock product that can be handled and processed more readily. Uncontrolled blasting gives uneven contours, overbreak, overhangs, and excessive shattering. Costs are increased because of the need for extra scaling and rock reinforcement, and long-term maintenance costs of poorly blasted rock are likely to be high.

2.6.2 Scaling and trimming

Scaling should be completed after blasting each bench and before excavating deeper. The wall is inspected to remove any remaining loose rock, installing rockbolts and other stabilizing measures as needed. The work has to be completed while access remains available; it is too late to recognize unstable wedges or incompetent, loose blocks after they are undercut and dangerous to stabilize. Close geotechnical su-

(a)

(b)

Figure 2.17 Smooth-wall blasting to achieve stable rock cuts with little or no reinforcement. (a) Powerhouse excavation (*photo courtesy Manitoba Hydro*); (b) Highway cut (*Photo courtesy E. Magni, Ministry of Transportation, Ontario*).

pervision guards against "runaway excavations" in which stabilization work lags far behind.

Inspections from a distance seldom give a reliable guide to the extent and best methods of scaling. Usually much more than the expected amount of rock comes down, and other apparently loose blocks prove difficult or impossible to remove. The scaling bar is used to "sound" the rock face, and the rock engineer makes an on-the-spot decision regarding which blocks are best removed and which are safe to drill and bolt. In mining work, this presents no problem, but in civil construction, this degree of flexibility calls for careful writing of specifications.

Access to low and intermediate-height slopes is possible with suspended power platforms (spiders) or hydraulic boom cranes (giraffes, Fig. 2.18). Scaling on the upper reaches of high faces is usually carried out by workers on ropes. Useful scaling tools include hand pry bars, bencher drills, jackhammers, air operated scaling tools, hydraulic splitters or jacks, and explosives.

In special cases where the risk is high, particularly along transportation routes in urban areas, the cure to a rock slope stability concern may be removal of the entire slope in small, carefully controlled stages. In areas with high risk of rock avalanches or rock-debris torrents, blasting or pressure injection of water is used to bring down dangerous rock when all people have been evacuated for safety.

Figure 2.18 Scaling a 10-m-high rock cut during construction of the Science North Museum, Sudbury, Ontario.

2.6.3 Stabilization or protection?

Methods of *stabilization* (Fig. 2.19) can be classified into the following four categories:

- Cut-and-fill methods that remove material at the crest or place it at the toe to reduce the overall slope angle
- Drainage systems that increase stability by reducing water pressure in the slope
- Erosion protection systems that reduce or prevent erosive action on the slope face or toe
- Reinforcing and retaining systems that provide positive support to prevent the initiation of falls and slides

Add to these three further methods of *protection* (Fig. 2.20) designed to avoid the consequences of instability, when prevention is impractical or uneconomic:

- Catch systems that act as a passive, second line of defense to prevent the falling rock from damaging downslope structures

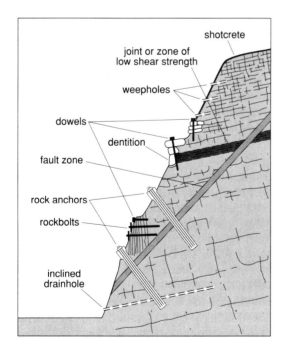

Figure 2.19 Various methods for stabilizing rock slopes. (*From Geotechnical Control Office, Hong Kong, 1984*)

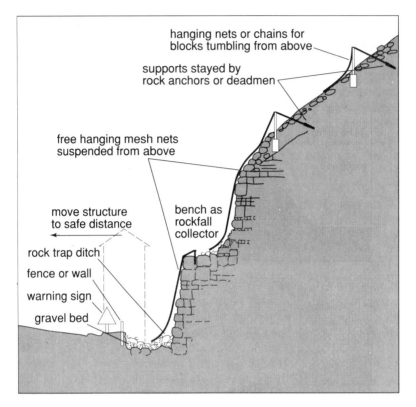

Figure 2.20 Rockfall control measures. (*From Geotechnical Control Office, Hong Kong, 1984*)

- Deflection systems that will either deflect a moving mass or cause it to lose enough of its energy so that the hazard is reduced
- Avoidance systems such as relocation, tunneling, or route diversion that bypass the slope stability problem entirely

Often these are combined. For example, Lumb (1975), Brand et al. (1983), and the Geotechnical Control Office Manual (1984) describe slope engineering in Hong Kong in massive granites and closely jointed volcanics, weathered to depths of up to 50 m. The slopes are stabilized using concrete buttresses, prestressed rock anchors, and rockbolts. Surface protection on the more weathered and broken areas is provided sometimes by shotcrete, but more often by "chunam," a mixture of cement and sand derived from the weathered granite, which is troweled onto the rock face. Drainage systems are effective and widely used.

In cases of active regional landsliding, avoidance is rarely possible, and conventional solutions such as bolting and drainage must give way to massive land reshaping, drainage, or living with slow movements.

2.6.4 Cut-and-fill methods

2.6.4.1 Recontouring. The stability of a slope can be enhanced by "recontouring," reducing the overall angle by unloading the slope crest, loading the toe, general excavation and benching, or by a combination of several methods. Often toe berms are combined with other gravity structures such as gabion walls or reinforced earth walls.

Slide debris has a stabilizing influence; because it comes from the crest and is deposited at the toe, it reduces the slope angle. However, positive effects can be offset by the interruption of natural drainage channels followed by loosening, softening, and reduction of shear strength. Slide debris often must be removed and replaced by free-draining material, or excavated and recompacted in thin lifts with drainage blankets.

Cutting into the toe of a slope can be dangerous. Hutchinson (1977) warns that "...when modifying a slope profile by excavation and filling one must ensure that neither the cuts nor the fills trigger the existing or potential slide that they are designed to stabilize, nor generate fresh slides local to themselves."

2.6.4.2 Toe mining. Stability can sometimes be maintained by excavating at the toe of the landslip, generally a slow-moving one in soft or broken rock or soil. Toe mining continues with careful monitoring until the whole sliding mass is removed or the average slope becomes gentle and can be maintained.

Evidently the method lends itself more to mining than to civil engineering applications. Being similar in concept to "block caving" in underground mining (Chap. 7), the method depends for its success on establishing steady, controlled movements, and on avoiding catastrophic accelerations. Monitoring of rock movements is the key to the success of this method. It is typically used where complete stabilization is out of the question because of cost. Removal of waste rock from the crest is much less attractive than mining ore at the toe.

Asbestos at the Jeffrey Mine in Quebec has been mined with little difficulty for many years in conditions of continuous sliding (Mamen, 1973; Brawner, 1975). The volume of moving rock has fluctuated, but typically has included more than 30 million tons in one wall of the pit. Rock creep of 1 to 2 mm/day has been monitored using a laser distance measuring system with reflective targets. Only occasionally have movements accelerated to values of 10 to 100 mm/hr, and the monitoring has continued to provide ample warning for safety. Mining rates and strategies have been adjusted in response to the measurements.

2.6.5 Drainage

The provision of adequate drainage including control of both groundwater and surface runoff is nearly always an important part of a stabilization program.

The most common groundwater-related slope stability problems are those associated with high water pressures that reduce shear strength of the joints or bedding planes. Lowering the hydraulic head in a rock slope by about 7 m in one project allowed the slope to be steepened by between 3° and 6° for the same safety factor (Brawner, 1968). Benefits in terms of reduced stripping and increased ore recovery can far outweigh drainage costs.

Water flowing over the crest and down the slope face should be diverted by a system of *collector ditches* to lessen erosion, and *vegetation* should be maintained or reestablished. *Trench drains* can be used in soft or weathered rocks to lower the water table directly beneath the slope face to counteract frost deterioration and shallow sliding. Deeper slides are stabilized by drilling horizontal *gravity-drain holes* to intersect as many water-conducting joints as possible. *Drainage galleries* can be effective when large-scale treatment is needed, and for stabilization of very large slides. *Vertical drains* are used mainly for short-term dewatering during construction (*Rock Engineering,* Chap. 18).

Sumer et al. (1988) present a case history on the effectiveness of dewatering in reducing both spoil pile and highwall stability problems in a coal strip mine in Alberta. Five alternatives were studied: highwall vertical wells, highwall horizontal drains, in-pit drainage ditches, pit floor vertical wells, and drainage galleries in the highwall. The goal was to lower the water table to the level of the coal seam. After regional groundwater studies and numerical modeling of the various alternatives, a combination of highwall wells and pit floor wells and ditches was chosen. Coal recovery increased from 65 to 87 percent because much less coal was sterilized by spoil pile slumps, and highwall instability disappeared. A further benefit was that coal quality improved; both the noncoal content and the moisture content dropped.

2.6.6 Erosion control

Toe erosion caused by wave action or river flow is most commonly controlled using *concrete or crib walls, gabion walls, riprap, groynes,* or *spur dikes*. Surface erosion is prevented by *surface drains, hydraulic seeding*, encouragement of *vegetation*, and sometimes by applying a layer of asphalt, soil-cement, or shotcrete.

Vegetation can bind surface materials, prevent erosion, reduce surface water velocity, and also reduce infiltration. Although most often used on soil slopes, vegetation can take root in softer rocks and even in the joints and weathered surface layers of harder rock types. Slope vegetation is only effective after major instabilities have been remedied by other means.

Deep-rooted plants such as crown vetch and bird's-foot trefoil are much more effective than shallow-rooted types such as grasses. Slopes can be sodded or seeded or directly planted with larger shrubs and trees. *Hydroseeding* is a useful method when the area to be treated is extensive. A mixture of seed together with fertilizer and a bonding agent is sprayed against the slope face by a mobile spray tanker.

2.6.7 Reinforcing and retaining systems

Rockbolts and dowels are used to stabilize small-scale slab, wedge, and toppling failures less than 3 to 4 m deep. Bolting is inappropriate where the joints are closely spaced (block size 200 mm or less) or the rock is loose, because of the risk of drilling vibrations causing disintegration and rockfalls. Before reinforcement, the rock face should be made safe by thorough scaling. Bolts can be installed on a regular grid or spot-located to anchor critical blocks. For permanent stabilization, steel components must be protected against corrosion.

High-capacity *soil and rock anchors* can be used alone or together with retaining structures to stabilize potentially unstable rock cliffs, translational rock slides, and large boulders (Figs. 2.21 and 2.22). Stabilization of open-pit mine slopes by anchoring is discussed by Seegmiller (1974). Steel plates or cast concrete blocks are used as bearing pads to distribute load to the rock. The pad size is increased when the anchor loads are high and the rock weak or closely jointed. If smaller anchor spacings become necessary, it becomes economical to replace the pads by beams or straps spanning between anchors. Anchors must in all cases be deep enough to be seated in firm rock that will not move.

Shotcreting, often in combination with bolting or anchoring, is an effective method for preventing slaking and raveling; it binds together loose blocks into a coherent skin, and seals the rock face to prevent wetting and drying. It offers a rapid, mechanized, and often simple solution to many rockfall problems. If it can be done safely, the slope should be scaled to remove loose blocks, and then subexcavated, usually with pneumatic tools, to remove seams of heavily weathered or

(a)

(b)

Figure 2.21 Stabilization of rock cliffs to protect the historic town of San Leo, Italy (Ribacchi and Tommasi, 1988). (a) The butte on which sits San Leo. Competent sandstones and calcarenites rest on a clay-shale substratum; (b) stabilization of the rock cliff by anchor bars and tendons. (*Photos courtesy P. Tommasi*)

Figure 2.22 Stabilizing the scenic Cabot Trail, Cape Breton Highlands National Park, Nova Scotia, Canada. (a) Airtrack rig drilling anchor holes at the base of a section of bin wall; (b) installing a 90-t cable anchor. (*Photo courtesy Geocon, Fredericton N.S., and Public Works Canada*)

weakened rock. Weep holes are drilled to relieve water pressures, and pipes installed behind the shotcrete to drain water from permeable strata. Without drainage, the problem of small pieces raveling from the face may be replaced by instability of much larger masses of coherent rock and shotcrete.

Gravity and cantilevered walls are rarely used to retain rock, except when they form part of a larger structure such as a tunnel portal or a bridge abutment. The alternative of an anchored or tied-back (*tieback*) wall is more attractive. Types include stone masonry, cast-in-place concrete, precast concrete panels with anchored ribs or waling (sometimes faced with rock slabs to enhance the natural appearance of the wall), and steel sheet piling or bin walls with anchored waling (Figs. 2.22 and 2.23). *Anchored beams* are a compromise between anchors with localized bearing pads and tied-back retaining walls that cover the entire rock face.

Anchored *cable nets* and mesh restrain masses of small loose rock or individual blocks as large as 2 or 3 m. Cable lashing is a variation in which large rock blocks are supported by cables anchored to the slope. Vertical concrete or steel ribs can be used to bridge between the cables. Anchored mesh is similar in concept to an anchored cable net but is much lighter and less expensive. Steel mesh, welded or woven, is pulled tight against the rock face using rockbolts.

Natural *buttresses* can be left behind as support, or artificial buttresses can be constructed of concrete (Fig. 3.5*b*). They are most often used to support overhangs in road cuts at highway level and undercuts in river banks. Patches of *dental masonry* are like small-scale buttresses, used to repair undercuts and sections of loose or deteriorated rock (Fig. 2.24).

2.6.8 Catch systems and deflection measures

Falling rocks can be intercepted by catch systems, diverted to different trajectories, or stopped or slowed down before they damage structures.

2.6.8.1 Catch systems.

The various catch systems augment more positive methods like reinforcement, support, or drainage, or can substitute for these methods when reinforcement is impractical or uneconomical. To design a catch system, the trajectory of falling rocks needs to be predicted, as well as the rate at which the debris is likely to accumulate. Storage capacity of ditches and fences should be sufficient to avoid overflow between periods of maintenance.

A *ditch* is cut down, whereas a *bench* is cut horizontally and a *berm* is built up. The functions of all three are similar, and all are usually inexpensive and easy to maintain.

Slope ditches intercept falls before they have a chance to gather momentum. They are often placed at the top of a scree slope to catch rock falling from a cliff before it starts to bounce and roll. *Toe ditches,* such as roadside ditches, are designed to catch rock that reaches the toe of the slope. Design criteria (Fig. 2.25) are given by Ritchie (1963).

(a)

(b)

Figure 2.23 Rock retention systems. (a) Anchored bin wall in horizontally bedded shale and limestone, Hamilton, Ontario (*Photo courtesy E. Magni, Ministry of Transportation, Ontario*); (b) masonry retaining wall in weathered rock, Hong Kong.

Figure 2.24 Boulder stabilized by masonry buttress, Cesnola, Dora Baltea Valley, Italy. (*Photo courtesy P. Grasso, Geodata, Torino*)

If lack of space prevents a ditch from being as wide as needed, it can be supplemented or replaced by a *catch net or wall* or by a *draped mesh curtain or blanket* (Fig. 2.26a). British Columbia contracts have specified galvanized 11-gauge hexagonal triple-twist gabion mesh suspended by 16-mm galvanized wire rope. Fences and walls increase the storage capacity of a ditch so that maintenance intervals can be extended. The system is designed for easy removal of accumulated rock

Landslides, Open-Pit Mines, and Rock Cuts

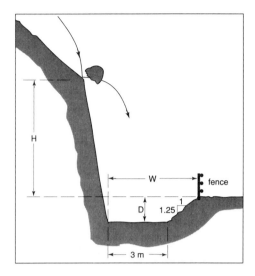

Rock slope angle	Height (m)	Fallout area width W (m)	Ditch depth D (m)
Near vertical	5 to 10	3.7	1.0
	10 to 20	4.6	1.2
	> 20	6.1	1.2
0.25 or 0.3:1	5 to 10	3.7	1.0
	10 to 20	4.6	1.2
	20 to 30	6.1	1.8*
	> 30	7.6	1.8*
0.5:1	5 to 10	3.7	1.2
	10 to 20	4.6	1.8*
	20 to 30	6.1	1.8*
	> 30	7.6	2.7*
0.75:1	0 to 10	3.7	1.0
	10 to 20	4.6	1.2
	> 20	4.6	1.8*
1:1	0 to 10	3.7	1.0
	10 to 20	3.7	1.5*
	> 20	4.6	1.8*

*May be 1.2 m if catch fence is used.

Figure 2.25 Design criteria for shaped ditches to catch falling rock (after Ritchie, 1963, and Piteau and Peckover, 1978).

debris. A low barrier can be formed of fill, standard highway guard rail, or gabions, but higher barriers must be specially constructed.

In contrast to anchored mesh which is designed to prevent rockfalls, the purpose of draped mesh is to prevent rocks from building up momentum, to reduce fall velocities and bounce heights, and to guide

(a)

(b)

Figure 2.26 Rock catch systems. (a) Draped mesh blanket to control raveling rock (*Photo courtesy E. Magni, Ministry of Transportation, Ontario*); (b) catch fences protecting apartment buildings below a 65-m-high rock cut, Tsuen Wan, Hong Kong (*Photo courtesy Dr. T.Y. Irfan, Geotechnical Control Office, Hong Kong*).

(c)

Figure 2.26 (*Continued*) Rock catch systems. (c) catch fences protecting apartment buildings below a 65-m-high rock cut, Tsuen Wan, Hong Kong (*Photo courtesy Dr. T.Y. Irfan, Geotechnical Control Office, Hong Kong*).

falling rock into a toe ditch where it will do no harm. Draped mesh works best when no individual blocks are larger than a meter, and when the slope is uniform enough for continuous contact between the mesh and the rock. The blanket is anchored at the crest by grouted rock dowels with a cable slung between them. It is rolled over the crest and the vertical seams are wired together. The bottom end should be left open, a meter or so above ditch level, and only a narrow ditch is required. The mesh usually is made of 9- or 11-gauge galvanized, chain link, or gabion wire. The gabion mesh alternative has a double twist hexagonal weave that does not unravel when broken.

Catch fences (Figs. 2.26*b* and 2.27) absorb more energy if made of steel cables or netting as opposed to the more rigid types of post-supported fencing. Cable-suspended net systems can catch flying

(a)

(b)

Figure 2.27 Catch fence for protection against rockfalls, Val Susa, Italy. (*a*) Tied-back fence soon after construction; (*b*) yielding fence absorbs the momentum of boulders. (*Photo courtesy P. Grasso, Geodata, Torino*)

blocks as large as a meter with little damage to the mesh. They are anchored to sound rock, if available, or to deadman anchors buried in soil or scree.

Rail fences consist of vertical posts or pieces of railing set in hand-dug or blasted holes that are backfilled with concrete, with further rails or cables strung horizontally between them, and with smaller cables or coarse wire woven between the main cables to form a crude net or mesh. They tend to be rigid and overdesigned, but if constructed of scrap materials can be inexpensive.

Catch walls are more sturdy than catch fences and can stop rolling or bouncing rocks as large as 2 m in diameter. Concrete catch walls are common in steep mountainous terrain, either cast in place or precast in short sections and assembled on site. Gabion catch walls are more flexible and absorb more energy. Gabions on impact tend to deflect and deform, but remain functional.

2.6.8.2 Diversion structures and relocation.

Relocation of highways, railways, power plants, or dam sites is expensive, but may be the only solution where rockslides are extensive and deep-seated. Protection against shallower movements such as avalanches and debris torrents can often be provided at less expense by placing sections of the route underground in a tunnel or rock shed, or by building diversion structures higher up the valley walls. The high capital cost is offset by the elimination of maintenance costs, and these alternatives can provide a high degree of security.

Large fast-moving avalanches and debris torrents are difficult to control. They should if possible be prevented from gathering momentum by a series of structures each of which will absorb a part of the energy (VanDine, 1985). Often their channels are easily identified. Small avalanches can be deflected by cutting *channel diversions* to guide them away from critical paths, or by installing strong *debris racks* in stream courses. Larger ones can be controlled using *check dams* at the lip of the flattest portions of the stream, where the mass will be moving the most slowly. They are made of massive rock blocks that are unlikely to be entrained in the flow. Their extremely high permeability allows water flow through the dam under normal conditions.

2.7 Monitoring and Maintenance

2.7.1 Monitoring objectives

Unstable natural slopes and those of doubtful stability are instrumented to warn of impending landslides and rockfalls, and to provide information for the design of stabilizing works. In open-pit mines, monitoring provides a means of checking the pit slope design, and permits safe and uninterrupted mining beneath the steepest possible slopes, so as to minimize stripping of waste rock and soil. Adjacent to highways and other civil engineering works, monitoring has the added role of safeguarding not only the construction crew but also the public (*Rock Engineering,* Chap. 12; Franklin and Denton, 1973; John, 1977; Wilson and Mikkelsen, 1978).

2.7.2 Inspection and maintenance

Slopes along transportation routes in mountainous terrain or valleys are often too extensive to be fully stabilized and trouble-free. Periodic inspection and scaling of loose, overhanging, or protruding blocks are basic maintenance requirements.

Rock faces are inspected from a moderate distance and from several angles using high-powered binoculars to detect loose material. Inspections are by helicopter when road access is difficult or nonexistent, for example, along pipe line routes and around reservoirs. Photographic records help decide priorities for remedial work. Telephoto lenses and stereo-pairs of oblique air photographs are particularly helpful to identify and locate loose rock. Other symptoms of instability, such as tilted trees and interrupted drainage, have been discussed in Sec. 2.4.1. The methods used for scaling and trimming during maintenance operations are the same as those used after rock blasting (Sec. 2.6.2).

2.7.3 Rockfall occurrence warnings

Along railways and highways through mountainous terrain there is a need for devices to inform that a rockfall or landslide has occurred. They give a "last ditch" line of defense when stabilizing and protective measures and before-the-fact warning systems are too expensive or impractical, and provide a check on the effectiveness of whatever stabilizing systems may be installed.

Patrols are a reliable and flexible method, but they are expensive and often possible only in good weather. Simple monitoring instruments can be used as rockfall indicating devices. Single-position extensometers can be connected to alarm systems, and electric fences and wires can be used on the principle that falling rock will break or pull out a wire and actuate a signal. The standard electric warning fence used along railways consists of wires strung between poles. Overhead wires can be used when rock faces are steep and close to the right-of-way.

Wire systems have the disadvantage of obstructing snow cleanup, and have been known to give false alarms. Other methods that have been considered or implemented on a small scale include seismometers to pick up ground vibrations from falling rocks, television monitoring, radar, and laser beams (Piteau and Peckover, 1978).

2.7.4 Landslide prediction

Once a zone of potential instability has been localized, it becomes possible to install a monitoring and warning system. The justifiable scope and cost of such a system depend on the significance of potential sliding. Localized high-risk slopes such as above a dam or the haul road in an open-pit mine justify a greater effort and expense than extensive stretches of highway cut.

Detection of movements provides the most direct and often the most useful information on stability. Commonly, geodetic surveying, extensometer, shear strip, and inclinometer observations are used to detect movements both at surface and at depth. Even tension crack gauges show patterns of movement (Yamaguchi and Shimotani, 1986). Other forms of observation such as of water pressures, support loads, and microseismic noise can assist as part of the overall instrumentation plan, depending on the situation.

Zavodni and Broadbent (1978) report cumulative displacements of 80 to 2500 mm at the time of failure, or "collapse," and suggest that displacement magnitudes give no useful warning. Usually, velocity and acceleration are the best indicators of landslide activity; a decelerating slide is likely to remain stable for the present, whereas an accelerating one continues to the point of failure, followed by postslide creep, usually at a decelerating rate. Opinions vary regarding the actual values to be used for prediction. Saito (1965) suggests imminent failure when slide velocities approach 5 to 100 mm/day, whereas Hungr (1981) reports a much broader range, from as little as 2 mm to more than 14 m during the day preceding failure.

Some empirical equations used to fit monitoring data and extract a prediction of failure time (t_f) are listed below:

$$\dot{d} = \frac{C}{(t_f - t)^n} \quad (Saito, 1980)$$

$$\dot{d} = K \cdot (e^{t/a} - 1) \quad (Varnes, 1983)$$

$$d - d_0 = \frac{d_1 t^{b+1}}{b + 1} \quad \text{with} \quad d = d_1 t^b \quad (Cruden\ and\ Masoumzadeh, 1987)$$

where time is t; displacement is d; d at $t = 0$ is d_0; d_1 is displacement at 1 min and the dot is time differentiation (displacement rate or velocity). Other symbols are constants fitted to the data records by statistical methods.

These equations were used in an attempt to statistically predict the time to failure t_f in an Alberta coal mine (Cruden and Masoumzadeh, 1987). Precise prediction proved impossible, although by assuming three separate accelerating creep stages, a critical velocity for pit evacuation was defined. Using the Zavodni and Broadbent (1980) equation, the critical velocity was reached on average 12 h before failure.

The time of sliding was forecast, with skill and undoubtedly a measure of luck, to within a few hours in a classic case reported by Kennedy and Niermeyer (1971). In December 1967, at the Chuquicamata open-pit copper mine in Chile, tension cracks began to open behind the highwall over a length of 200 m after a magnitude 5 earthquake. Various monitoring devices were installed, including tension crack stakes, crack extensometers, linear extensometers, tape and transit lines, and a seismograph. From the displacement information obtained, the final slope collapse on February 18, 1967, was accurately predicted some 5 weeks beforehand. This enabled the mine to reroute truck and rail access with a resultant shutdown of only 2½ days, and to avoid personal injuries and equipment damage.

In 1979 a rock slide at the King Beaver asbestos mine in Quebec interrupted traffic on the haulage ramp (Fig. 2.28). Avoiding major stabilization work, the roadway was reinstated by some regrading, steel mesh, and rockbolts (Bullock, Underwood and Franklin, 1980; Franklin, 1990). Only with monitoring could site traffic be allowed to flow again. A fail-safe system was installed consisting of seven independent cable extensometers passing beneath and across the ramp, each connected to an electric limit switch. Movement at any one station would extinguish a green light and activate a flashing red light and alarm bell in a control cabin. The controller, on duty day and night and in radio contact with every truck, could at once halt and redirect the traffic. Backup measurements were made by inclinometer and geodetic surveying. The haul road was reopened to traffic soon after testing of the warning system. During the initial weeks of monitoring it rained heavily, and several of the alarms were activated by movements of up to 20 mm. Movements have continued at seasonally varying rates but the alarm system has allowed the haul road to remain in service.

At the Premier Diamond Mine in South Africa, a 350-m-high 85° gloryhole has been experiencing rim slides that dilute the ore, and block drawpoints and ventilation raises (Bartlett and Raubenheimer, 1988). Monitoring using surveying methods in one case gave 6 week's warning of wedge failure.

(a)

(b)

Figure 2.28 King Beaver Mine haul road instability and monitoring, Asbestos, Quebec, Canada. (*a*) and (*b*) Views of the rockslide instrumented with a fail-safe warning alarm system.

(c)

Figure 2.28 (*Continued*) King Beaver Mine haul road instability and monitoring, Asbestos, Quebec, Canada (c) supplementary in-depth monitoring using a probe inclinometer.

A difficulty with predictions of unlikely events, even using real-time monitoring data, is the possibility of too many false alarms. The slope may accelerate during a period of heavy rain, then restabilize. Warning criteria must therefore be developed carefully and adapted to suit conditions at the mine or construction site. Monitoring should be used as part of an integrated geomechanics program, not just as a "crystal ball" for predicting slides.

References

Banks, D. C., 1972. "Study of Clay Slopes." In: *Stability of Rock Slopes. Proc. 13th Symp. Rock Mech.*, U. of Illinois, Urbana, Ill., pp. 303–328.

Bartlett, P. J., and Raubenheimer, M., 1988. "Rim Failure at Premier Mine." *Proc. 1st Reg. Conf. Rock Mech.*, Africa, pp. 49–52.

Barton, N., 1972. "Progressive Failure of Excavated Rock Slopes. In: *Stability of Rock Slopes. Proc. 13th U.S. Symp. Rock Mech.*, U. of Illinois, Urbana, Ill., pp. 139–170.

Baumer, A., Searle, P. J., and Tillmann, V. H., 1973. "Slope Stability on the Bougainville Copper Project." *Q. J. Eng. Geol.*, vol. 6, nos. 3 and 4, pp. 303–314.

Bjerrum, L., 1967. "Progressive Failure in Slopes of Overconsolidated Plastic Clay and Clay Shales." *J. Soil Mech. and Found. Div.*, ASCE, vol. 93, no. SM 5, pp. 1–49.

Brand, E. W., Hencher, S. R., and Youdan, D. G. 1983. "Rock Slope Engineering in Hong Kong." *Proc. 5th Int. Cong. Rock Mech.*, Melbourne, Australia, vol. C, pp. 17–24.

Brawner, C. O., 1968. "The Influence and Control of Groundwater in Open-Pit Mining."

Proc. 5th Can. Symp. Rock Mech., Toronto, Dec. 1968; also *Western Miner* vol. 42, no. 4, 1969, pp. 42–55.

——, 1974. "Rock Mechanics in Open-Pit Mining." *Proc. 3d Int. Cong. Rock Mech.*, Denver, vol. 1A, pp. 755–773.

——, 1975. "Case Examples of Instability of Rock Slopes." *B.C. Professional Engineer*, Feb., pp. 6–13.

Broili, L., 1977. "Relations between Scree Slope Morphometry and Dynamics of Accumulation Processes." *Proc. Mtg. on Rockfall Dynamics and Protective Works Effectiveness.* ISMES Bul. no. 90, Bergamo, Italy, pp. 11–24.

Brunsden, D., and Prior, D. B., (eds.) 1984. *Slope Instability.* John Wiley, Chichester, 620 pp.

Bullock, W. D., Underwood, A. H., and Franklin, J. A., 1980. "Rehabilitation of a Failed Haul Road in an Asbestos Open Pit." *Proc. American Inst. Min. Eng. Ann. Mtg.*, Las Vegas, Nev.

CANMET (Canada Centre for Mineral and Energy Technology), 1976. *Pit Slope Manual* (10 chapters plus supplements), CANMET Report 76–22, Supply and Services Canada, Ottawa.

Cestelli-Guidi, C., Croci, G., and Ventura, P., 1983. "The Stability of the Orvieto Rock." *Proc. 5th Int. Cong. Rock Mech.*, Melbourne, Australia, vol. C, pp. 31–41.

Charles J. A., and Soares, M. M., 1984. "The Stability of Slopes in Soils with Nonlinear Failure Envelopes." *Can. Geotech. J.*, 21, pp. 397–406.

Choi, Y. L., 1974. "Design of Horizontal Drains." *J. Eng. Soc. Hong Kong*, Dec. 1974, pp. 37–49.

Choquet, P., and Tanon, D. D. B., 1985. "Nomograms for Assessment of Toppling Failure in Rock Slopes." *Proc. 26th U.S. Symp. Rock Mech.*, Rapid City, S. Dakota, pp. 19–30.

Chowdhury, R. N., 1986. "Geomechanics Risk Model for Multiple Failures Along Rock Discontinuities." *Int. J. Rock Mech. Min. Sci. and Geomech. Abstr.*, vol. 23, pp. 337–346.

Christiansen, E. A., and Whitaker, S. H., 1976. "Glacial Thrusting of Drift and Bedrock." In: *Glacial Till*, Legget, R. F. (ed.), Royal Soc. of Can. Spec. Pub. no. 12, pp. 121–230.

Coulthard, M. A., 1979. *Back-Analysis of Observed Slope Failures Using a Two-Wedge Method.* CSIRO Div. of Applied Geomech., Melbourne, Australia, Tech Rept 83.

Cruden, D. M., 1985: "Rock Slope Movements in the Canadian Cordillera." *Can. Geotech. J.*, vol. 22, pp. 528–540.

——1976. "Major Rock Slides in the Rockies." *Can. Geotech. J.*, vol. 13, pp. 8–20.

——and Eaton, T. M., 1987. "Reconnaissance of Rockslide Hazards in Kananaskis Country, Alberta." *Can. Geotech. J.*, vol. 24, pp. 414–429.

——and Masoumzadeh, S., 1987. "Accelerating Creep of the Slopes of a Coal Mine." *Rock Mech. and Rock Eng.*, vol 20, pp. 123–135.

Cundall, P. A., Voegele, M., and Fairhurst, C., 1975. "Computerized Design of Rock Slopes Using Interactive Graphics for the Input and Output of Geometrical Data." *Proc. 16th U.S. Symp. on Rock Mech.*, pp. 5–14.

Davies, T. R. H., 1982. "Spreading of Rock Avalanche Debris by Mechanical Fluidization." *Rock Mech. and Rock Eng.*, vol. 15, pp 9–24.

Dearman, W. R., 1974. "Weathering Classification in the Characterization of Rock for Engineering Purposes in British Practice." *Bull. Int. Assoc. Eng. Geol.*, vol. 9, pp. 33–42.

De Freitas, M. H., and Watters, R. J., 1973. "Some Field Examples of Toppling Failure." *Geotechnique*, vol. 23, pp. 495–514.

Dolezalová, M., Zemanová, V., and Horeni, A., 1985. "Finite Elements for Open Pit Mine Stability Analysis." Preprints, *Int. Symp. on the Role of Rock Mech. in Excavations for Mining and Civil Works*, Zacatecas, Mexico, pp. 113–122.

Dusseault, M. B., and Morgenstern, N. R., 1978. "Characteristics of Natural Slopes in the Athabasca Oil Sands." *Can. Geotech. J.*, vol. 15, pp. 202–215.

Eisbacher, G. H., and Clague, J. J., 1984. "Destructive Mass Movements in High Mountains: Hazard and Management." *Geol. Sur. of Canada*, Paper 84-16, 230 pp.

Falcetta, J. L., 1985. "Un Nouveau Modèle de Calcul de Trajectoires de Blocs Rocheux." *Rev. Fran. de Géotech.*, vol. 30, pp. 11–18.

———(ed.), 1990. *Mine Monitoring Manual.* Canadian Institute of Mining & Metallurgy, Montreal, Quebec, Canada.

Franklin, J. A., and Denton, P. E., 1973. "The Monitoring of Rock Slopes." *Q. J. Eng. Geol.,* vol. 6, nos. 3 and 4, pp. 259-286.

Geotechnical Control Office, Hong Kong, 1984. *Geotechnical Manual for Slopes.* 2d ed., Govt. Pubn. Centre, Hong Kong, 295 pp.

Ghosh, A., and Haupt, W., 1989. "Computation of the Seismic Stability of Rock Wedges." *Rock Mech. and Rock Eng.,* vol. 22, pp. 109–125.

Golosinski, T. S., 1989. "Selective Continuous Surface Miners and Their Possible Applications in Oil Sands Mining." *Bul. Canad. Inst. Min. Metall.,* vol. 82, no. 925, pp. 51–57.

Goodman, R. E., and Bray, J. W., 1976. "Toppling of Rock Slopes." In: *Rock Engineering for Foundations and Slopes, ASCE Specialty Conf.,* Boulder, Colo., vol. 2, pp. 201–234.

———and John, K. W., 1983. "Surface and Near-Surface Excavations." *General Report, Theme B, Proc. 5th Int. Cong. Rock Mech.,* Melbourne, Australia, 12 pp.

Heim, A., 1932. "Bergsturz und Menchenleben." Zurich, *Vierteljahrsschrift 77,* No. 20, Beiblatt, p. 218.

Hocking, G., 1978. "Analysis of Toppling-Sliding Mechanisms for Rock Slopes." *Proc. 19th U.S. Symp. Rock Mech.,* Reno, Nev., pp. 288–295.

Hoek, E., 1974. "Methods for the Rapid Assessment of the Stability of Three-Dimensional Rock Slopes." *Q. J. Eng. Geol.,* vol. 6, nos. 3 and 4, pp. 243–255.

———, and Bray J. W., 1973. "Rock Slope Engineering." *Inst. Mining and Metal.,* London, 309 pp.

———, and Londe, P., 1974. "The Design of Rock Slopes and Foundations." *Gen. Rept. 3rd Cong. Int. Soc. Rock Mech.,* Denver, Colo., 40 pp.

Hungr, O., 1981. "Dynamics of Rock Avalanches and Other Types of Slope Movement." Ph.D. thesis, University of Alberta.

———, 1987. "An Extension to the Bishop's Simplified Method of Slope Stability Analysis to Three Dimensions." *Géotechnique,* 37, pp. 113–117.

———, Morgan, G. C., and Kellerhals, R., 1984. "Quantitative Analysis of Debris Torrent Hazards for Design of Remedial Measures." *Can. Geotech. J.,* vol. 21, no. 4, pp. 663–677.

Hutchinson, J. N., 1977. "Assessment of the Effectiveness of Corrective Measures in Relation to Geological Conditions and Types of Slope Movement." *Gen. Rept., Theme 3, Bul. Int. Assoc. Eng. Geol.,* no. 16, pp. 131–155.

Ikeya, H., 1989. "Debris Flow and its Countermeasures in Japan." *Bul. Int. Assoc. Eng. Geol.,* no. 40, pp. 15–33.

Ishida, T., Chigira, M., and Hibino, S., 1987. "Application of the Distinct Element Method for Analysis of Toppling Observed on a Fissured Rock Slope." *Rock Mech. and Rock Eng.,* vol. 20, pp. 277–283.

Jackson, L. E., Hungr, O., Gardner, J. S., and Mackay, C., 1989. "Cathedral Mountain Debris Flows, Canada." *Bul. Int. Assoc. Eng. Geol.,* no. 40, pp. 35–54.

John, K. W., 1977. "Monitoring of the Performance of Rock Slopes as Related to Design Considerations." *Proc. Field Measurements in Rock Mechanics,* Balkema, Rotterdam, pp. 735–755.

Kalkani, E. C., and Piteau, D. A., 1976. "Finite Element Analysis of Toppling Failure at Hell's Gate Bluffs, British Columbia." *Bull. Assoc. Eng. Geol.,* vol. 13, pp. 315–327.

Kennedy, B. A., and Niermeyer, K. E., 1971. "Slope Monitoring Systems Used in the Prediction of Major Slope Failure at the Chuquicamata Mine, Chile." *Symp. on Planning in Open Pit Mines,* Johannesburg, South Africa, pp. 215–225.

Kerisel, J., 1987. *Down to Earth. Foundations Past and Present: the Invisible Art of the Builder.* A. A. Balkema, Rotterdam, 149 pp.

Kullmann, D., and Barron, K., 1989. "Most Likely Tension Crack Positions in Planar Shear Slope Failure." *Bul. Canad. Inst. Min. Metall.,* vol. 82, no. 927, pp. 46–54.

Lee, C. F., 1985. "Stability of Pit Slopes in Brittle Rock Strata." Preprints, *Int. Symp. on the Role of Rock Mech. in Excavations for Mining and Civil Works,* Zacatecas, Mexico, pp. 40–43.

Lembo-Fazio, A., Manfredini, G., Ribacchi, R., and Sciotti, M., 1984. "Slope Failures and Cliff Instability in the Orvieto Hill." *Proc. 4th Int. Symp. on Landslides,* Toronto, vol. 2, pp. 115–120.

Lumb, P., 1975. "Slope Failures in Hong-Kong." *Q. J. Eng. Geol.,* vol. 8, pp. 31–65.

McMahon, B. K., 1971. "A Statistical Method for the Design of Rock Slopes." *Proc. 1st Australian–New Zealand Conf. on Geomech.,* Melbourne, Australia, vol. 1, pp. 314–321.

———, 1975. "Probability of Failure and Expected Volume of Failure in High Rock Slopes." *Proc. 2d Australian–New Zealand Conf. on Geomech.,* Brisbane, Australia, pub. Inst. Engrs., Aust., pp. 308–313.

Mamen, C., 1973. "Geotechnical Study Indispensable for Economical Slope Stability." *Can. Mining J.,* November.

Mollard, J. D., 1977. "Regional Landslide Types in Western Canada." In: *Landslides. Reviews in Eng. Geol.,* vol. 3, Geol. Soc. Amer., pp. 29–56.

Morgenstern, N. R., 1977. "Slopes and Excavations in Heavily Over-Consolidated Clay Shales." *Proc. 7th Int. Conf. Soil Mech. and Found. Eng.,* Tokyo, vol. 2, pp. 567–581.

Morris, P., and Stoter, H. J., 1983. "Open-Cut Slope Design Using Probabilistic Methods." *Proc. 5th Int. Cong. Rock Mech.,* Melbourne, Australia, vol. C, pp. 107–113.

Müller, L., 1968. "New Considerations on the Vaiont Slide." *Felsmechanik und Ing. Geol.,* vol. 6, pp. 1–91.

Norman, J. W., 1970. "The Photogeological Detection of Unstable Ground." *J. Inst. Highway Engrs.* (London), vol. 17.

Panet, M., and Rotheval, J.-P., 1976. "Stabilité des Talus Rocheux." *Bul. des Labo. Ponts et Chausées,* N° Spécial 3, pp. 171–186.

Peckover, F. L., 1975. "Treatment of Rock Falls on Railway Lines." *American Railway Eng. Assoc. Bul. 653,* Chicago, pp. 471–503.

——— and Kerr, J. W. G., 1977. "Treatment and Maintenance of Rock Falls on Transportation Routes." *Can. Geotech. J.,* vol. 14, pp. 487–507.

Perri, A. G., 1983. "Graphical Method for the Analysis of Rock Slopes in Urban Areas." *Proc. 5th Int. Cong. Rock Mech.,* Melbourne, Australia, vol. C, pp. 65–71.

Pfleider, E. P., (ed.), 1968. *Surface Mining.* Am. Inst. Min., Metall., Petrl. Engrs., New York.

Piteau, D. R., and Peckover, F. L., 1978. "Rock-Slope Engineering." Chap. 9 in *Landslides: Analysis and Control.* Special Rept. 176, Transp. Res. Bd., Nat. Acad. of Sci., Washington, D.C. pp. 192–228.

Popescu, M., Craciun, F., and Balasescu, A., 1985. "Stability Analysis of Lignite Highwalls." Preprints, *Int. Symp. on the Role of Rock Mech. in Excavations for Mining and Civil Works,* Zacatecas, Mexico, pp. 69–88.

Priest, S. D., 1980. "The Use of Inclined Hemisphere Projection Methods for the Determination of Kinematic Feasibility, Slide Direction and Volume or Rock Blocks." *Int. J. Rock Mech. Min. Sci. and Geomech. Abstr.,* vol. 17, pp. 1–23.

Ribacchi, R., and Tommasi, P., 1988. "Preservation and Protection of the Historical Town of San Leo, (Italy)." *Proc. IAEG Conf. Eng. Geol. of Ancient Works, Monuments and Historical Sites,* Balkema, Rotterdam, pp. 55–64.

Richards, B. G., Coulthard, M. A., and Toh, C. T., 1981. "Analysis of Slope Stability at Goonyella Mine." *Can. Geotech. J.,* vol. 18, pp. 179–194.

Ritchie, A. M., 1963. *Evaluation of Rockfall and its Control.* Highway Res. Bd., Highway Res. Record 17, pp. 13–28.

Ross-Brown, D. M., 1973. "Design Considerations for Excavated Mine Slopes in Hard Rock." *Q. J. Eng. Geol.,* vol. 6, nos. 3 and 4, pp. 315–334.

Saito, M., 1965. "Forecasting the Time of Occurrence of a Slope Failure." *Proc. Int. Conf. Soil Mech. Found. Eng.,* Montreal, vol. 6, pp. 537–541.

———, 1980. "Semi-Logarithmic Representation for Forecasting Slope Failure." *Proc. 3d Int. Symp. on Landslides.,* New Delhi, vol. 1, pp. 321–324.

Salcedo, D. A., and Tinoco, F. H., 1983. "A Rock Slide in an Urban Area: a Case History." *Proc. 5th Int. Cong. Rock Mech.,* Melbourne, Australia, vol. C, pp. 73–81.

Schuster, R. L., 1986. "Landslide Dams, Processes, Risk and Mitigation." *Proc. Am. Soc. Civ. Eng. Convention," Seattle, Wash., ASCE Geotechnical Special Publication no. 3,* ASCE New York, 164 pp.

Seegmiller, B. L., 1974. "Artificial Stabilization in Open Pit Mines: a New Concept to Achieve Slope Stability." *Ann. Mtg. American Inst. Min., Metal. Petrol. Engrs.*, Vail, Colo.

Selby, M. J., 1982. "Controls on the Stability and Inclinations of Hillslopes Formed on Hard Rock." *Earth Surface Processes and Landforms*, vol. 7, pp. 449–467, John Wiley, London.

Sharp, J. C., 1970. "Drainage Characteristics of Subsurface Galleries." *Proc. 2d Cong. Int. Soc. Rock Mech.*, Belgrade, vol. 3, pp. 197–204.

Skempton, A. W., 1964. "Long Term Stability of Clay Slopes." *Geotechnique*, vol. 14, no. 2, pp. 77–102.

Stead, D., and Singh, R., 1989. "Loosewall Stability in United Kingdom Surface Coal Mines." *Can. Geotech. J.*, 26, pp. 235–245.

Stimpson, B., 1978. "Failure of Slopes Containing Discontinuous Planar Joints." *Proc. 19th U.S. Symp. Rock Mech.*, Reno, Nev., pp. 296–300.

———, Barron, K., and Kosar, K., 1987. "Multiple-Block Plane Shear Slope Failure." *Can. Geotech. J.*, vol. 24, pp. 479–489.

Styles, K. A., Hansen, A., Dale, M. J., and Burnett, A. D., 1984. "Terrain Classification Methods for Development, Planning and Geotechnical Appraisal: a Hong Kong Case." *Proc. 4th Int. Symp. Landslides*, Toronto, vol. 2, pp. 561–568.

Sumer, S. M., Elton, J. J., and Tapics, J. A., 1988. "Dewatering Optimization Using a Groundwater Flow Model at the Whitewood Open-Pit Coal Mine, Alberta." *Can. Geotech. J.*, vol. 25, pp. 684–693.

VanDine, D. F., 1985. "Debris Flows and Debris Torrents in the Southern Canadian Cordillera." *Can. Geotech. J.*, vol. 22, pp. 44–68.

Varnes, D. J., 1978. "Slope Movement Types and Processes." Chap. 2, *Landslides: Analysis and Control*. Special Report 176, Transportation Research Board, U.S.A.

———, 1983. "Time-Deformation Relationships in Creep to Failure of Earth Materials." *Proc. 7th SE Asian Geotech. Conf.*, vol. 2, pp. 107–130.

Voight, B. (ed.), 1978. *Rockslides and Avalanches*. Elsevier, vol. 1.

———, Calvache, M. L., and Herrera, O. O., 1987. "High-Altitude Monitoring of Rock Mass Stability Near the Summit of Volcano Nevada del Ruiz, Colombia." *Proc. 6th Cong. Int. Soc. Rock Mech.*, Montreal, vol. 1, pp. 275–279.

Wilson, S. D., and Mikkelsen, P. E., 1978. "Field Instrumentation." Chap. 5 *in Landslides: analysis and control*. Special Rept. 176, Transp. Res. Bd., Nat. Acad. of Sci., Washington, D.C. pp. 112–138.

Yamaguchi, U., and Shimotani, T., 1986. "A Case Study of a Failure in a Limestone Quarry." *Int. J. Rock Mech. Min. Sci. and Geomech. Abstr.*, vol. 23, pp. 95–104.

Young, D. S., 1985. "A Generalised Probabilistic Approach for Slope Analysis." *J. of Min. Eng.*, vol. 3, pp. 215–228.

Zaruba, Q., and Mencl, V., 1982. *Landslides and Their Control*. Elsevier Scientific Publishing, Amsterdam, 324 pp.

Zavodni, Z. M., and Broadbent, C. D., 1978. "Slope Failure Kinematics." *Proc. 19th U.S. Symp. Rock Mech.*, Reno, Nev., vol. 2, pp. 86–94.

———and ———, 1980. "Slope Failure Kinematics." *Can. Institute of Mining Bulletin*, vol. 73, no 816, pp. 69–74.

Chapter

3

Rock Foundations

3.1 Introduction

The presence of rock guarantees a reliable foundation, but only if it is in sound condition and not undermined or decomposed by weathering, broken by jointing or faulting, or shattered by blasting. Most types of rock are adequate for most types of foundation; only for the most heavily loaded foundations, such as beneath an arch dam, are the softer rocks, such as shale, unacceptable. The three main concerns in foundation engineering are to reliably establish (1) the depth to "sound" rock, often buried beneath soil and weathered rock, (2) the costs of excavating to remove unwanted rock, and (3) the requirements for maintaining stability of surrounding buildings, services, natural slopes, and basement excavations.

Of all rock foundations, the greatest risks are associated with foundations for dams, and the greatest expenditures are allocated for their investigation, design, and treatment. Because of their importance, dam foundations are discussed in greater detail in Chap. 4.

3.2 Types of Foundation

3.2.1 Foundation extent and loading

Foundations for buildings, bridges, and dams occupy a limited area, in contrast to those for highways, railways, and oil and gas pipelines that extend for long distances. The more localized a foundation, the more easily it can be explored, tested, and mapped in detail, and the less variable is the nature and behavior of the ground.

From another perspective, foundations differ according to whether the superstructure is heavy or light and sensitive or insensitive to settlements. In the "localized, heavy, sensitive" category, for example, are the foundations of arch dams and power plants, which require very careful and detailed testing and design. In contrast, the foundations for pavements and for most types of buildings are lightly loaded and settlement tolerances are less stringent.

3.2.2 Shallow and deep foundations

Foundations are usually classified into "shallow" and "deep" categories according to the depth and type of the *footing* which transmits load from the structure to the ground (Fig. 3.1). *Slab-on-grade* construction and *spread and strip footings* fall into the shallow category, and *deep basement excavations, caissons,* and *piles* into the deep category.

Slab-on-grade construction (Fig. 3.1a) is used for roads and other types of pavement and for lightly loaded buildings where a suitable bearing stratum is close to the surface. The weight of the superstruc-

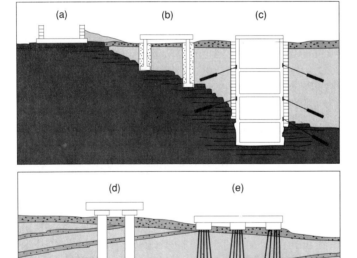

Figure 3.1 Types of footing. (a) Slab on grade; (b) spread or strip footing; (c) deep basement excavation; (d) caissons; (e) end-bearing piles.

ture is distributed over the complete foundation, or is transferred to the subgrade as a moving wheel load in the case of a pavement.

Shallow spread footings of various shapes and strip footings (Fig. 3.1b) carry loads from columns or from the perimeter and interior bearing walls of buildings.

Basement excavations (Fig. 3.1c) are convenient when competent bearing strata are deep, and provide underground space for shopping, parking, or storage. Usually soil must be supported while excavating to reach rock, and often the basement penetrates rock to some substantial depth. Using the most common *H-piles and lagging* method of support, steel H-piles are inserted in holes predrilled around the perimeter of the building. The base of each hole is socketed into rock and filled with concrete. As excavation proceeds downward, timber lagging boards are inserted between the pile flanges. At appropriate depths, rows of downward-raked holes are drilled and *tieback anchors* are cemented into the surrounding soil or rock. Horizontal *walers* (steel beams) transfer load from the tensioned tiebacks to the piles. Slab, spread, or strip footings are constructed in the floor of the completed excavation.

Caissons (Fig. 3.1d) can be used in place of pile and lagging support for deep basements, in which case they are installed in touching or overlapping rows, usually with every second or third caisson reinforced by an H-pile, supporting the filler piles in between (Fig. 3.2). Caissons or piles are used individually or in pile clusters when the shallowest satisfactory bearing stratum is below the depth of economical basement excavation (Fig. 3.1e). Caissons and bored piles are constructed by augering vertical holes into which concrete is pumped. Driven piles are of steel with an "H" or circular cross section (pipe piles) or of precast concrete or timber.

3.3 Shallow Footings

3.3.1 Failure mechanisms

Foundations can fail by a variety of mechanisms such as shown in Fig. 3.3. Shallow foundations can fail by shearing and punching (bearing capacity failure), sliding instability, differential settlement, sinkhole collapse, fault displacement, frost heave, or swelling of the foundation soils or rocks. Deep basement foundations can, in addition, suffer damage from lateral squeeze of basement walls, base heave and buckling of excavation floors, and buoyancy uplift of the building from its foundation. Piled foundations can suffer from defects in the piles, bending and misalignment, or failure to reach a satisfactory bearing stratum.

Bearing capacity failures in which plastic shearing predominates

Figure 3.2 Caisson wall (interlocking pile wall). View of basement excavation for the Admiral Hotel in Toronto, Ontario. In this case the wall had a triple function, to act as shoring, as a cut-off wall keeping the lake out of the basement, and as the structural basement wall. (*Photo, W. Lardner, Deep Foundations Contractors Inc.*)

(Fig. 3.3a and b) are associated with the overloading of a soil foundation, but are uncommon for structures on rock. This mode of foundation deformation can occur beneath heavily loaded footings on weak clay shales.

Consolidation failures are quite common in weathered rocks where the footing is placed at too high a level in the weathering profile (Fig. 3.3c and e). Unweathered rock corestones are pushed downward under the thrust imposed by the footing, because of a combination of low shear strength along clay-coated lateral joints and voids or compressible fillings in the horizontal joints.

A few porous rock types can yield by a *punching* mechanism (Fig. 3.3d) in which the pores collapse, leaving the grains in much closer contact (Ladanyi, 1972). Porous limestones, typically chalk, are particularly prone to this behavior, and have been known to settle 100 to 200 mm under applied pressures of only 30 to 100 kPa. Tests on chalk reported by Burland and Lord (1970) gave plastic yield at applied pressures in the range 400 to 600 kPa. Tests on porous limestone at the Bermudan National Sports Center resulted in a 300-mm-diameter plate punching continuously into the rock when the applied pressure

Rock Foundations 115

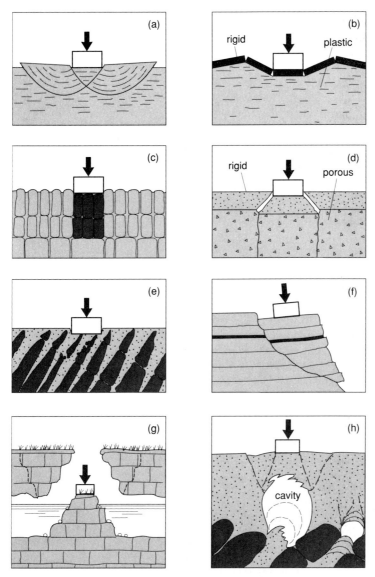

Figure 3.3 Mechanisms of foundation failure (adapted from Sowers, 1976). (a) Prandtl-type shearing in weak rock; (b) shearing with superimposed brittle crust; (c) compression of weathered joints; (d) compression and punching of porous rock (rigid crust); (e) breaking of pinnacles of weathered rock surface; (f) slope failure caused by superimposed loading; (g) collapse of shallow cave; (h) sinkhole caused by soil erosion into solution cavities.

reached about 4 MPa (Fig. 3.4).The bearing pressure normally permitted for Bermudan foundations on these very porous limestones is 300 kPa (giving a safety factor against punching of about 10).

Intensely weathered igneous rocks can have collapsible structures with intergranular porosities as high as 50 percent. Scoria, vesicular basalt, or layers of pumice (all materials found in volcanic environments) can also have high porosities.

3.3.2 Sliding instability

3.3.2.1 Undermining of foundations. If a footing is located on or near the crest of a natural slope, rock cut, or trench, there is some risk of slope instability by any of the several mechanisms discussed in Chap. 2 (Fig. 3.3*f* and 3.5). The investigation for a radar station foundation on a rocky headland in Labrador had been limited to a walking tour of the site until a photograph of the 300-m-high cliff came to light. The cliff face (which could not be seen from on top) contained wedge-shaped indentations and an accumulation of talus at the toe (Fig. 3.6). This discovery prompted extensive investigation by helicopter, and stability analyses.

The problem of a footing near a slope crest reduces to one of slope stability with a superimposed loading (Chap. 2). Factors of safety are calculated using the method of limiting equilibrium, which requires information on the inclinations and shear strengths of joints daylighting in the slope face, and on water pressures in the rock mass. If the computed factors of safety are inadequate, the footing can be moved further from the crest, or the slope can be stabilized by tensioned rock anchors or other means. Often, however, calculations demonstrate that the load imposed by a footing is insignificant compared with that imposed by the dead weight of rock. The analysis resolves itself into a simple assessment of slope stability.

Numerical stress-strain analysis or discrete block methods (*Rock Engineering,* Chap. 7) can be used to supplement an equilibrium analysis. Guenot and Panet (1983) employed a finite element model to study the behavior of a footing at the crest of a stratified rock slope, and were able to simulate the development of plastic zones and to predict bearing capacity and differential settlement.

Sliding stability must be considered in the case of laterally loaded footings. An extreme case is a dam loaded by upstream water pressure, which may tend to displace downstream by sliding along weakness planes in the foundation and abutments (Chap. 4).

3.3.2.2 Hazards from above. Another potential risk is that a footing at the base of a slope may be endangered by rockfalls, landslides, or slope creep, even though it has a stable foundation.

Rock Foundations 117

(a)

(b)

Figure 3.4 Foundation tests on chalk, Bermuda National Sports Center. (a) Loading beam and jack; (b) plate resting on chalk before loading. (*Photos courtesy Public Works Department, Bermuda*)

Figure 3.4 (*Continued*) Foundation tests on chalk, Bermuda National Sports Center. (*c*) Punching failure with compression and collapse of the porous chalk. (*Photos courtesy Public Works Department, Bermuda*)

Figure 3.7 shows an example of such a situation, the Ramat Viaduct under construction in northern Italy. This crosses a rock talus that, from investigations and monitoring, appears to be creeping. The piers are carried to stable rock beneath the zone of instability. Shafts have been excavated to depths of up to 20 m and with a diameter of 16 m to give room for lateral creep of soils, avoiding lateral thrusts on the piers.

3.3.3 Foundations over sinkholes and mine openings

Problems caused by cavities beneath a foundation (Fig. 3.3*g* and *h*), whether of geological or mining origin, are obvious. The cavities themselves, however, can often pass undetected until revealed by subsidence or even complete collapse of the structure. The causes and prevention of mining-induced subsidence are discussed in Chap. 7. This section will focus on the effects of mine openings and naturally occurring cavities beneath foundations.

3.3.3.1 Solution pipes and caves. Mechanisms of rock solution (*Rock Engineering,* Sec. 4.1.4) rarely act rapidly enough to undermine a

(a)   (b)

Figure 3.5 Foundations in Hong Kong. (*a*) Footings for high-rise apartments notched into the rock outcrop; (*b*) concrete buttresses beneath a similar building.

foundation during the few decades of concern to most engineering projects. Toulemont (1970) reports cases where gypsum strata have dissolved very quickly because of the pumping of groundwater, leading to collapse of structures. Gypsum (sodium sulfate) and rock salt (sodium chloride) are the most soluble of rock types, and where present at shallow depth are often associated with problems of subsidence (Terzaghi, 1970).

A more common type of sinkhole problem is associated with limestone terrain in which cavities have formed by solution acting over hundreds of years (Beck, 1984). Solution enlarges some of the intersecting vertical joints into chimney-like openings in the rock called *pipes* that can extend downward from the base of the soil overburden many tens of meters deep, and are sometimes as large as 15 m in diameter (Sowers, 1976). These link with subhorizontal systems of tubes and caves. Over these openings, the rock progressively collapses and "caves" upward until, on reaching surface, it forms a *rock collapse sink* or *sinkhole* (Fig. 3.3*g*). The landscape of some *karstic regions* is dotted with thousands of sinkholes, from a few meters to more than 100 m in diameter.

When pipes are covered by granular soils, the finer components (silt and sand) wash downward into the pipes, leaving a metastable arch of coarse sand and gravel (Fig. 3.3*h*).

Figure 3.6 Site for a radar station, perched atop a 300-m-high cliff in Labrador, Canada. A substantial talus mound has accumulated in the sea beneath the cliff, derived from rock slides to the right and left of the main rock buttress.

Structures and roads can be built over cavities or soil arches that later collapse. The collapse can be triggered by increased stress caused by lowering of the water table, by sudden erosion and loss of loose infilling materials, or by the weight of the structure itself. Sowers (1976) describes an erosion-induced sinkhole more than 100 m in diameter and 25 m deep that occurred overnight near Birmingham, Ala. Similar sinkholes in central Florida have consumed houses and streets.

Several years ago at Welwyn Garden City in England, sinkholes had been appearing one by one beneath partially completed houses on a new estate. Investigations revealed 7 m of granular soils over chalk, with vertical solution pipes at the intersections of an orthogonal jointing pattern. Over the years, the silts and sands in the granular soils had been washed downwards into the pipes, leaving an arch of gravel bridging each pipe. In preparation for building the houses, the topsoil had been stripped. This had allowed rainwater infiltration sufficient to collapse the arches, creating elliptical cones of subsidence at the surface, some measuring several meters in diameter (Fig. 3.8).

At this site, geophysical methods and a heavy vibrating roller were unsuccessful in detecting or collapsing the pipes that remained hidden beneath the soil. Detection was achieved by injection of copious volumes of water through pipes jetted into the subsidence cracks. The

Rock Foundations 121

(a)

(b)

Figure 3.7 Motorway from Turin (Italy) to Frejus (France). (*a*) Venaus Viaduct; (*b*) Ramat Viaduct, length 920 m, span 100 m, and maximum height 95 m. (*Photos courtesy Prof. Sebastiano Pelizza, Politecnico di Torino*)

Figure 3.7 *(Continued)* Motorway from Turin (Italy) to Frejus (France). *(c)* A foundation shaft for the Ramat Viaduct, depth 20 m and diameter 16 m to give room for lateral creep of soils. *(Photo courtesy Prof. Sebastiano Pelizza, Politecnico di Torino)*

water was left running overnight, and by morning, sinkholes appeared along orthogonal lines radiating from the jet point. Treatment by backfilling would have been inexpensive and effective had it been carried out before construction. With construction well under way, expensive underpinning and piling operations were necessary. The developer might have been forewarned of the problem by the name of the road running alongside the property: Moneyhole Lane. Old newspaper reports also described cows and tractors disappearing into sinkholes.

3.3.3.2 Construction on cavernous ground. The existence of cavities must first be recognized as a problem. Then, the extent and size of individual cavities must if possible be determined. Geophysical methods such as ground-probing radar are the most promising in this application *(Rock Engineering,* Sec. 6.4). Special designs and remedies are possible in situations where buildings must be built over sinkholes, mining excavations, or areas of anticipated subsidence.

Bhattacharya et al. (1984), from a world-wide survey, give tolerances to ground movements in areas of subsidence (Table 3.1). These are expressed in terms of building categories 1 (brick and masonry

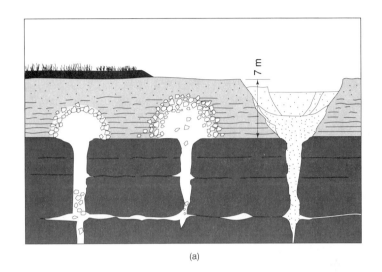

Figure 3.8 Sinkholes at Welwyn Garden City, England. (*a*) Cross section; (*b*) typical sinkhole depression in soil. (*Photo courtesy Rock Mechanics Ltd., U.K.*)

structures), 2 (steel and reinforced concrete frame structures), 3 (timber frame structures), and 4 (massive structures of considerable rigidity).

Where individual cavities can be located, they can be plugged or bridged by structural concrete, or grouted to backfill the entire open-

TABLE 3.1 Ground Movement Limits
Criteria for subsidence damage to buildings (abbreviated from Bhattacharya, Singh and Chen, 1984). Blanks in the table indicate where no data are available.

Building category	Level of damage	Angular distortion ($\times 10^{-3}$)	Horizontal strain ($\times 10^{-3}$)	Deflection ratio ($\times 10^{-3}$)	Radius of curvature (km)
1	Architectural	1.0	0.5	0.3	—
	Functional	2.5–3.0	1.5–2.0	0.5	20
	Structural	7.0	3.0	—	—
2	Architectural	1.3	—	—	—
	Functional	3.3	—	—	—
	Structural	—	—	—	—
3	Architectural	1.5	1.0	—	—
	Functional	3.3–5.0	—	—	—
	Structural	—	—	—	—
4	Architectural	—	—	—	—
	Functional	—	—	—	—
	Structural	—	—	—	—

ing and overlying zones of loosened, caved rock. Otherwise, some accommodation must be made in the structure or foundation. Flexible structures are designed for areas where there is a high risk of settlement, and services such as pipelines can also be designed for flexibility. An alternative is to provide jacking facilities for the releveling of a more rigid but lightweight building. Heavier structures over shallow cavities can be founded on deep caissons or piles reaching below the level of the suspected problem.

3.3.4 Foundations on faults

Permanent, rigid buildings of any type usually should not be built on historically active fault surface traces. Structures of great flexibility, such as suspension bridges, can be permitted to span faults because lateral movement along the fault will not cause collapse. Highways and pipelines over faults may have to be periodically rebuilt as episodic or creep displacements cause surface distress.

Geological faults often must be accommodated beneath the foundations of dams (Chap. 4). This is because dams are located on rivers, and rivers follow geological planes of weakness. A fault beneath a dam or a major structure such as a power plant must be proven inactive, and even then has to be treated to bring the foundation within the fault zone to a level of competence similar to that of the surrounding, more massive and stronger rock.

A fault was encountered during investigations for the Science North

Museum in Sudbury, Ontario. This major geological feature had a mapped length of 56 km, a throw of over 600 m (measured in a nearby mine), and a shear zone 30 m wide with seams of soft clay gouge that passed beneath and between the heavily loaded piers of the building (Fig. 3.9). The presence of the fault was confirmed from a study of air photographs, and its character was investigated by inclined core drilling. Evidence relating to the fault's potential activity was obtained from test pits excavated across the fault plane, which revealed unbroken beds of varved clay. The fault was classified as "inactive." Nevertheless, structural precautions were taken, such as "toughening up" the joints in the structure, and providing extra doweling in the foundations (Franklin and Pearson, 1985).

The probability of the fault becoming active was in this case assessed on the basis of the U.S. Bureau of Reclamation criteria, which state that a fault is designated as "inactive" if it can be proven not to have displaced during the preceding 10,000-year period (*Rock Engineering*, Sec. 6.3.3). Dating methods are discussed by Murphy et al. (1979). Alternatively, using criteria of the nuclear power industry, the potential for movement to recur along a fault has been based on the history of faulting within the past 500,000 years. This is called the "capability" of the fault; U.S. Nuclear Regulatory Commission 1978 criteria define a *capable fault* as one that has exhibited one or more of the following characteristics:

> Movement at or near the ground surface at least once within the past 35,000 years or movement of a recurring nature within the past 500,000 years
>
> Macroseismicity instrumentally determined with records of sufficient precision to demonstrate a direct relationship with the fault
>
> A structural relationship between a fault and a capable fault according to the above characteristics such that movement on the capable fault could be reasonably expected to be accompanied by movement on the other fault.

If a fault is active, but construction is essential, careful design will be required to accommodate the anticipated displacements without impairing the integrity or safety of the structure.

3.3.5 Frost heave

Frost heave in rock occurs by ice lensing within joints close to foundation level. For frost heave to be possible, the foundation must be either weak or jointed, it must have access to water, and its temperature must drop periodically below freezing.

Figure 3.9 Fault in the foundation at Science North museum, Sudbury, Ontario. (a) Creighton fault exposed during stripping of soil; (b) principal shear with wedged glacial boulders, seen from within the glazed spiral staircase; (c) view of staircase (on the fault) with associated black diabase dyke, and limestone masonry.

Dense, strong, and massive rock formations, including most igneous rocks, are unlikely to suffer frost heave. Foundations above the water table and out of the reach of capillary water migration should also be problem-free. Precautions may be needed, however, for closely jointed or weak rocks at or close to the water table. In such cases, frost action can be minimized or eliminated by drainage, provided that the drains can be relied upon to function effectively for as long as needed. For example, drain outlets must themselves be protected from freezing.

Depths of frost penetration vary with latitude and location and can be assessed from local experience or by using the isothermal maps that are published for northern latitudes. Surface frost penetration can be reduced or eliminated by insulation, for example, by constructing earth embankments above and around the perimeter of a building, or by using artificial insulating materials. In permafrost areas, surface insulation may cause the permafrost level to rise, triggering frost heave. Freezing of the foundation is an alternative, either by direct refrigeration or by using heat wicks, such as those used for the foundations of the Alaska pipeline.

3.3.6 Swelling

Swelling is a time-dependent volume increase involving physicochemical reaction with water. It results from a variety of processes including the adsorption of water by clay minerals, oxidation of anhydrite into gypsum, and oxidation of pyrite and marcasite when in contact with air, water, and the "sulphur oxidans" bacterium. Swelling mechanisms and tests are described more fully in *Rock Engineering,* Chap. 10.

The water-sensitive minerals may be the principal constituent of a rock, as in the case of smectitic clay shales, or may cause problems even when present in minor amounts. Often, swelling materials are concentrated along beds, as a gouge or infilling in faults, or along joints or veins, in which case their swelling can be disruptive to the rock mass as a whole.

The most expansion-prone shales are those containing substantial amounts of the smectite family of clay minerals, which have the ability to adsorb considerable amounts of water. The best known member of this family is montmorillonite. Swelling is most pronounced when the rock is first dried and then rewetted, because changes rather than absolute water content are the cause of swelling and shrinkage, and because the cyclic expansion helps to destroy the rock fabric.

Certain black shales such as those in Oslo, Ottawa, and Cleveland are well known for foundation swelling problems resulting from the oxidation of pyrite in the shale. Sulfuric acid is also produced and can attack concrete in the foundation.

Various measures are used to mitigate swelling problems (Linder, 1976). Some involve direct treating of the foundation soil or rock. The reactions can be stopped by preventing drying out, for example, by spraying exposed surfaces immediately with asphalt (Grattan-Bellew and Eden, 1975). Waterproofing below and around the foundation area is best done immediately after excavation to prevent drying and subsequent large expansions on rewetting. Flooding the foundation with water for a period before construction to trigger any swell can be effective. Chemical stabilization using salt or lime is used with varying degrees of success, particularly in road construction. Breaking up, saturating, and recompacting the swell-prone strata can make the problem worse.

Structural solutions are generally more expensive but more effective. A "rigid box" design is used to withstand the predicted swelling stresses; a flexible construction approach can limit damage while allowing differential heave to occur. The rock engineer may specify sheathed end-bearing caissons, piers, or piles carried below the deepest elevation of swell-prone strata, or to sufficient depths to ensure that stresses will not permit swell to occur at the bearing tip. As a last resort, complete excavation of the swell-prone strata and replacement with a nonswelling material may be considered.

3.4 Deep Foundations

3.4.1 Foundations on piles and caissons

When driving a pile to rock, the main consideration is the depth at which rock will be encountered. Irregularities in the topography of buried bedrock can be as dramatic as the cliffs and gulleys of exposed rock outcrops, but being buried beneath soil, they may not be fully revealed even by closely spaced drillholes. Driven piles are often deflected by irregular or sloping bedrock, or damaged by driving against hard protrusions. Some piles break, and others bend so badly that they never reach the required driving resistance. A driving shoe with a special carbide tip can help a driven pile bite into a steep rock surface, rather than be deflected by it.

3.4.2 Basement excavations

3.4.2.1 Wall squeeze mechanism. Because high horizontal compressive stresses are common, the walls of trenches and basement excavations in bedded rocks have a tendency to move inward in a mechanism called *bedding slip squeeze*. Blasting releases locked-in stresses and allows the horizontal beds to slide inward over one another like a deck of cards (Fig. 3.10 and *Rock Engineering,* Sec. 5.1.3.3).

Some movement occurs immediately after excavation, but movements can continue for years or even decades by sliding and creep of bedding planes along which shear strength is exceeded. Bedding slip squeeze can cause extensive damage to concrete structures poured directly against the rock walls. Shear displacement can occur for distances as great as several hundred meters from the excavation. Wall displacements can reach magnitudes of 200 mm or more, and displacements of 50 mm are commonplace (Franklin and Hungr, 1978; Quigley et al., 1978).

The "wheel pit" excavations for hydroelectric turbines in Niagara Falls, New York, provide a well-documented example where wall convergence has been going on since 1905 when the wheel pits were excavated (*Rock Engineering,* Sec. 5, Fig. 5.7). The movement has been

Figure 3.10 Bedding plane slip as a result of stress relief after blasting, dolomitic limestones, Ontario.

monitored and fluctuates annually but continues nevertheless. Horizontal convergences of about 50 mm during construction have been followed by an additional 1 mm each year, amounting to 70 mm during the period 1905 to 1970. "Rigid" concrete and masonry members spanning between the wheel pit walls have buckled. The turbines have been closed down because of consequent misalignment (Lo et al., 1975).

A similar problem arose at the primary crusher plant of Dufferin Quarry at Milton, Ontario. Cracks were developing in the concrete of the crusher building, which was constructed in a 20-m-deep slot in the quarry floor. The problem was diagnosed as differential movement between the base of the slot (the crusher foundation level) and the upper rock beds, which were also in contact with the concrete structure. The solution was to cut the upper beds with a pneumatic impact breaker, to isolate them from the building. The cracks, monitored by a sensitive measuring device, were observed to close, and the problem did not recur.

Various amounts and rates of squeeze have been monitored by inclinometer measurements alongside cut-and-cover sewer and water main excavations, for example, in the vicinity of Hamilton, Ontario, where precautions are now routine.

The same problem does not seem to occur when excavating into hard rocks with more or less random jointing. The normal pattern of behavior in such cases is for elastic inward wall movement of a few millimeters almost instantaneously during blasting. The rock walls and floor are usually stable by the time the basement concrete is poured.

3.4.2.2 Counteracting squeeze. Attempts have been made to counteract wall squeeze by a variety of measures including anchoring, time delays, and placing deformable buffers between the rock and the structure. Only the latter is reliable, as discussed below. The problem should be viewed in terms of displacements, because pressures generated by stress release are usually too great to be resisted by the building itself or by any feasible system of rock anchors. The solution is to accommodate rock movement, not to resist it.

A simple calculation demonstrates the futility of installing *tensioned anchors* to stop stress relief squeezing. Back-calculation shows that in reaching a wall displacement of 50 mm (such as often recorded), the adjacent rocks must have been stress-relieved for a distance of some 400 m (this is based on the limestone rocks of Southern Ontario with typical Young's modulus of 40 GPa, and a horizontal ground stress of 10 MPa). To reduce the movement by 50 percent (to 25 mm from the expected 50 mm), 200-t anchors, each 420 m long, would have to be installed horizontally at 900-mm centers. This is hardly worthwhile. Even vertical anchors installed from the crest of the excavation (if space were available) would have similarly insignificant results. The only useful role of rockbolts and anchors in the excavation walls is to protect against local gravitational failures, caused by wedge slides backed by steeply dipping joints.

Some reduction in the amount of stress transfer can be expected if there is a *time delay* between excavation and construction of the building. However, stress-release deformation is slow, and may not be complete even after many years. Wheel pit excavations in Niagara as described previously have been converging for the past 80 years (Sec. 3.5.1). Construction delays are therefore not to be relied upon, and more positive steps such as buffers or gaps are needed.

The suggested course of action is to assume that movements will occur and will continue, and for open excavations in horizontally bedded stressed rock, could amount to as much as 100 mm. An allowance of 50 mm has proven insufficient in some cases, notably in a Hamilton, Ontario, cut-and-cover trench where after detection of springline cracking in a concrete pipe, the trench backfill was excavated to find that the Styrofoam buffer had been compressed from 50 mm to about 10 mm.

Buffer design is based on a boundary displacement (rather than boundary stress) assumption. A gap equal to the assumed rock displacement (e.g., 100 mm) may be left open if the rock walls are stable or can be stabilized by bolting or shotcreting. Where necessary the gap is bridged across the top with a sliding slab or plate. Alternatively, the gap can be filled with a compressible buffer layer consisting, for example, of uncompacted soil, expanded polystyrene sheeting, or sprayed polyurethane foam. The buffer is designed by calculating its required thickness in terms of stress-strain characteristics of the buffer material and the predicted rock displacement. The layer must accommodate the maximum expected movements without unacceptable transfer of stress to the structure.

3.4.2.3 Effects on surrounding buildings.
When basement walls move inward by typically 50 mm, the surrounding rock also expands and moves toward the basement, carrying with it a burden of superimposed structures and services. However, ground strains in this example are not very great. The 50-mm displacement being spread over 400 m, an adjacent building measuring 50 m along its foundation wall would be subjected to a differential horizontal displacement of only about 6 mm. The authors know of no cases of reported damage attributed to this stress-relief expansion of ground adjacent to an excavation.

3.4.3 Base heave

3.4.3.1 Heave mechanisms. The same high horizontal stresses that cause inward squeeze of walls create high compressive stresses in the floor of a rock excavation. The floor tends to buckle (heave upward) when the constraining effect of overlying soil and rock is removed. The tendency for buckling is most pronounced when the rock beds are thin and horizontal and the stresses high.

Buckling and heave can cause severe settlement of footings; joints opened by heave will close when the weight of the building is applied.

Buckling and heave can be explosively violent or may pass unnoticed, but even the quiet type can cause severe settlement of footings. Ductile rocks such as shales will, like soils, undergo a generalized heaving with closely spaced shears and distortion of the overstressed rock. Brittle and dense limestones and igneous rocks in contrast tend to buckle, sometimes violently in a rockburst-like manner. Buckles in foundations are similar to the "pop-ups" now widely reported in the floors of quarries throughout southern Ontario (*Rock Engineering*, Sec. 5.1.3.3). Quarry workers have noted the appearance of these ridge-like features overnight or even during working hours, with "bumps" or more violent noises. Heave was observed in the invert of the Chippewa Canal when it was dewatered for maintenance some years ago.

The tendency for base heave is greater if *water pressures* are acting in horizontal joints. Heave can even occur under the action of water pressure alone, irrespective of the horizontal stress level. The hydraulic conductivity of horizontally layered rock is greater parallel to the layering, and without drain holes or open vertical joints, the vertical drainage tends to be poor. A stress-and-water type of base heave occurred in limestone-shale foundations for the Admiral Hotel in Toronto. A 10-mm bedding plane separation was observed between the limestone and the underlying shale in a foundation sump. The limestone layer appeared sound from above, but had to be removed to avoid settlement.

Base heave problems developed at the Morwell thermal power project in Australia, where coal is extracted from a seam 110 to 140 m thick. When the pit had reached 50 m in depth, extensive heaving occurred. High water pressures were temporarily relieved by drilling wells, but after deepening a further 30 m, heaving started again. Further well drilling and pumping brought the problem under control. Deep drainage continues, with the cone of depression detectable 50 km away and with the whole region displaying a significant, but uniform, subsidence because of consolidation.

3.4.3.2 Remedies for base heave. Base heave problems can be avoided by using vertical rockbolts to pin down the floor, which has the effect of increasing the rigidity, fixity, and weight of the foundation rock slab. The bolts should be installed a few at a time during staged excavation so that buckling is prevented by the weight of material yet to be removed.

Water pressures can be relieved by drilling an adequate number of vertical drain holes to several meters below the foundation level, again before the foundation is fully excavated.

3.4.4 Buoyancy uplift

A structure founded within an excavation that can become flooded by groundwater or surface runoff can be floated from its foundations if not securely anchored. In one such case, a sewage tank was pulled from its foundation and tilted to one side. It had been sitting in a recess blasted into the local granite hillside, anchored to the floor by rockbolts only 150 mm long. Drainage provisions had been ineffective, allowing groundwater to rise around the empty tank. The buoyant uplift was sufficient to pull out the anchors, which in the closely jointed granite should have been at least 10 times as long. Archimedes' principle can be used to compute the unbalanced buoyancy forces, which can be counteracted by bolting.

3.5 Foundation Investigations

The main purpose of a site investigation is to produce maps and sections that show variations in soil, rock, and groundwater characteristics. With the aid of computer graphics, the results can be compiled into a three-dimensional model of the site. Information to be gathered includes contours of pre- and postconstruction topography and of the soil-bedrock and weathered-fresh rock contacts, boundaries of geotechnical mapping units (GMUs) within both soil and rock, locations and dimensions of cavities, if present, and the groundwater table elevation. In addition, characteristics of each GMU have to be determined by a combination of visual observations and index testing (*Rock Engineering,* Chap. 6). Key issues to be addressed by an investigation include:

- Feasibility of construction
- Preferred site or route
- Most appropriate type of foundation
- Depths to reach satisfactory bearing strata
- Difficulties and costs of excavating to these depths
- Bearing pressures, stability, and settlements

- Merits and costs of grouting, dewatering, and other treatments
- Precautions needed during construction
- Requirements for monitoring and maintenance

Geophysical methods and drilling can be particularly useful for rapid bedrock profiling and water table delineation. For GMU characterization, however, sampled soil borings and rock diamond drillholes are needed unless the depths of interest are such that they can be reached with test pits and outcrop measurements alone. The depths explored have to be sufficient to determine the effects of elevated ground stresses beneath the loaded area, usually several footing diameters below the deepest proposed footing. This is because the bearing horizon and footing type may need to be modified depending on the quality of strata found at that depth, and because the bearing capacity and settlement depend on the ground below, not at, footing level. An approximate guide to the depths and spacings of boreholes is given in Table 3.2.

3.6 Foundation Design

3.6.1 Design methods

For routine construction of residential, commercial, and light industrial buildings, foundation design is often guided by local building codes, which incorporate conservative factors of safety against excessive settlement or rupture of rock beneath the footing. These codes establish permissible foundation bearing pressures according to local practices and experience.

Foundation behavior is modeled numerically only if the rock is soft

TABLE 3.2

Typical depths and spacings of boreholes for foundation investigations (courtesy Trow Ltd. Consulting Engineers, Toronto, Canada). If adverse ground conditions are encountered, the numbers and depths of borings are increased.

Types of Project	Depth (m)	Spacing (m)
Roads	1.5–2.0	50
Sewers	1.5 below invert.	30–50
Parking lots	1.5–2.0	50
Light buildings	3–6	30
Medium buildings	7–12	25–30
Heavy buildings	12–20	15–25
Bridges	10–20	At least 1 at each abutment and pier
Dams	At least 10	1.5 times the height of the dam

or closely jointed, the loading unusually high, or the structure unusually sensitive. In these cases, predictions of settlement are compared with structural tolerances.

The designer has to choose an appropriate foundation type and depth, and has to assess and guard against each possible type of foundation failure (discussed in Secs. 3.3 and 3.4). Failure is defined as a movement sufficient to damage the structure. It is not necessary that the rock rupture entirely, because damage usually ensues from excessive deformations. Peck (1976) points out that settlement is a much more pertinent design criterion than the classical concept of bearing failure, and this is particularly valid for rocks.

3.6.2 Choice of site or route

When constructing on the small and highly valued sites that become available in cities, the geotechnical study is limited to determining costs and solving problems particular to that location. The only choices available are the depth and type of footing, and the extent and type of foundation preparation and treatment.

Away from congested urban centers, particularly for transportation routes, several alternative sites, designs, and depths are generally available. Highways and pipelines can be rerouted or placed on fills or viaducts, or in cuts or tunnels (Fig. 3.7 and Sec. 2.4.2). Buildings can be sited to take advantage of the easiest possible excavating conditions, and sometimes the option is available to orient them with respect to the directions of jointing for optimum stability of the basement excavation. Rejection of one site in favor of another is usually possible in a rural environment.

3.6.3 Choice of foundation type and depth

Questions of foundation elevation, type, and treatment have to be considered together. Often alternatives range from slab-on-grade to footings on rock with or without grouting to piled foundations taken to deeper bearing strata. Rational selection is possible only when construction costs are calculated and weighed against adequacy of the foundation. This requires a good site model and reliable cost information for the various options. Foundations partly on rock and partly on soil are usually avoided because of the possibility of differential settlement.

Piles and caissons are driven or augered to "refusal" in the rock. Foundations of other types are most often placed below the level of the rock-soil interface, and penetrate the layer of weathered and loosened rock. Foundation strata are compressible and weak when they contain closely spaced joints with large apertures, filled with substantial amounts of compressible material such as clay. These conditions are found closest to surface, so bearing capacity can be increased and settlement reduced if necessary by carrying the foundation deeper.

However, the cost of carrying a foundation to a competent bearing stratum increases with depth, and even rock-bearing piles become impractical when they are longer than about 50 m.

3.6.4 Allowable bearing pressures

The bearing pressures permitted by typical building codes are only a small fraction of the strength of the weakest rock. Plate load and punching tests show that stresses of 10 to 100 times those prescribed as "allowable" can often be applied before the rock actually yields or ruptures. For most buildings this is of little consequence, because the cost of a high factor of safety is only a small fraction of the total cost of construction. Such a degree of conservatism is unacceptable for transportation routes, where foundation costs are a significant part of total costs.

Large safety factors can be justified as a precaution against inadequate foundation preparation. The bearing capacity of a foundation containing loose rock, rubble, mud, or other debris is evidently much lower than that of a clean, hard rock surface. Under the normal conditions of construction and inspection it is often difficult to guarantee the quality and completeness of foundation cleaning and scaling.

Because settlements are governed by the intensity of jointing, allowable bearing pressures can be assessed as a function of RQD (Fig. 3.11). The Canadian Foundation Engineering Manual (1985) offers a

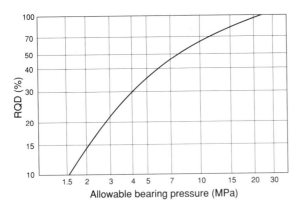

Figure 3.11 A prediction of allowable bearing pressure from Rock Quality Designation (Peck, 1976).

table of "presumed allowable bearing pressures" based on a combination of strength and joint spacing (Table 3.3) together with a formula that gives allowable bearing pressure as one-tenth of uniaxial compressive strength for joint spacings in the range 0.3 to 1.0 m, and one-quarter of uniaxial compressive strength for joint spacings of 1 to 3 m.

3.6.5 Differential settlements

3.6.5.1 Consequences of settlement. Uniform settlement is usually of much less consequence than differential movement. The Palace of Fine Arts in Mexico City, founded on silt and clay, has "settled" uniformly by about 3 m since construction, and the building is still intact although the second floor is now the first. The Italian-American architect spurned engineering advice, and is on record as saying that "if the structure is pleasant to my eye, it is structurally sound" (Leggett, 1962).

Differential settlements are of greater importance, and structures such as arch dams and nuclear reactors are particularly vulnerable

TABLE 3.3
Allowable bearing pressures as defined in the *Canadian Foundation Engineering Manual* (1985). "Sound condition" here means "minor cracks spaced no closer than 1 m." The tabulated values do not apply for stratified and foliated rocks with inclined discontinuities, which require evaluation by a rock engineer.

Types and conditions of rocks	Uniaxial compressive strength of rock material	Presumed allowable bearing pressure (MPa)
Massive igneous and metamorphic rocks (granite, diorite, basalt, gneiss) in sound condition	High to very high (50 to 200 MPa)	10
Foliated metamorphic rocks (slate, schist) in sound condition	Medium to high (15 to 50 MPa)	3
Sedimentary rocks: cemented shale, siltstone, sandstone, limestone without cavities, thoroughly cemented conglomerates, all in sound condition	Medium to high (15 to 50 MPa)	1 to 4
Compaction shale and other argillaceous rocks in sound condition	Low to medium (4 to 15 MPa)	0.5
Broken rocks of any kind with moderately close spacings of discontinuities (0.3 m or greater), except argillaceous rocks (shale)		1
Limestone, sandstone, shale with closely spaced bedding		Assess in situ
Heavily shattered or weathered rocks		Assess in situ

(Geddes, 1977). Settlements that continue long after construction cause more damage than short-term ones because they cannot be accommodated before the structure is in service. Peck (1976) points out that in nuclear power plants, until the individual heavy components are joined structurally, the magnitudes of settlement are largely irrelevant. By delaying connection until most of the loading has been imposed, differential settlements can be rendered inconsequential.

At the site of a proposed nuclear particle accelerator at Mundford, England (Ward et al., 1968), even very small differential settlements could not be tolerated. The chalk rocks of the foundation were classified into five grades according to block size, and a single large-diameter loading test was conducted with a tank that was 18 m in diameter and capable of exerting a load of up to 180 kPa when filled with water. Plate tests were also conducted at the bottom of auger holes at various elevations in the chalk of various grades. The exhaustive field examination and testing program, aided by back-analysis, led to a reliable understanding of in situ chalk rock deformation. The European community nevertheless decided to locate the accelerator in Geneva.

3.6.5.2 **Settlement prediction.** In rock, elastic displacements usually amount to a few millimeters and are inconsequential except for the most sensitive of structures. Calculations that assume elastic behavior therefore give only optimistic predictions. Settlements can become much more substantial when the rock behaves inelastically by sliding or yielding. Foundations on weak rocks can exhibit viscous flow under high contact stresses and may require a study of long-term (time-dependent) behavior. By far the majority of serious problems are caused by cavities in the rock, resulting from solution or mining. Nevertheless, elastic calculations are easy and continue to be made, if only as a guide to optimum behavior. If settlements are monitored during construction, a comparison of measured and predicted settlements will demonstrate if behavior is other than elastic.

Analysis of foundation behavior can make use of analytical elastic solutions, such as those of Boussinesq and Cerutti, or a numerical method such as finite element analysis, which may include inelastic behavior and inhomogeneity in the foundation rocks. The rock mass may be considered as an equivalent continuous medium, at least as a first approximation (Hobbs, 1975). D'Appolonia et al. (1976) describe equivalent continuum finite element analysis applied to the arch foundations for the Olympic Sports Complex in Montreal, in which horizontal displacement of each abutment had to be less than 12 mm. After grouting and anchoring the limestone bedrock, actual displacements measured during construction amounted to less than 7.5 mm.

In situ dilatometer or plate loading tests (*Rock Engineering,* Sec. 9.2) are used to determine the elastic parameters for equivalent continuum analyses (Fig. 3.12). The influence of scale and loading duration must always be assessed in these tests. Load path and stress history effects are important, because irreversible behavior of the rock mass results from the closing of fissures. The modulus for first loading is lower than for subsequent cycles. Foundation problems can result if the low modulus at first loading is ignored or averaged, when in reality it may create the most critical conditions. Load cycling can occur at different rates in different parts of the foundation area, leading to differential settlements.

To interpret plate-bearing tests, it is common to use an equation derived from elastic theory, and a value of Poisson's ratio is assumed. This may be valid in an unjointed continuum, but it is a dubious practice in a jointed medium where the effective Poisson's ratio can be variable and difficult to predict. The use of small-scale laboratory tests on intact rock is questionable in application to foundation analysis. Perhaps the best approach is to use several plate-bearing tests at as large a scale as possible, and then to derive the equivalent continuum parameters that most closely predict observations.

Analyzing the data is easier if deformation data are obtained from a multipoint extensometer located centrally beneath the plate, as well as from several surface deformation gauges away from the bearing plate. For vertical loads, a Boussinesq-type solution for the layered case can be used, but no similar equivalent exists for the Cerruti horizontal force solution.

Often, an assumption of isotropy is unrealistic. Gaziev and Erlikhman (1971) have shown graphically the extreme differences in stress distributions that can occur beneath a loaded foundation as a result of variations in rock jointing when the rock is layered and anisotropic (Fig. 3.13). Compression tends to be guided along deep and narrow bands within the rock mass and to close the fissures within these bands. In a heavily loaded foundation, this can have a substantial effect on the hydraulic conductivity of the rock mass, and therefore on water flows and pressures (Sec. 4.4.4).

3.6.6 Design of piled foundations

Piles carry their load by a combination of side shear and end bearing. In soft rock and soils, the contribution of end bearing is often insignificant, particularly when the piles are long and may not be straight (Horvath et al., 1981). Also, piles that have to resist uplift forces rely entirely on side shear. End bearing can be unreliable because of debris accumulations that are difficult to clean from the face of a pile socket.

140 Rock Engineering at Surface

(a)

(b)

Figure 3.12 Plate bearing test on a deep chalk foundation, Littlebrook power station, England. (*a*) Anchors and reaction beam at the head of the shaft; (*b*) reaction transferred through a steel column. (*Photos courtesy Soil Mechanics Ltd., U.K.*)

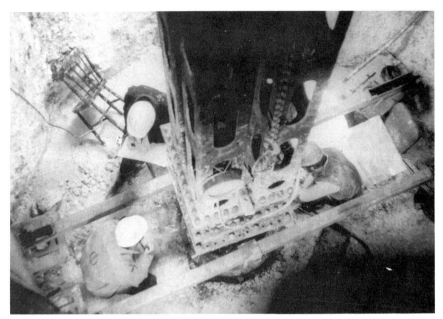

Figure 3.12 (*Continued*) Plate bearing test on a deep chalk foundation, Littlebrook power station, England. (c) Testing in progress at the base of the shaft. (*Photo courtesy Soil Mechanics Ltd., U.K.*)

In the case of a rock pile through extremely soft soils such as sensitive clays, only end bearing can be relied upon.

Finite element calculations can be used to predict pile settlements and stress distributions from estimated or measured values for shear strength and elastic properties of the concrete and the rock. Chiu and Donald (1983) describe such an analysis using a variable modulus strain-softening model applied to piles in the Melbourne mudstone of Australia.

Pile loading tests give a check on the assumptions of design and on the adequacy of pile installation. A concrete plug is cast into the pile socket on top of a deformable sheet of expanded polystyrene so that load bearing is entirely by side shear. The plug can be loaded either downward or upward, but in the latter case, check calculations are needed to ensure that the rock surrounding the pile is not being fractured. Load and corresponding displacement are measured.

3.6.7 Nuclear power plant foundations

Foundation design for a nuclear power station provides an example where the most rigorous methods must be applied, both for site investigation and design, in order to satisfy the standards set by regulatory authorities and to meet public concerns. Detailed reports are required

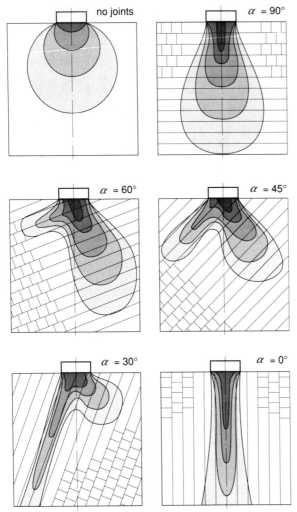

Figure 3.13 Stresses in layered foundations as a function of the angle α between the primary joints and the direction of loading (Gaziev and Erlikhman, 1971).

to demonstrate that there are no geological problems associated with active faults, solution or mining cavities, or landslides (Swiger, 1976; McClure and Hatheway, 1979).

The containment vessel for a pressurized water reactor, the largest and heaviest structure on the site, typically imposes a foundation pressure in the range 330 to 380 kPa, at a depth of 10 to 20 m below surface. Under the worst scenario of a maximum design earthquake at the same time as failure of the primary coolant system, the containment vessel becomes pressurized to about 345 kPa, which causes a radial expansion of about 18 mm. To permit this expansion, the containment walls must be separated from the rock.

All structures are designed to resist large earthquakes loads. The reactor structure and its foundations are numerically analyzed to simulate two levels of artificial earthquake: the *operational basis earthquake* (OBE) for which all nuclear systems must remain operational and the more severe *safe shut-down earthquake* (SSE) for which only those systems must remain functional whose failure could release radioactivity.

3.6.8 The Toronto CN Tower

The CN Tower in Toronto at a height of 550 m is the world's tallest freestanding structure. Although founded on a shale bedrock of only moderate strength, stresses imposed on the foundation are modest, and the design relied more on a thorough investigation than on rock mechanics calculations (Robinsky and Morton, 1973). The concrete tower broadens at the base forming a tripod with legs 19 m wide by 33 m long measured from the center of the tower (Fig. 3.14). The investigation made use of 760-mm-diameter, 21-m-deep observation wells into the shale bedrock beneath the bearing area, in addition to five NX exploratory core holes at the perimeter of the proposed excavation. Piezometers were installed at various depths. Additional NX holes were drilled at the bottom of each of the observation wells. The deformability modulus was estimated from in situ sound velocity and other index measurements combined with results of previous plate loading tests in adjacent parts of downtown Toronto.

Beneath 10 m of sandy fill and beach deposits was shale with a uniaxial compressive strength of 9 to 25 MPa, with some weak rubble zones and clay seams, and with some stronger interlayers of siltstone and limestone. RQD averaged 50 to 80 percent, dropping to zero in the rubble zones. The foundation elevation was chosen at about 6.7 m below bedrock surface to avoid one of the thickest and most persistent clay seams.

Design calculations made use of elastic solutions (Boussinesq equations and Newmark charts), and were based on an assumed deformability modulus of 3.7 GPa, an average for the foundation rocks. The weight of the CN Tower was estimated at 107,200 t. Wind loading for a 500-year return period amounted to 12,900 MN · m of moment, combined with a horizontal force of 5230 t. The foundation was dimensioned to ensure that the peak resultant force would pass within the middle third of the base.

The final design resulted in a foundation bearing stress of 580 kPa under no-wind conditions, and a maximum peak bearing stress of 2.59

(a)

(b)

Figure 3.14 CN Tower, Toronto, Ontario. (a) Model of Toronto and the CN Tower before construction; (b) foundation excavations into the Georgian Bay shale formation. The overlying clay-silt till is protected by plastic sheeting weighed down with car tires. (*Photos courtesy E.I. Robinsky Associates Ltd.*)

MPa, which compares with typical design values of 7.2 MPa for caissons in the Toronto shales. These conservatively low bearing stresses were recommended to preclude the possibility of fatigue breakdown of the softer rock strata as the result of cyclic wind loading. A total heave of about 6 mm was predicted, followed by a settlement of 6 mm comprising about 1 mm of elastic compression with the balance in creep and consolidation of clay layers.

3.7 Foundation Construction

3.7.1 Piles and caissons

Bored piles, piers, and caissons are made by pouring concrete into a rock socket. In caisson construction, a large-diameter auger penetrates through soil and for a limited distance through soft or weathered rock until a satisfactory bearing horizon is reached. Irregular hard rock at the foot of a caisson is cut out with a chopping bit that is dropped repeatedly onto the rock surface to provide satisfactory seating.

When the water table is high, a water seal must be obtained at the base of the casing to prevent inflows of groundwater and soil during the placing of concrete. In deep caisson holes, the concrete must be *tremied* (pumped down a pipe to the base of the hole) to prevent aggregate segregation or dilution of the concrete by groundwater. A cage of reinforcing steel is usually placed in the caisson hole before concrete placement.

Caisson sockets must be inspected and probed in situations where pockets of weathered rock could extend beneath the apparently hard rock surface. Probe drilling with recording of penetration rates is one way of doing this. If a soft seam is found, the caisson is excavated deeper or the rock is reinforced by grouted steel dowels.

Difficulties were experienced in the foundations for an office building near Toronto, where more than a hundred caisson sockets had to be augered into bedrock through 7 m of water-bearing sands and gravels. The rock surface, described as shale in the site investigation reports, was found to be a hard limestone. A satisfactory water seal was difficult to obtain. Inspections of the caisson holes as they neared completion uncovered a buried river channel crossing the site, along which the rock surface had been incised into a pattern of waterfall-like steps, with rounded, gravel-filled clefts (*clints*) more than a meter deep and as wide as the inspector's arm. The auger was replaced by a chopping bit to remove the hard limestone by percussion to a depth of about a meter, sufficient to cut off water inflow through the clints. A dispute regarding whether these conditions could have been foreseen was resolved in favor of the contractor.

3.7.2 Control of groundwater

Dewatering may be required to give temporary increased stability to the walls of a deep basement or caisson excavation and to avoid the hazards of working in excessively wet conditions (*Rock Engineering*, Sec. 18.1). Small inflows can be pumped from sumps within the excavation. When quantities are expected to be too great for pumping, inflows can be reduced by grouting around the excavation perimeter, or by pumping from perimeter wells. Hydrogeological methods are used to design these systems, with piezometers to assess and monitor performance.

3.7.3 Precautions during excavation

Foundations require stripping of weak overburden, and deep basements are excavated further into sound rock. Most excavations into rock require blasting, but rippers and toothed mechanical excavators can be less disruptive and cheaper in soft or weathered formations (Fig. 3.14*b* and *Rock Engineering*, Chaps. 13 and 14).

In deep foundations, geometry of the *excavation perimeter* must be kept simple. Attempts to save concrete by leaving noses of rock often fail because of the high cost of careful blasting and bolting. Rock cannot be cut like cheese except in favorable situations (e.g., Fig. 2.17) and an elaborate rock perimeter that looks simple on paper may turn out to be difficult or impossible to excavate.

Allowance is made for *blast overbreak* when estimating and writing specifications for a contract. Overbreak can undermine adjacent buildings and services and may necessitate expensive backfilling with concrete. Inaccurate blasthole drilling is the main cause. Also, blasthole depth must be monitored to avoid excessive subdrilling beneath foundation elevations, which will loosen rock beneath the footings and increase scaling requirements. Smooth-wall blasting methods are often used to minimize disturbance. The extra cost of closely spaced and lightly loaded drillholes is offset by reduced costs for wall scaling and perimeter support.

Because of the proximity of buildings and services to the perimeters of most urban foundation excavations, blasting must be very carefully controlled. *Vibration levels* are minimized by limiting the charge weight per round and by the use of delayed detonation. Blasting mats are placed over the rock surface before detonation to limit fly rock.

3.7.4 Scaling and support of excavation walls

Depending on their height and on the quality of the rock, foundation excavation walls may stand unsupported or may require shoring or anchoring. Rockbolts are used for stabilization, supplemented by mesh or shotcrete when the rock is expected to deteriorate on exposure (*Rock Engineering,* Chaps. 16 and 17). Small loose material must be scaled or bolted in place, and potentially unstable larger rock wedges and slabs need to be identified and anchored. Deep foundation walls are scaled during excavation, and should be bolted from the top down while access is easily available. This is safer, more convenient, and less expensive than scaling and bolting the completed high rock faces.

Faces are best inspected after washing, because joints are most visible when partly wet, and because exposed faces soon become covered with a coating of dust raised by blasting and construction.

3.7.5 Scaling of footings

Rock exposed after overburden stripping requires local trimming at locations of footings or slabs, and may need extensive scaling to remove weathered, jointed, or blast-shattered materials (Fig. 3.15). Large foundations almost always are traversed by one or several shear zones or pockets and seams of infilling or weathering which may have to be removed or grouted.

Excavation and replacement by concrete (*dental work*) is the most economical alternative for thin layers of poor quality rock, but this technique becomes prohibitively expensive when the rocks are deeply

Figure 3.15 Air jet cleaning of footing, Science North.

weathered or closely jointed to substantial depths. In such cases it is cheaper to grout the rock.

The excavation can be done by pneumatic tools or high-pressure water jets to avoid possible blast damage to the foundation. Jackhammer drills or mobile, truck-mounted impact hammers on articulated hydraulic arms may be used to remove shattered material, underbreak, and loose grout, and to bring the rock surface to the required elevation and degree of competence. The rock surface must be sound, relatively flat, and free from overhangs; otherwise, voids will be formed when concrete is poured. These voids are sources of weakness and future settlement when the foundation is for a building, but more importantly, are potential leakage paths when the foundation is for a dam.

Soft seams should be uncovered, excavated, and backfilled with concrete. Weathered or sheared rock can be subexcavated to a depth of 1.5 to 2 times the slot width, and backfilled with concrete. For example, 20,000 m^3 of concrete were required to thoroughly fill several fault seams in the Nagawado dam abutments (Fujii, 1970). Extra reinforcing steel may be needed in the footings to bridge across foundation soft spots.

3.7.6 Rock protection and improvement

A shotcrete or troweled concrete sealing coat is often applied immediately after excavation to protect shales and weathered rocks that are sensitive to moisture and thermal variations.

Consolidation grouting (*Rock Engineering,* Sec. 18.2.2) can be used to increase the stiffness and strength of a foundation. By filling open fissures with a grout whose properties are similar to those of the rock, the bulk mechanical characteristics are upgraded nearly to those of the intact rock material. In particular, there is a reduction in the irreversible component of rock mass deformation and an increase in the rigidity. Consolidation grouting is applied to the near-surface rock of almost all heavily loaded foundations and sensitive structures where foundation deformations must be minimized. Typical applications are foundations for arch dams and nuclear power plants, and those for heavy machinery where alignment of components must be maintained to within close tolerances.

High-strength grouts are essential for consolidation purposes. Portland cement is preferred to chemical grouts that are usually too weak (the silicates) or too expensive (the epoxies and other polymers). Neat cement-water grouts are the most common, although sand-cement and flyash-cement are less expensive alternatives for filling larger cavities. Often it is better to grout a foundation before, rather than after, the stripping of overburden, to permit the use of higher grouting pressures without hydraulically fracturing the rock. Secondary grouting may be needed as a precaution after the foundation becomes exposed.

3.7.7 Foundation anchoring

Foundation anchors and dowels may be required to resist uplift forces, for example, cable anchors for guyed towers and suspension bridges, and doweling systems to resist groundwater buoyancy (Fig. 3.16). These must resist external forces, in contrast to the anchors for rock reinforcement which are subjected only to internal tensions and shears generated by movements within the rock mass (*Rock Engineering*, Sec. 16.2; Pease and Kulhawy, 1984; Callanan and Kulhawy, 1985).

Anchors must be designed with a sufficient factor of safety against failure of the anchor steel, of the bond between grout and steel, and of the bond between grout and rock. For economy, the factors of safety should be similar in each of the three cases. Anchors subjected to external loading can fail in a fourth mode, by conical rupturing or lifting of beds. The factor of safety against rock mass failure is then determined using the limiting equilibrium method, by comparing uplift forces with the gravitational and frictional forces resisting uplift. Unless the rock is weak or the anchor shallow, shearing will occur along rock joints rather than through intact rock material. Sometimes joints are conservatively assumed to have little or no strength, so the entire uplift force must be resisted by the dead weight of a rock cone with its apex near the base of the anchor. The cone angle is often assumed to be 45°.

As a check, pullout tests can be conducted in the field. Low-capacity foundation dowels, with a holding capacity in the order of 10 to 20 t, are easy to test, but the higher-capacity tendons are more difficult. Heavy steel beams are needed to carry reactions to areas of rock remote from the anchor to allow freedom of rock uplift. The rock directly surrounding the anchor cannot be used for reaction as in testing for the strength of bond or of anchor steel. As an alternative, a high-capacity crane can be used to provide the pull force.

3.7.8 Sounding, probing, inspection, and testing

Foundations are usually inspected and approved just before the footing concrete is poured. Thorough cleaning is essential, using water or air jets, and preferably both. Standing water is removed by brushing

(a)

(b)

Figure 3.16 Foundation anchoring at Science North Sudbury (Franklin and Pearson, 1985). (a) Testing an inclined anchor for rock reinforcement; (b) one of the heavily loaded pier foundations with anchors installed, ready for concreting. As a precaution against blast damage to the fault-broken rock, the front face of this foundation was protected by line-drilling.

or pumping to expose the complete foundation together with any jointing and fracturing, and to allow the inspector to distinguish between rock, concrete, and compacted debris.

Scaling bars and sledgehammers are used to sound the rock surface. Competent rock gives a sharp rebound and a ringing sound, whereas loose or slabby materials are detected by their "drummy" sound. Pneumatic air track drills can probe beneath the exposed surface, and drill penetration rates indicate rock quality and reveal cavities and hidden soft spots. Rock quality can be confirmed by index testing using point load, Schmidt rebound hammer, or cartridge gun test techniques (*Rock Engineering,* Sec. 2.3).

Inspection of rock conditions goes hand in hand with quality control of foundation anchors and dowels, and sampling of foundation concrete for testing in the laboratory. Grouting is monitored to check the quantities, consistencies, and pressures of grout injected. Rock surfaces and structures adjacent to the work are instrumented to check for possible damage by grout heave or blasting vibrations.

Monitoring of a foundation (*Rock Engineering,* Chap. 12) is an important extension of design, and often warrants the installation of instruments, particularly when the rock is of doubtful quality or difficult to explore or to model by numerical methods.

The instrumentation should be installed as early as possible so that it can be used to monitor behavior continuously, both during and after construction. Blast vibrations are monitored as a precaution against damage to the rock and to surrounding buildings and services. Piezometers measure recovery of the water table and the efficiency of foundation drainage systems. Settlement monitoring instrumentation and pressure cells may be needed to confirm the displacements and pressures predicted by the design, and to check the long-term performance of the foundation.

References

Amir, J. M., 1985. *Piling in Rock.* A. A. Balkema, Rotterdam, Netherlands, 112 pp.

Beck, B., (ed.), 1984. *Proc. 1st Multidisciplinary Conf. on Sinkholes,* Orlando, Fla., Balkema, Rotterdam, 429 pp.

Bhattacharya, S., Singh, M. M., and Chen, C. Y., 1984. "Proposed Criteria for Subsidence Damage to Buildings." *Proc. 25th U.S. Symp. Rock Mech.,* Northwestern Univ., Evanston, Ill., pp. 747–755.

Brawner, C. O., 1968. "The Influence and Control of Groundwater in Open Pit Mining." *Proc. 5 Canad. Symp. Rock Mech.,* Toronto; also *Western Miner,* vol. 42, no. 4, 1969, pp. 42–55.

Burland, J. B., and Lord, J. A., 1970. "The Load-Deformation Behavior of the Middle Chalk at Mundford, Norfolk: A Comparison between Full-Scale Performance and In Situ and Laboratory Measurements." *Proc. Conf. on In Situ Investigations in Soils and Rocks,* London, England, May 1969, Publ. British Geotech. Soc., pp. 3–15.

Callanan, J. F., and Kulhawy, F. H., 1985. "Evaluation of Procedures for Predicting Foundation Uplift Movements." Report EL-4107 by Cornell University Geotechnical Engineering Group for Electric Power Research Inst., Palo Alto, Calif.

Canadian Foundation Engineering Manual, 1985. 2d. Edition, Can. Geotech. Soc., Tech. Committee on Foundations. Bitech Publishers, Vancouver, 456 p.

Chiu, H. K., and Donald, I. B., 1983. "Prediction of the Performance of Side Resistance Piles Socketed in Melbourne Mudstone." *Proc. 5th Int. Cong. Rock Mech.*, Melbourne, Australia, vol. C, pp. 235–243.

D'Appolonia, E. F., Shaw, D. E., Richard, J., and Raynaud, D. A., 1976. "Finite Element Analyses of Arch Abutments." In: *Rock Engineering for Foundations and Slopes. Proc. ASCE Specialty Conf.*, Boulder, Colo., vol. 1, pp. 55–81.

Franklin, J. A., and Hungr, O., 1978. "Rock Stresses in Canada, Their Relevance to Engineering Projects," *Rock Mech.*, Springer, New York, suppl. 6, pp. 25–46.

Franklin, J. A., and Pearson, D., 1985. "Rock Engineering for Construction of Science North, Sudbury, Ontario." *Canad. Geotech. J.*, vol. 22, pp. 443–455.

Fujii, T., 1970. "Fault Treatment at Nagawado Dam." *ICOLD. 10th Int. Cong.*, Montreal, Report O37-R59.

Gaziev, E. G., and Erlikhman, S. A., 1971. "Stresses and Strains in Anisotropic Rock Foundation (Model Studies)." *Rock Fracture, Proc. Int. Symp. Rock Mech.* ISRM, Nancy, France. Report II-1.

Geddes, J. D. (ed.), 1977. *Proc. Conf. on Large Ground Movements and Structures*, John Wiley, Toronto, 1064 pp.

Grattan-Bellew, P. E., and Eden, W. J., 1975. "Concrete Deterioration and Floor Heave due to Biogeochemical Weathering of the Underlying Shale." *Canadian Geotech. J.*, vol. 12, no. 3, pp. 372–378.

Guenot, A., and Panet, M., 1983. "Numerical Analysis of Rock Masses with Planar Structure" (in French). *Proc. 5th Int. Cong. Rock Mech.*, Melbourne, Australia, vol. F, pp. 181–185.

Hobbs, N. B., 1975. "Factors Affecting the Prediction of Settlement of Structures on Rock: With Particular Reference to the Chalk and Trias." *Proc. Conf. on Settlement of Structures*, Cambridge, England, Pentech Press, London, pp. 579–610.

Horvath, R. G., Kenney, T. C., and Kozicki, P., 1981. "Influence of Shaft Roughness on Field Performance of Drilled Piers Socketed into Weak Shale." *Proc. 34th Canad. Geotech. Conf.*, Fredericton, New Brunswick, 13 pp.

Ladanyi, B., 1972. "Rock Failure under Concentrated Loading." In: *Basic and Applied Rock Mechanics, Proc. 10th Symp. Rock Mech., Austin, Texas (1968).* SME-AIME, New York. Chap. 13, pp. 363–387.

Legget, R. F., 1962. *Geology and Engineering*, 2d. ed., McGraw-Hill, New York, 884 p.

Linder, E., 1976. "Swelling Rock: A Review." In: *Rock Engineering for Foundations and Slopes. Proc. ASCE Specialty Conf.*, Boulder, Colo., vol. 1, pp. 141–181.

Lo, K. Y., Lee, D. F., Palmer, J. H. L., and Quigley, R. M., 1975. "Report on Stress Relief and Time-Dependent Deformation of Rocks," Can. Nat. Res. Council Spec. Project 7307, Univ. Western Ontario, London, Ont., 320 pp.

McClure, C. R., Jr., and Hatheway, A. W., 1979. "An Overview of Nuclear Power Plant Siting and Licensing." In: *Geology in the Siting of Nuclear Power Plants. Reviews in Eng. Geol.*, vol. 4., Geol. Soc. Am., pp. 3–12.

Murphy, P. J., Briedis, J., and Peck, J. H., 1979. "Dating Techniques in Fault Investigations." In: *Geology in the Siting of Nuclear Power Plants. Reviews in Eng. Geol.*, vol. 4., Geol. Soc. Am., pp. 153–168.

Pease, K. A., and Kulhawy, F. H., 1984. "Load Transfer Mechanisms in Rock Sockets and Anchors." Report EL-3777 by Cornell University Geotechnical Engineering Group for Electric Power Research Inst., Palo Alto, Calif.

Peck, R. B., Hanson, W. E., and Thornburn, T. H., 1974. *Foundation Engineering.* 2d Ed., John Wiley, New York, 514 pp.

———, 1976. "Rock Foundations for Structures." In: *Rock Engineering for Foundations and Slopes. Proc. ASCE Specialty Conf.*, Boulder, Colo., vol. 2, pp. 1–21.

Quigley, R. M., Thompson, C. D., and Fedorkiw, J. P., "A Pictorial Case History of Lateral Rock Creep in an Open Cut into the Niagara Escarpment Rocks at Hamilton, Ontario," *Can. Geotech. J.*, **15**(1), 128–133.

Robinsky, E. I., and Morton, J. D., 1973. "Foundation Investigations for CN Tower, Toronto." *Proc. 26th Canad. Geotech. Conf.,* Toronto. 23 pp.

Sowers, G. F., 1976. "Foundation Bearing in Weathered Rock." In: *Rock Engineering for Foundations and Slopes. Proc. ASCE Specialty Conf.,* Boulder, Colo., vol. 2, pp. 32–41.

Swiger, W. F., 1976. "Foundations in Rock for Nuclear Power Stations." In: *Rock Engineering for Foundations and Slopes. Proc. ASCE Specialty Conf.,* Boulder, Colo., vol. 2, pp. 43–54.

Terzaghi, R. D., 1970. "Brinefield Subsidence at Windsor, Ontario." *Proc. 3d Symp. on Salt,* vol. 1, pp. 298–207.

Toulemont, M., 1970. "Geological Observations of the Dissolution of Gypsum in the Paris Area" (in French). *Proc. 1 Int. Cong., Int. Assoc. Eng. Geol.,* Paris, vol. 1, pp. 62–73.

Ward, W. H., Burland, J. B., and Gallois, R. W., 1968. "Geotechnical Assessment of a Site at Mundford, Norfolk, for a Large Proton Accelerator." *Geotechnique,* vol. 18, pp. 399–431.

Chapter 4

Dams and Reservoirs

4.1 Overview of Dam Construction

Dams and reservoirs are constructed for flood control, water supply, irrigation, hydroelectric generation, or pumped storage, or for recreation or navigation. The first dams appear to have been built in Egypt and Iraq around 2900 B.C. The Sadd el Kafara dam 30 km south of Cairo, circa 2500 B.C., was constructed from a 36-m-thick earth core between rubble masonry walls (Schnitter, 1967). Having no spillway, it was overtopped and destroyed soon after completion. A 100-km dike impounding Lake Hungtze in China, built in the sixteenth to seventeenth centuries, was one of the longest of early dams. The highest of the known Roman dams (24 m) was the Cornalbo earth dam near Marida in southwest Spain.

The arch, introduced by the Romans for construction of bridges, was first used in dam engineering during the sixth century. The oldest arch dam still preserved was built in the fourteenth century on the Kebar River, 150 km southwest of Tehran, Iran. It is a 26-m-high 55-m-long masonry structure with a uniform wall thickness of less than 5 m (Goblot, 1965; Mary, 1968). Arched masonry dams were constructed in Spain and Italy in the sixteenth and early seventeenth centuries (Mary, 1968). The height of arch dam did not surpass 50 m until after World War I, but reached 261 m with Vaiont in Italy in 1961. The Marèges dam, built in 1935, was a thin cupola, with a crest of 247 m and a height of 90 m (Walters, 1962). The Glen Canyon arch dam in Arizona (Fig. 4.1), built in 1964, is not one of the highest (216 m) but has an exceptional crest length (475 m) and volume of concrete (3.75 million cubic meters). The Inguri arch dam in Georgia, U.S.S.R., rises to a majestic 271 m from a jointed rock foundation that some regarded as too deformable for construction of a dam of such magnitude (Fig. 4.2).

Figure 4.1 Glen Canyon arch dam on the Colorado River, Arizona, constructed by the U.S. Bureau of Reclamation. (*Photo courtesy USBR*)

Figure 4.2 The 271-m-high Inguri arch dam, Georgia, U.S.S.R.

Between 1900 and 1960, 1200 dams more than 15 m high were built in western Europe, and 1650 in the United States. Most dams are constructed by utility companies or public agencies. In the United States, for example, the main dam builders have been the U.S. Army Corps of Engineers, the U.S. Bureau of Reclamation, and the Tennessee Valley Authority. Many more dams are built by private-sector utility companies. In Canada, provincial authorities are in the process of developing one of the nation's major resources, water power. The 240-m-high Mica embankment dam in British Columbia is one of the largest in North America. Earth and rockfill dams and some concrete gravity structures dot northern Manitoba on the Churchill and Nelson River systems, and include some on permafrost terrain. The Prairie Farm Rehabilitation Association (PFRA) has built many dams on smectitic clay shales, one of the weakest rock types. The Hydro Quebec James Bay Project, one of the largest rock engineering works in history, is expected to be followed by a second development phase between 1990 and 2010.

National electrical agencies in Austria, France, Italy, and Switzerland in the first half of the twentieth century constructed many thin arch dams in steep mountain valleys. The Snowy Mountain Authority was created to develop the hydroelectric potential of southeastern Australia through a series of dams, tunnels, and rock works. Hydropower developments in the U.S.S.R. are proceeding at an accelerating pace. Today, the focus of hydropower construction has shifted to developing countries, notably in South America, China, and India where construction in the Himalayas presents many challenges.

4.2 Types of Dams

The six main types of dams (Fig. 4.3) represent a spectrum with a progressive decrease in the area of the foundations from the earth dam (largest foundation area, smallest stress) to the double curvature arch dam (smallest area, greatest foundation stress):

- Earth embankment dams (Fig. 4.4)
- Rockfill dams (Fig. 4.5)
- Concrete gravity dams (Figs. 4.6 and 4.7*b*)
- Buttress dams
- Cylindrical arch dams
- Double curvature arch dams (Figs. 4.1 and 4.2)

Components of a typical earth embankment dam are shown in Fig. 4.8. All dams are, in one way or another, gravity structures. The thrust of water on the dam (which comes from gravity) is borne by the foundations

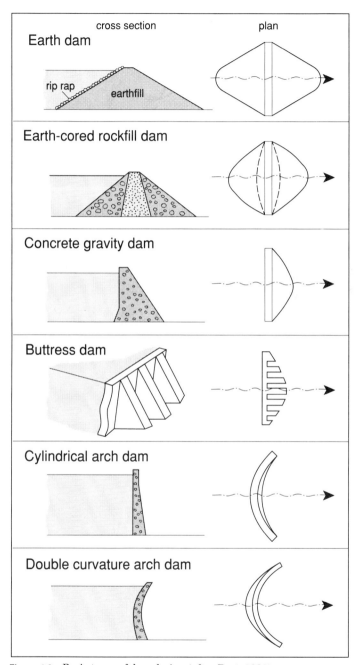

Figure 4.3 Basic types of dam design (after Best, 1981).

(a)

(b)

Figure 4.4 Sugar Pine earthfill dam, Auburn, California. (a) Laborer using high-pressure water to wash down the rock foundation in the river bed; (b) view during dam construction showing clay core and concrete spillway. (*Photos courtesy J. Santa, Water & Power Resources, USBR*)

Figure 4.4 (*Continued*) Sugar Pine earthfill dam, Auburn, California. (c) Downstream face as dam nears completion. (*Photo, courtsey J. Santa, Water & Power Resources, USBR*)

and abutments. The abutments themselves are gravitating masses, stable under their own loads and geometry, and potentially destabilized by new thrusts and increased water pressures in the earth and rock.

4.3 Benefits and Risks

Dams provide irrigation for crops, water and hydroelectric power for developing urban areas, and regulation of river flows to prevent flooding. In arid regions they also provide habitat for wildlife and reservoirs for recreation. These benefits have to be weighed against the costs and risks. As the height of the dam and the head and volume of stored water increase, so do the challenges of constructing a safe and durable dam and reservoir system (Deere, 1976; Deere et al., 1967).

In spite of greatly improved technology, dams remain the most accident-prone of all engineered structures, and when they fail, the consequences are often severe. Between 1900 and 1965, about 1 percent of the 9000 large dams in service throughout the world failed, and another 2 percent suffered serious accidents (Stapledon, 1976). Between 1900 and 1960, 18 dam failures and the ensuing floods accounted for 1690 deaths (Schnitter, 1967).

With each failure, however, has come improved understanding, and today's mistakes are fewer and different from yesterday's. A study of

(a)

(b)

Figure 4.5 Chiew Larn dam, Electricity Generating Authority of Thailand. This is a clay core rockfill dam 94 m high with a 761-m crest length, for hydropower, irrigation, flood control, fishery, and recreation purposes. (a) Downstream face and penstocks; (b) upstream face protected by riprap.

Figure 4.6 Alum Creek concrete gravity dam, Huntington District, U.S. Bureau of Reclamation. (*Photo courtesy USBR*)

historic dam failures serves is useful. As in medical pathology, when an accident occurs, the remains must be dissected and conclusions drawn. For this reason, and not from any morbid curiosity, the following account is presented of dam and reservoir problems before proceeding on a more positive note with a discussion of investigation, design, and construction.

The authors agree with John (1988), who concluded that "few real problems but also a limited need of rock mechanics are encountered when dealing with simple, reasonably dimensioned hydroelectric power projects, well adapted to well explored natural conditions. High technology is no remedy for poor, unrealistic basic designs."

4.4 Case Histories of Foundation Instability

4.4.1 Sources of instability

Legget (1962) remarks that

> the construction of a dam and the subsequent impounding of water cause more interference with natural conditions than do almost all other works of the civil engineer. The elevated water table tends to find an escape

Figure 4.7 Revelstoke concrete gravity dam, British Columbia, Canada. (*a*) Preparation of foundations; (*b*) completed dam. (*Photo courtesy BC Hydro*)

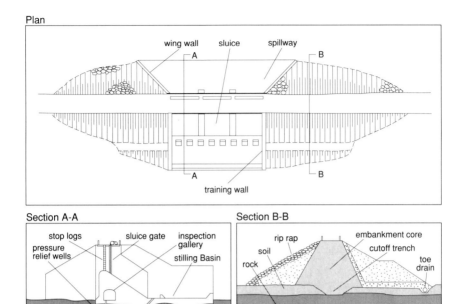

Figure 4.8 Typical components of an embankment dam. (*From Ontario Ministry of Natural Resources, Dam Operations and Maintenance Guidelines, 1983*)

route; the materials that become saturated for some considerable distance beyond the limits of actual impounded water may tend to dissolve or become unstable.

Harza (1974) calls the foundation of a dam the "...least knowable in advance, the least visible after project completion, and often the most expensive and dangerous part of the dam as a whole."

Most dam failures have been caused by inadequate spillway capacity or defective foundations, about 10 percent by foundation seepage (Clevenger, 1974). There is a lesser incidence of problems caused by differential settlements or movements along active faults.

4.4.2 Erosion and solution within the rock

Although most rocks are resistant to erosion and solution, some contain seams or pockets of erodible or soluble materials. Rock in a foundation, despite being strong when dry, can behave differently when wet. Tests should always be conducted for durability and for solubility (*Rock Engineering,* Sec. 2.3.3). Materials that require close investigation as potentially erodible include highly weathered and porous rocks, sedimentary rocks with soluble cements such as calcite or gypsum, shales and other clay-bearing rocks susceptible to swelling or pronounced slaking, and fault gouges and joint fillings of all types.

Water-sensitive conglomerates and schists were responsible for the rupture, in 1928, of the 63-m-high St. Francis concrete dam near Los Angeles. Water rushed 80 km downstream killing 400 people and destroying much valuable property. A block of the dam weighing more than 2000 t was carried 800 m downstream. A similar block toppled over, although the central section stood without much movement. No geological investigation had been made at the dam site, and no crushing or immersion tests were made on the conglomerate. Testing after the collapse showed the conglomerate to have a dry strength of 340 kPa, and to disintegrate on immersion.

> A startling change takes place...air bubbles are given off, flakes and particles begin to fall, and the water becomes turbid with suspended clay. After 15 min to an hour the rock has disintegrated into a deposit of loose sand and small fragments covered by muddy water (Ransom, 1928; Walters, 1962).

A 24-m-high earth dam in western Colorado was built on shale with about 20 percent by volume of gypsum in the joints and bedding planes. Underseepage dissolved the gypsum, the reservoir emptied by leakage, and a 15-m section of the dam dropped by a meter. Similarly at the Malad reservoir in Idaho, work had to be stopped in 1917 when the water had risen to within 3 m of final level, because of leakage through a bed of limestone 150 m long and 8 m thick, affected by solution, that had been buried beneath a talus slope (Legget, 1962).

4.4.3 Piping at foundation level

Leakage between an earth embankment and its rock foundation, or through the earth embankment itself, is potentially more dangerous than leakage beneath a concrete dam. This is because of the possibility of scour and piping (internal erosion) of the soil from which the dam is made.

On June 5, 1976, the 93-m-high Teton Dam in Idaho collapsed as its reservoir was being filled for the first time. In spite of downstream warnings, the failure killed 14 people and caused property damage estimated at between $400 million and $1 billion. An independent review panel (U.S. Department of the Interior, 1976) attributed the failure to erosion caused by flow of water along open joints in the volcanic rocks of the foundation. Incomplete compaction of soils into these joints and into the cutoff trench, and incomplete grouting of the rock may have contributed to the failure. Piping in the fill of the key trench soon led to the formation of an erosion tunnel and to the collapse of the entire dam, which was documented on film.

The critical *hydraulic gradient* for piping to occur is about 1:1, although backward erosion can occur at lesser hydraulic gradients. To provide an adequate factor of safety, the gradient is usually kept to a maximum of 1:4 by constructing an impervious barrier beneath the dam, either a grout curtain or cutoff wall, or by placing a compacted clay, concrete, or asphalt apron on the upstream floor of the reservoir.

4.4.4 High pressures and blowouts—the Malpasset incident

Malpasset may be the single most studied dam failure in history. It was the first known case of failure of the arch itself, although this was later shown to be the consequence of the foundation movements and load transfer. The disaster called into question most aspects of rock mass stability analysis. Only after this incident and follow-up studies did engineers begin to appreciate the interaction of rock stresses, joints, and groundwater (Londe and Sabarly, 1966; Wittke and Leonards, 1985; Erban and Gell, 1988).

On December 2, 1959, the 62-m-high Malpasset double curvature arch dam collapsed while its reservoir was being filled. Because of torrential rainfalls, the final 4 m of filling took just 3 days to complete (Bellier, 1967). A fissure appeared along the heel of the dam, followed by sliding and rotation in the foundation along an upstream foliation and a downstream fault. This was followed by blowout of the heavily fissured foundation rock on the downstream lower left abutment, and 340 people were killed when the released water traveled 24 km downstream at an average speed of 22 km/h to the Mediterranean, destroying part of the town of Fréjus (Jaeger, 1979; Mary, 1968).

The foundation was gneiss with pegmatite intrusions, shears, and many closely spaced clay-filled joints. The rock, because of its jointing, was estimated to be on average 100 times less permeable in compression than in tension, a condition that encouraged the buildup of substantial water pressures in the highly compressed toe of the foundation. There were no foundation drainage provisions such as might have relieved water pressures and prevented toe blowout and failure (Carlier, 1974).

Surveying of the dam and surrounding rock in July 1959, 5 months before the rupture, indicated that deformations had begun to accelerate (Bellier, 1967) and that an unexpected heave of 15 mm had developed in the central part of the foundation in front of the arch. These data became known to the consulting engineer only after the disaster.

Increased seepage from the right abutment joints was observed as much as 15 days before the failure; however, because of the heavy rainfall, this was not taken as a sign of danger. Also, cracking was observed in the stilling basin several days before rupture. Nevertheless, there were no obvious signs of distress ½ h before the foundations gave way; the watchman left for his home shortly before midnight, and later reported nothing unusual.

4.4.5 Earthquakes and active faults

Faults are found in many dam foundations, because many faults exist in the mountainous terrain most suitable for dams, and because river channels often follow fault lines which are easily eroded.

Faults may be active or inactive. Activity may be renewed by the reservoir water pressure, which decreases the effective stresses across the faults. Movements may be of the stick-slip type, occurring only during earthquake events, or characterized by creep, perhaps at rates slow enough to escape notice in the investigation stage (2 to 5 years), but important in the life span of the dam.

In the last 150 years, geologists have noted about 200 surface fault breaks during earthquakes. Displacements along preexisting as opposed to newly created fault planes have ranged from a few millimeters to more than 10 m. There are no known failures of dams caused by such displacements but there have been several "near misses" including one at the San Andreas dam near San Francisco where the 1906 earthquake fault break passed through the abutment without damaging the dam itself (Sherard, 1974).

If found or suspected in a foundation, faults must be assessed for activity before a dam of any size and importance can be built (*Rock Engineering*, Sec. 6.3.3). The local and regional geology are studied to determine the degree of movements and activity over a specified time period, such as 10,000 years. With suitable precautions and a conservative design approach, construction can proceed across active faults. The Coyote and Cedar Springs dams in California have been built on faults known to be active; both are conservatively proportioned earth embankments. Foundation faults in rock usually require treatment by grouting or subexcavation and filling with concrete, to stiffen the foundation, reduce leakage, and prevent erosion of fault gouge (Fujii, 1970).

Of greater concern than fault displacements is the possibility of foundation liquefaction under the prolonged shaking that often accompanies an earthquake. This problem is restricted to dams on earth foundations as opposed to rock, and is therefore beyond the scope of present discussions.

4.5 Reservoir Engineering

Challenges of a different sort center on the reservoir. They include flooding and loss of valley lands; excessive leakage or siltation, sometimes enough to leave the dam high and dry; landslides around the perimeter slopes, sometimes large enough to generate waves that overtop the dam and inundate towns downstream; earthquakes, often sizeable, induced by the weight of water; and modifications to the groundwater regime that can have a profound and not always favorable effect on agriculture.

4.5.1 Land acquisition, rehousing, and social issues

An unavoidable and nontechnical problem of constructing a reservoir is the cost of flooding land within the valley. This land has to be purchased, the population compensated and rehoused, and irreplaceable buildings of historic importance may have to be relocated. Roads and railways must be diverted because a reservoir is difficult to bridge, except at the dam itself or far upstream.

Historic and religious sites, such as ancient burial grounds, present difficult social issues to dam site planning. The filling of the reservoir (now Lake Nasser) at the Aswan High Dam in Egypt caused the flooding of major artifacts dating from the period of the Pharaohs. An international effort was mounted to move or preserve the major artifacts. Of those remaining, some were inundated by the reservoir, and others were affected by raising of the water table, which greatly accelerated rock deterioration, by evaporation and deposition of soluble salts.

4.5.2 Leakage

The purpose of a reservoir is to contain water, so the hydrological balance of the catchment, including inflows and leakages, requires careful evaluation when designing the reservoir-dam system.

The greatest potential for leakage occurs where the water table is far below the bed of the reservoir, as is often the case for reservoirs in mountainous terrain. Impounding will then bring about *groundwater recharge* (downward flow) and create a new hydrogeological environment as the rocks become saturated. In contrast, reservoirs fortunate enough to have a high water table in rocks around the reservoir perimeter (a *groundwater discharge* condition with springs or seepages emerging in the valley walls) will keep leakage to a minimum even if the rocks are permeable.

An unusual case of water loss occurred near Mansfield, England, where subsidence above an underground coal mine reversed the flow of a stream that had been feeding a 90 million gallon reservoir (Walters, 1962).

More often, leakages occur through ancient *river channels,* cut into the bedrock or clay soils, and filled with gravel. These can often be detected in aerial photography, but may be covered by landslides or glacial deposits. Across central Alberta and Saskatchewan, extending to North Dakota, there exists a major preglacial drainage network, obscured in most places by glacial till. This huge channel is the major potable water aquifer in western Canada, and at many locations lies across or near potential dam sites.

In the days before groundwater studies became routine, rock reservoirs that leaked as fast as they collected water were not uncommon. Dams were constructed for the Jerome reservoir in Idaho and the Hondo reservoir in New Mexico, only to find that the valleys remained dry. A concrete buttress dam 540 m long built in 1910 across a boulder moraine never retained water. The owner paved a large part of the reservoir with concrete and asphalt and constructed a cutoff wall to a depth of 24 m, all to no avail—the penstock and powerhouse had to be dismantled (Legget, 1962).

By far the worst problems of leakage occur in *karstic limestone* terrain *(Rock Engineering,* Sec. 4.1.4). The 20-m-high Hales Bar Dam near Chattanooga, Tennessee, was built in 1910 on limestone and shale (Lundin, 1974). Leakage was so severe that construction costs escalated from $3 to $11.3 million, and a 2-year construction period became 8. Four visible boils below the dam in 1914 had increased to more than fifteen by 1919. Inlets in the river bed were treated with dumped rock, gravel, and clay, which were washed out and replaced by rags, old carpet, sacking, cinder concrete, and wire mesh. As soon as one source of leakage became sealed another would develop nearby. In 1919 to 1920, 11,000 barrels of asphalt were pumped into the foundation but leakage increased again and by 1939, flows of 54 m^3/s were measured. By 1944, more than 90 percent of the leakage had been stopped by installing a cutoff wall at a cost of $2.75 million. However, the leakage again gradually increased and by 1964 was about the same as in 1940. The dam was eventually replaced by another 10 km downstream.

Karst drainage can bypass the dam, or the water can be diverted through the solution cavity network into an adjacent valley. Karst channels can act as syphons, draining the reservoir far below the point at which sudden leakage began. A particular danger is that as hydraulic gradients are increased during filling of the reservoir, karst channels that had been naturally silted may suddenly unplug.

A thin or discontinuous clay blanket may conceal a highly permeable aquifer such as fractured rock or karstic limestone. Hydraulic gradients generated by reservoir filling can be sufficient to puncture the blanket, starting leakages that are difficult to control. The May reservoir in Turkey and the Hawthorne reservoir in Nevada, both on limestone, failed abruptly when a natural "lining" was punctured after partial filling of the reservoirs (Knill, 1974).

Erguvanli (1979) presents an extensive bibliography of engineering geological problems at karstic dam sites. Seven dams with dried reservoirs and a further six on karstic limestone formations with heavy leakage (more than 10 m^3/s) are listed. Therond (1972), in a study for Electricité de France, details 17 karstic damsites and gives a comprehensive introduction to karst and its impact on reservoir and dam design.

The Camarasa dam in Spain required the injection of 190,000 dry tons of grout over a curtain 1100 m long and up to 360 m deep. The grout curtain, which reduced karstic leakage by 80 percent, required the injection of 650 dry kilograms of grout for each square meter of its designed surface area, and the corrective measures lasted from 1927 to 1931. The Bouvante dam in France underwent repeated sealing efforts, concentrating on the placements of walls and aprons on the immediate upstream portion of the abutments. The 73.5-m-high Montejaque arch dam in southern Spain was abandoned because the reservoir drained faster than it could be filled; corrective efforts failed to solve the problem.

4.5.3 Instability of perimeter slopes

Water level fluctuations in a reservoir are the main reason for landslides (Chap. 3). The greatest risks coincide with *rapid drawdown* (emptying of the reservoir), when steep hydraulic gradients and large seepage forces are generated by water trying to escape from the rock and soil. The slope face acts as a dam, trying to retain water. Instability develops during filling only if the rocks are weakened or dissolved when newly saturated. Once the water level has stabilized, the perimeter slopes are usually as stable as when the reservoir was empty, because water pressures on the slope face and in rock joints are self-balancing.

In the most catastrophic recorded case of reservoir perimeter instability, on the night of October 9th, 1963, close to 300 million cubic meters of rock broke loose from Mount Toc overhanging the Vaiont reservoir in Italy (Müller, 1968; Jaeger, 1979; Hendron and Patton, 1986). The sliding mass accelerated to a speed of between 90 and 110 km/h, jumped an 80-m-wide gorge, and climbed 140 m up the opposite valley wall. The resulting wave overtopped the 262-m-high dam and flooded into the valley, destroying the town of Longarone where 2500 lives were lost. The dam itself, built on jointed limestone, still stands undamaged, a testament to good dam design overshadowed by reservoir slope instability.

The suddenness and violence of the slide were unexpected although cumulative movements of nearly 4 m had been measured during three attempts at raising the level in the reservoir. On the first two occasions the movement stopped as soon as the water level was lowered, but on the third it did not. Recent geological observations indicate that this was a reactivation of an old slide, and that movement occurred along continuous clay layers with an angle of shearing resistance of about 12° (Hendron and Patton, 1985).

4.5.4 Perimeter erosion

Perimeter erosion is caused by wind-generated waves and fluctuations in reservoir water level such as occur in pumped storage reservoirs that are drawn down and refilled each day. Erosion can undercut slopes and trigger larger slides. Erosion problems are accentuated by harvesting of timber before filling of the reservoir and by vegetation being killed by a rise or fall in the water table. With protective vegetation gone, slopes erode more rapidly. Small reservoirs can be protected by installing riprap between high and low water levels, but for larger reservoirs this solution is impractical and the perimeter must be left to find a new equilibrium.

4.5.5 Induced seismicity

Reservoir filling is often accompanied by small to moderate or even major earthquakes (Gough, 1978). This phenomenon, known as *reservoir-induced seismicity,* is the combined result of the surface load of water and the increased groundwater pressures at depth. Increased water pressures reduce the effective stress on geological faults along which shear stresses exist, and so trigger sliding and release of the stored strain energy in the form of earthquakes. Reservoir planning should be preceded by a study of faulting, including estimation of the seismogenic history and potential activity of the faults.

Settlements that accompanied the formation of Lake Mead behind the Hoover dam in Nevada amounted to nearly 130 mm during 15 years, and were accompanied by several hundred minor earthquakes in an area with no previous seismic history. The strongest earthquake was of magnitude 5, and it occurred 10 months after first maximum reservoir filling to a depth of 221 m.

Seismic tremors greater than magnitude 5 were recorded at the

Kariba reservoir in southern Africa and of magnitude up to 6.3 at the Kremasta reservoir in Greece (Olivier 1961; Gough and Gough, 1970). The 103-m-deep Konya reservoir in India experienced many small shocks during impoundment, and 3 years later, a major shock of magnitude 6.5. In contrast to this delayed effect, the Kremasta earthquake occurred 3 months before reservoir-full conditions.

Filling of the Monticello reservoir in South Carolina in 1977 to 1978 was accompanied by considerable induced seismic activity. Tremors were detected soon after impoundment began and reached a peak of about 100 events per day. The activity lasted for more than a month, but earthquake magnitudes were less than 2.9 on the Mercalli scale (Haimson and Lee, 1984).

4.5.6 Hazards of raising the water table

Raising the water table in the soil adjacent to the reservoir often will improve the agricultural potential of the region. However, there can be negative effects on the environment and on farming. Slopes may become unstable, well water supplies may be affected, building foundations may settle or rise, and soils around the reservoir may become more saline.

Forested areas are often harvested for timber, then cleared and grubbed before they become flooded by the reservoir. Without extensive stripping, vegetation becomes submerged when the reservoir level is raised. Chemical and bacteriological decay of the remaining plants can generate toxic and explosive gases such as hydrogen sulphide and methane. These modify the water chemistry, which can kill fish, and also may damage electrical power generating equipment.

4.6 Site Investigation

4.6.1 Scope of investigations

Methods of site investigation are reviewed in *Rock Engineering,* Chap. 6. During a dam and reservoir project, perhaps more than in any other type of investigation, rock engineers form part of a team that may include hydrologists, hydrogeological, civil, mechanical, and electrical engineers, as well as specialists in agriculture and ecology. The project benefits greatly if the team can remain associated with the work through investigation, design, and construction phases.

The site investigation must answer questions in the following six categories:

> *Storage capacity and head:* Watertightness of the reservoir and dam foundation, and available head and volume in relation to hy-

droelectric generating and other requirements. Can this head be increased by constructing saddle dams?

Effects of impoundment: Land acquisition costs, the risks of unstable valley walls and of induced seismicity, and the ecological effects of an elevated water table.

Conceptual design: Type of dam, its location in the valley, spillway type and location, and method for diverting the river during construction.

Construction materials: Alternative sources, amounts and quality of rock, sand, gravel, and clay, and the most suitable quarrying methods and transportation routes to the dam site.

Foundation conditions: Required depth and extent of the foundation excavation; digging, blasting, and dewatering requirements; competence of the foundation; and the requirements for rock reinforcement and grouting to reduce leakage and erosion.

Conveyances: Routes to convey water from the dam to the place of use; choice among canals, pipes, or tunnels; and anticipated problems and methods of construction.

In recognition of the importance of a thorough knowledge of ground conditions, the budget for investigating a dam site and reservoir is substantial, often 7 to 10 percent of the cost of construction. This cost is defrayed by improved forecasts and superior design (Van Schalkwyk, 1983). Inaccurate forecasts usually precipitate gross escalations in construction costs. For example, sealing of a reservoir in the Italian Apennines cost over 6 times as much as the gravity-arch concrete dam itself (Legget, 1962). About 5 times the estimated quantity of weathered rock had to be removed from the foundation of the Aswan dam on the Nile in Egypt. Adequate site investigation should have identified these conditions.

Rock mechanics specialists should insist on the best possible exploration at dam sites. Often low-quality drilling is combined with an excessive amount of drilling. John (1988) remarks that "the output of an exploration always depends more on the effort and ingenuity that went into planning and evaluation than on the extent of holes drilled, adits driven, or tests performed." A geomechanics consultant encounters

> ...too many major projects in which large to huge amounts of extra funds had to be spent due to unforeseen conditions...in most cases these expenditures could have been drastically reduced by better exploration, better communication between exploration and design, and better reflection of results of exploration in contract documents.

Geological detail is critical to successful dam engineering. The rock engineer must never be tempted to ignore qualitative geological information in favor of the more quantitative data from measurements and tests. Dams are located on narrow portions of river courses. Does the narrow section exist because rocks are more resistant and therefore more suitable for a dam site, or in contrast, because of landslides? Are there buried river beds within the valley, perhaps at higher elevations than the present bed? Does the river follow a fault or an erodible and permeable stratum or an aligned karstic feature in soluble limestones or gypsum? The answers to these and similar questions will dictate the site investigation strategy and guide subsequent design and construction work.

4.6.2 Sequence of investigation and reporting

Site investigations are usually phased, starting with regional studies and becoming more detailed later. Geological conditions progressively unfold with each phase, and require continual review right up to final inspection and approval of the prepared rock foundation surface. Only during construction can the full extent of unusual features be appreciated.

The *feasibility investigation* relies mainly on outcrop mapping and air or satellite photographs, assisted by test pits, trenches, and geophysical methods. Limited core drilling may be used to establish regional geological conditions. The report compares potential sites with the aid of geological plans and cross sections, and assesses foundation conditions and the availability of construction materials. Foundation treatment requirements and suitability for different dam types are evaluated.

Groundwater conditions must be thoroughly known; investigations start early in the preliminary design phase. Piezometers are installed and monitored beneath the proposed dam and abutments, and at saddle locations around the reservoir, to determine hydraulic pressures and gradients. Tests are conducted to measure the hydraulic conductivities of the strata and to identify major conductive zones such as faults or karst. Investigations may include flow studies using dyes and isotopes. Determining hydraulic properties for abutments, foundations, and reservoir strata requires care and attention to detail, as well as good techniques (de Andrade, 1987; Freitas de Quadros et al., 1988).

In the *preliminary and final design* phases, exploratory drilling is a major component. Abutments are explored by fan drilling and by excavation of trenches and exploratory adits and shafts that permit visual inspection, sampling, and in situ testing (Fig. 4.9). The adits also serve for monitoring, grouting, and drainage while the dam is being constructed, and may be established as permanent drains or monitoring sites for the life of the dam. Extensive index testing assists in rock mass classification and mapping. Rock mass deformability and shear strength are measured by large-scale in situ testing to provide data for stability calculations.

Seismic velocity measurements and *tomography* are often used to map rock quality variations in the foundations and abutments, and to identify zones of weak or closely jointed rock to be excavated or grouted. Bonaldi et al. (1983) describe a characterization program for the foundations of a 71-m-high concrete gravity dam on weak schists in Southern Italy. Permanent cross-hole seismic drillhole installations were placed under the dam along three cross sections to permit initial mapping and subsequent monitoring of improvements in rock quality during grouting, and changes in velocity during impoundment.

Seismic tomography can help to detect karstic fractures, caves, and

Figure 4.9 Engineer lowered into 1-m-diameter drilled shaft to inspect rock conditions, Site C, Peace River, British Columbia, Canada. (*Photo courtesy BC Hydro*)

conduits. The Storglomvatn rockfill dam is to be constructed by the State Power Board of Norway 50 km north of the Arctic Circle, on a foundation of calcareous schist, marble, and quartzite with nearly vertical bedding (By et al., 1988). Tomography was successful in detecting low-velocity zones along bedding planes at the marble-schist contact, correlating with karstification visible in outcrop.

The aim of the investigation is to produce a reliable set of contract documents, technical descriptions, specifications, and contract drawings that faithfully reflect the geotechnical conditions and identify potential problems. The site investigation report should include detailed geotechnical maps (typically 1:200 scale) and three-dimensional diagrams or models in which the dam foundation is zoned according to rock quality, and major weaknesses such as voids, faults, and weathered dikes are identified and evaluated (ASCE-USCOLD, 1967).

John (1988) remarks that information released to contractors

> is sometimes kept to a minimum to keep the bid price low, forgetting that the extra costs that generally come due later, such as for settlements of claims based on changed conditions, are likely to by far exceed the savings initially hoped for.

4.7 Design

In no other type of construction is the continued contribution of the geotechnical specialist so important to design. Deere (1974) remarked that

> the layout and design of nearly every element of a hydroelectric project require close cooperation between the design engineer and the engineering geologist. This is true for the coffer dams, diversion tunnel, power tunnel, and surface or underground powerhouse and even more so for the dam and its appurtenant structures, the power intake and the spillway. Decisions must be made which will seriously affect not only the time and the cost of construction but also the safety of the structures themselves.

Simple and conservative designs are needed particularly in developing countries, with fairly easy constructability and not more high technology than really required to accomplish design objectives (John, 1988). Low demands for monitoring and maintenance are preferable, with a reasonable balance between the cost of construction now versus delayed costs of maintenance and repair later on.

4.7.1 Choosing the type of dam

4.7.1.1 Dam alternatives.
The six main types of dam defined in Sec. 4.2 and Fig. 4.3 represent a spectrum with a progressive decrease in

the area of the foundations from the earth dam to the double curvature arch dam. The single biggest mistake that can be made in dam design is to select the wrong type of dam for a given site (John, 1988).

Composite dams may be designed to accommodate geological and topographical features of the site (Fig. 4.10). For example, a buttress dam may be more economical in the central portion of a valley, with mass concrete gravity or earth embankment sections in the flanks. An arch-gravity design may be chosen, combining the two approaches to obtain some of the advantages of each (Oberti et al., 1986). The multiple arch dam is a variety of buttress dam and uses even less concrete, although it requires more skilled labor.

Three key factors control the choice of dam type: topography of the site, strength and uniformity of the foundation, and availability of construction materials. Selection is also influenced by the presence or absence of suitable sites for spillways and outlet structures, and by the local cost of labor. The cost of the dam itself is typically about one tenth of the total project cost.

4.7.1.2 Valley topography. Valleys are termed *gorges* if the ratio of valley width to height is less than 3, *narrow valleys* if this ratio is 3 to 6, and *wide valleys* if greater than 6. Typical width-height ratios for economic dam construction are less than 3 for thin arch dams, 3 to 5 for thick arch dams, and greater than 5 for concrete gravity dams.

4.7.1.3 Quality of foundation. Approximate foundation bearing capacity requirements for the various dam alternatives may be broadly summarized as follows:

Earth and rockfill dams	Not critical
Concrete gravity dams	1 MPa
Buttress dams	2 MPa
Thick arch dam	3 MPa
Thin arch dam	5 MPa

Earth and rockfill dams are most suitable for weak foundations because their broad base spreads thrust over a large area, and because they can tolerate large differential settlements. In contrast, a concrete gravity dam requires competent rock at a depth usually no greater than 6 to 9 m.

A buttress dam gives a 30 to 50 percent saving in concrete costs for heights greater than 12 m compared with a concrete gravity dam. Arch dams are the most sensitive to differential movements and

Dams and Reservoirs 179

(a)

(b)

Figure 4.10 La Grande LG-3 dam, James Bay, Quebec, Canada. (*a*) View of dam and spillway in operation; (*b*) spillway excavation and stilling basin. (*Photo courtesy Hydro Quebec*)

therefore require stiff foundation and abutments, typically with a rock mass modulus of at least 3.5 GPa.

Closely jointed foundation rocks are avoided by excavating deeper, although moderately close jointing can be treated by careful *consolidation or blanket grouting* to make the rock mass stiffer, stronger, and less permeable. Grouting is commonly used for the upstream portion of the foundation of concrete dams and also sometimes for earth dams. The pattern consists of closely spaced shallow holes, typically 3 to 6 m spacing and 3 to 12 m depth (discussed further below).

4.7.1.4 Availability of materials. Earth and rockfill dams can be constructed only when suitable materials are available from quarries and pits near the site. Concrete aggregates, being required in lesser quantity, can be transported for moderate distances but the quality requirements are more stringent. Rock for a rockfill dam must be obtainable in the required gradation and must be durable. Clay for a clay core must be of a nonshrinking type. An earthfill dam requires both free-draining and impervious materials and also rock for riprap protection. The requirements and tests to determine quality of these materials of construction are discussed in Chap. 1.

4.7.2 Analyzing stability of the dam

4.7.2.1 Analysis requirements. Analyses of *embankment dams* should consider liquefaction, shear failure, conditions of sudden drawdown, steady state seepage at maximum pool (spillway crest elevation), earthquakes, and potential for overtopping. Factors of safety in the range 1.0 to 1.5 are recommended in the U.S. Department of the Army guidelines, depending on which case is being considered. The potential for erosion and piping are evaluated by considering the water velocities, exit gradients, erodibility, and solubility of the joint filling materials and of the rock.

Analyses of *concrete dams* should consider overturning, sliding, and overstressing in the foundations, abutments, and concrete, as well as the durability of the concrete and the potential for long-term alkali-aggregate reactions (Sec. 1.4.2.4). Stability calculations must include loads imposed by water, weight of structure, uplift, soil (and silt), earthquakes, and temperature changes. Where applicable, up to 600 mm of ice should be assumed to act at normal pool elevation, equivalent to a pressure of not more than 240 kPa. Stresses in the concrete should not exceed one-third of the in situ concrete strength. Tensile stresses in unreinforced concrete can be accepted only where cracks will not adversely affect overall performance and stability. Foundation stresses should not exceed the bearing capacity of the foundation material, although this is rarely an issue in rocks.

Typical design sequences for hydroelectric structures together with the testing requirements are reviewed by Camargo et al. (1988). An important objective is to prepare a geomechanical model idealizing the soil, rock, groundwater, ground stress, topographic, and structural features to form the interface between exploration and design (John, 1988, Wittke, 1990, Chap. 14, and *Rock Engineering,* Sec. 7.1.2)

4.7.2.2 Sliding stability. The sliding stability of a dam is determined by comparing *disturbing forces* and *resisting forces* in a limiting equilibrium analysis (*Rock Engineering,* Sec. 7.2.6). The following example illustrates use of the limiting equilibrium method for computing the factor of safety (F) in a simple case of sliding on a horizontal contact between dam and rock (Fig. 4.11):

$$F = \frac{T}{H} = \frac{CA + V \tan \phi}{H}$$

Where, considering a unit length of dam

V = total vertical force, including the weight of dam and rock above the side plane v_1, plus water above the sloping upstream face v_2, plus any vertical component of anchor force

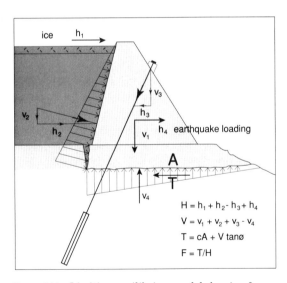

Figure 4.11 Limiting equilibrium model showing forces acting on a simple gravity dam resting on a horizontally stratified foundation.

v_3, minus the uplift force caused by water pressure in the slide plane v_4

H = horizontal disturbing force, equal to the thrust of ice h_1 (if any), reservoir water and silt h_2, and earthquake loadings h_4, less the horizontal component of any anchor force h_3

ϕ = angle of shearing resistance along the slide plane

c = cohesion along the slide plane

A = area of the slide plane (alternatives must be considered)

T = resisting force ($cA + V\tan\phi$) resulting from shear strength of the slide plane

Analyses must include jointing, search for the weakest potential surface of sliding beneath the dam (often clay seams give the lowest factors of safety), and assume that rock is unlikely to carry tension. Link (1969) provides an extensive review of direct shear testing and limiting equilibrium analysis applied to concrete dam foundations.

Shearing resistance depends on the normal stress, which is governed by dam self-weight, water pressure, thrust of water from the reservoir, and the forces imposed by any anchors. A linear Mohr-Coulomb criterion is usually employed, taking the residual (lowest) strength of the rock as a conservative assumption (*Rock Engineering,* Sec. 11.2). The value of cohesion usually is assumed to be low, often zero unless the designer can rely on interlocking along the potential slide plane. Cohesion can deteriorate with time, but the frictional component of shear strength can usually be relied on. Estimates of shear strength are critical to limiting equilibrium analyses. John (1988) points to the importance of appropriate tests on representative rock discontinuities rather than placing large investments in a few, possibly unrepresentative tests. Faced with the variability of rock, methods for estimating direct shear strength may often be preferred to in situ testing (*Rock Engineering,* Sec. 11.2).

Particular care is needed when designing dams on clay shales (Saint Simon et al., 1979). Peak strength has often been destroyed by shearing along thin bands, producing *clay mylonites,* which have little or no cohesion and low angles of shearing resistance. Clay mylonite seams may control the design of the dam, and cannot be washed out and grouted. At the Gardiner and Fort Peck Dams, the residual angle of shearing resistance was found to be as low as $\phi_r' = 5°$ to $12°$ in the more smectitic bands. Embankments of the Gardiner Dam, built in the 1950s, were planned at slopes of 1:4.5 to 1:6, but as investigations proceeded, the design was altered progressively to give final slopes of 1:11, no steeper than the flat surrounding valley walls. After impoundment, the dam started to creep downstream. The creep declined progressively to about 10 to 15 mm/year by the 1980s, partly as the result of the dam being in a valley that narrows downstream—displacements promoted an increase of abutment reaction through arching.

The *water pressure distribution* on the slide plane can range from the simple triangular case shown in Fig. 4.11 to more complex distributions modified by grout curtains, drainage, and thrusts from the dam, as described in the Malpasset example cited earlier. Depending on the type of dam, simple assumptions may suffice, or a more rigorous design may require a full hydrogeological analysis of water pressure distributions using numerical methods (*Rock Engineering*, Sec. 4.4). Rocha and Costa Filho (1988) review uplift pressures measured in the foundations of several Brazilian dams, at both the rock concrete contact and within the rock joints. The highly beneficial effects of drainage galleries in the foundations are clearly seen from the measurements.

A *factor of safety* of 4.0 is commonly required for a simple analysis, although if the strength and other characteristics are chosen conservatively and the design is a realistic one, this factor may be reduced to between 1.5 and 2.0 (Deere et al., 1967; Chitale, 1983). Two different design philosophies are common: high safety factors applied to average expected conditions and low safety factors applied to improbable conditions such as the weakest possible material, highest possible water pressure, and largest earthquake. A more rigorous approach would apply probability concepts to each parameter in the analysis.

Earthquake forces include inertia caused by the horizontal acceleration of the dam, and hydrodynamic forces from the reservoir water (using the Westergaard formula). The equivalent static load method is often sufficient. More sophisticated forms of analysis, however, can be based on either time-history or response spectra techniques (Seed, 1979). Dynamic testing of earth embankment and foundation materials may be needed. Scott and Dreher (1983) describe techniques for computing time-varying foundation loads from dynamic analyses, using a three-dimensional finite element model. These are combined with static and inertial forces, and are compared with dynamic shear strength measured using special equipment. The potential mode of instability and factor of safety are determined at each time step during the earthquake, and cumulative permanent displacements can be estimated.

Concrete can be saved by *anchoring* a dam to its foundation. A metric ton (10 kN) of anchoring force can replace up to 2 metric tons of concrete deadweight, depending on the angle of installation and the friction angle at the foundation contact (*Rock Engineering*, Sec. 16.2). Thus, anchored dams can be made thinner. The cost of anchoring is typically about one-third to one-half that of concrete used to apply the same force. With suitable design, up to one-half of the mass of the structure can be replaced by anchor forces. At the St. Michel dam in France, anchoring resulted in a 20 percent reduction in cost.

4.7.2.3 Overturning stability. Overturning stability for a gravity structure is checked by ensuring that the resultant of all forces, excluding those of earthquakes, acts through the middle third of any given horizontal section (including the plane of the foundation). When the earthquake component is included, the resultant should fall within the limits of the horizontal section. Fishman (1979) points out that planar sliding is seldom an issue in concrete dams, but that the combined effects of rotation and upstream joint opening can create overloading and compressive yield in rock or concrete in the toe. In these cases, stability analyses accounting for rotational load transfer and yield lead to lower safety factors than those calculated by simple rotational stability criteria.

4.7.2.4 Prevention of overstressing. To check overstressing in arch dams, which are the most sensitive to foundation conditions, three-dimensional *physical models* have been extensively used, with modeling materials selected to simulate variations in stiffness and jointing of the foundation rocks. The 102-m-high Ridracoli arch-gravity dam, built on sediments with weak bedding ($\phi = 13°$ to $14°$), was studied using a 1:100 physical model. The resulting rupture mechanism showed stepwise sliding as the shear plane moved upward through the sedimentary layers (Fumagalli, 1973; Oberti and Fumagalli, 1979).

Physical models are now often replaced by *numerical models;* an example relating to finite element modeling of the Auburn Dam in California is given in *Rock Engineering,* Sec. 7.2.3. A numerical model can also be "tested" to failure, but because the failure criterion and the geometry are preselected and simplified, results may not be realistic. A physical model often shows trends that cannot be quantified using numerical approaches, and a combination of the two methods is best if it can be justified for a large and complex structure (Paes de Barros and Colman, 1982; Weiyuan et al., 1987a, b).

Stresses in a thin concrete dam depend on the ratio of moduli, Ec/Er, concrete to rock. Variations in foundation stiffness are more important than absolute values. Even a five-fold overall change in the ratio may produce only a 20 percent change in the critical stresses in the concrete. John (1988) remarks that although measurements of in situ rock deformability are performed at most major dam sites to give data for numerical modeling of stresses and displacements (Fig. 4.12), in

(a)

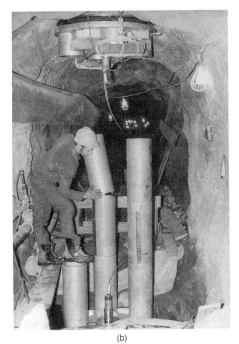

(b)

Figure 4.12 Testing to measure modulus of deformability of rock in a dam abutment (a) Equipment assembled for a horizontal test; (b) preparation for a vertical test at the same location. (*Photos courtesy Hydro Quebec & Roctest Inc., Montreal*)

many cases estimates of deformability appear to be good enough.

Differential settlements are reduced by thoroughly scaling loose rock and grouting the remaining rock to give a more uniform foundation. Often foundations and abutments, particularly of arch dams, benefit from reinforcement to improve and safeguard their strength and their stiffness.

4.7.3 Stability of abutments and the reservoir perimeter

Abutment thrusts are best directed normal to the planes of schistosity, foliation, or bedding to increase the strength of the rock mass. Rotation of a dam axis by even a few degrees in a narrow valley can increase stability greatly, and even slight alignment changes are worth investigating. Rockbolting and anchoring are often beneficial in the abutments, where the reservoir slopes are steepest and the consequences of slope failure are most severe.

Reservoir perimeter and abutment slopes should be analyzed for simulated earthquake loading and for conditions of rapid drawdown. Complete saturation of the slope is normally assumed, with no water pressure acting on the slope face. Perimeter slopes are usually too extensive for economical remedial treatments. All that can be done is to estimate risks and consequences, whether slides will occur, and if so, whether the sliding rock or the water wave could affect the dam. Model studies and hydraulic calculations help to determine whether a landslide is likely to generate a damaging wave. Filling and emptying of the reservoir are initially carried out at controlled and slow rates, while carefully monitoring the critical slopes.

4.7.4 Predicting and controlling leakage and uplift

4.7.4.1 Predicting leakage. The steps in planning are to outline the reservoir and catchment areas from the topographic contours, mark the river systems and divides, and identify potential leakages. These may develop through granular soils, dipping rock aquifers, zones of open jointing, individual conductive faults or karstic pipes, or as a result of erosion or solution of one or several strata.

Rocks usually become less permeable with depth because they are less weathered, and because the joints are tighter under the increased weight of overburden. The primary leakage paths are nearly always at shallow depths through the foundation and abutments of the dam, where the flow path is the shortest, hence the hydraulic gradient is the greatest. Some water loss is inevitable. Experience of the Tennessee Valley Authority indicates that seepages of between 0.1 and 0.6

L/min per lineal meter of dam can be expected after a normal program of grouting (Lundin, 1974). Usually it is uneconomical to reduce such seepages still further by additional grouting.

There may be secondary leakage through low points, or *saddles*, around the perimeter of the reservoir. Here the reservoir high-water level is limited not by the topographic elevations of the divides and saddles, but by the existing and potential levels of mounded groundwater. As the reservoir level rises toward the top of the groundwater mound, water starts to escape by percolation toward adjacent catchments. The rate of leakage is proportional to the hydraulic gradient and flow area in accordance with Darcy's law. Storage capacity can be increased by building *saddle dams* or long, low perimeter dikes, perhaps with grouted or excavated cutoffs to inhibit seepage losses.

Prognosis for a watertight reservoir is most favorable when the pre-existing water table is higher than the proposed high-water level in the reservoir. In this case, the water table mound will act as a barrier to leakage, and inflow will be supplemented by groundwater as it percolates inward from the divides. As a further advantage, only the superficial materials will become newly saturated, so the potential destabilizing effects of wetting will be confined to near-surface layers.

4.7.4.2 Design of grouting systems. Grouting methods and materials are described in *Rock Engineering,* Chap. 18. Rock treatments at dam sites include consolidation grouting to increase the stiffness and strength of the jointed foundation and abutments, and curtain grouting which, together with drainage provisions, limits water losses and reduces uplift pressures in the foundation. Further grouting may be needed at reservoir perimeter saddles and within the reservoir basin.

For dams less than 30 m high, curtain grouting is often omitted. For higher dams, flows and pressures are analyzed for various configurations and combinations of grout curtain, cutoff wall, and drainage system, the object being to achieve the maximum reduction in uplift pressures at the least cost.

A grout curtain may be injected through a single line of closely spaced holes, or up to five parallel lines about 3 to 15 m apart. Typically the maximum depth of grouting is in the range 0.5 to 0.8 times the maximum reservoir head and the grout depth decreases as the curtain extends up the abutments. Grout curtains can sometimes with advantage be inclined at 15° to 20° in the upstream direction to reduce hydraulic gradients between the grout curtain and the downstream drainholes.

4.7.4.3 Drainage curtains. Grouting and drainage are complementary and they combine to modify seepage forces within the rock mass. Hydraulic gradient and erosion potential are reduced on the downstream side of both a grout curtain and a drainage curtain. In rock of low conductivity (less than 5 lugeon units) drainage is essential and grouting is ineffective. In rock of high conductivity (more than 50 lugeon units), grouting is required for leakage control, whereas drainage may be needed only directly downstream of the grout curtain. For rock of intermediate conductivity, drainage is always useful, and a decision on whether to grout can be guided by comparing the costs of leakage with those of grouting.

A typical drainage curtain consists of a line of 75-mm-diameter drain holes at 3-m spacing. It is positioned wherever detrimental high-water pressures may develop, usually in the downstream portion of the foundation. In many concrete dams a gallery for drainage and grouting is provided near the upstream heel of the dam. Drain holes are drilled vertically beneath the dam to discharge into the drainage gallery; alternatively, they can be drilled obliquely from the downstream toe.

Placing drainage galleries as far upstream as possible, even in and underneath the heel, reduces uplift pressures under the main body of the dam, and permits lighter concrete structures and more efficient use of anchoring (Sapeguine et al., 1979). However, it may increase gradients and require impermeable bituminous or concrete aprons upstream from the dam heel.

Drainage blankets in earth and rockfill dams are provided by a layer of free-draining gravel or crushed stone beneath the downstream portion of the embankment.

4.7.4.4 Cutoff walls. Cutoff walls beneath an earth or rockfill dam provide continuity of the water barrier, acting to link the grout curtain in the foundation to the clay core in the embankment. Thus they protect the soil-rock contact, often the weakest link in an earth dam, against erosion and piping. The wall is made of concrete or compacted clay. It extends down in a cutoff trench excavated in the rock and up into the embankment for a total height of from 1 m to 7 m depending on the height of dam. It continues laterally up the abutments to the high-water elevation (ASTM, 1984).

The cutoff trench has the further benefit of giving direct access to cavities, faults, and open joints so that they can be treated. It permits intensive exploration and positive sealing of these features.

4.7.5 Design of forebay and spillway structures

For embankment dams in particular, spillway sizing is important to prevent overtopping, erosion, and failure. Embankments can be designed for overtopping if slope protection can be adequately provided, but the state of the art in erosion protection is not well developed.

Critical stability problems are often associated with concrete structures for water intakes and spillways, which may be perched high on the abutment alongside the dam where the rock is steeply sloping and deeply weathered. Stability of the structure depends on stability of the slope itself.

The design and construction of open excavations that form part of a dam and reservoir project follow much the same procedures as those for rock cuts in general (Chap. 2). The main difference is that spillways are subjected to water moving at high velocity and therefore to severe conditions of *erosion and uplift* (Cameron et al., 1988). Cuts into the rock are designed to resist scour, and are often reinforced with a dense pattern of rockbolts. The weaker and more jointed areas of rock are protected by durable reinforced concrete, anchored securely to the underlying rock.

4.8 Construction

4.8.1 Overburden stripping, rock blasting, and trimming

Soft rocks are trimmed to a uniform level using earth moving equipment. Final trimming is delayed until just before construction if the rock is susceptible to weathering, swelling, slaking, or distressing. In these cases it is important to immediately seal the exposed rock with concrete or with the first layers of fill.

Harder rocks often must be blasted in large quantities to prepare the foundation, abutment keyways, spillway, intake, and other structures. Perimeter blastholes are closely spaced and lightly loaded to reduce damage. Careful blasting minimizes loosening and reduces the requirements for grouting and reinforcement. Accurate blasthole drilling is a key factor. Techniques of controlled perimeter blasting such as presplitting, cushion blasting, and line drilling are discussed in *Rock Engineering,* Sec. 13.4.4.

Close to the final foundation level, the rock surface is scaled to remove loose material by light blasting or with jackhammers. Pockets of sand are removed by shoveling or jetting, and loose, weathered rock or weak grout are removed by scaling and wedging. Cleaning is completed by jetting with water or air (Fig. 4.4a).

Faults, clay-filled joints, and weathered zones are given special and individual treatment by excavating inferior materials and backfilling cavities with concrete. Thin seams are removed manually or by jetting, and thicker ones are removed by trenching or by surface or underground mining methods. Usually only the upper 5 to 10 m of foundation are so treated, although in unusual conditions, this may extend to depths greater than half the height of dam.

Foundations and abutments for the 123-m-high Feitsui arch dam 30 km upstream of the city of Taipei in Taiwan contained clay seams that had to be individually jetted clean and filled with mortar (Cheng, 1987). Even seams less than 10 mm thick were replaced, and many adits were used to gain access to all the seams. Initially, the seams had a shear strength of $c' = 0.03$ to 0.2 MPa and $\phi' = 17.5°$ to $30°$; after treatment, properties in all cases were estimated to be better than the design target of $c' = 1.1$ MPa and $\phi' = 38°$. The cost of this exceptional treatment amounted to 4.1 percent of the total cost of the project.

4.8.2 Foundation grouting

Water testing of grout holes, although expensive, provides a further test of foundation quality. For dam foundations, the test pressure should be at least equal to and preferably 50 percent more than the maximum reservoir head. Pressure is maintained for 10 or 15 min. Sections of hole that take less than about 1.25 L/min of water per meter length will take little grout, and can be omitted from the grouting program.

Most grouting uses a straight mixture of portland cement and water, usually with a 5:1 to 3:1 water to cement ratio, but sometimes as thick as 1:1 for rock with open fissures (*Rock Engineering,* Chap. 18). Often, progressively thicker grouts are injected in the same hole. Watery mixes at a ratio of 5:1 are likely to be weak and erodible, but thicker and stronger mixes are more difficult to inject into finer fissures, and require use of pressures high enough to open joints in the rock mass. Grout pressures vary depending on the depth, rock type, and grouting sequence. In the United States, pressures have often been limited to 22-kPa/m of overburden (1 psi/ft), whereas in Europe pressures of up to 4 times this value have employed successfully with thicker grout mixes, although usually with pretesting to determine hydrofracture pressure, and careful monitoring to detect and prevent heave.

Foundations can be pregrouted before the weathered rock or soil overburden has been excavated down to foundation level. The superincumbent weight of strata to be excavated keeps the joints relatively tight and allows the use of higher grouting pressures than would otherwise be possible.

Grout consumption (known as *grout take*), expressed as kilograms of cement per meter length of grout hole, can be designated as follows (Deere, 1976):

Very low	0.0 to 12.5 kg/m
Low	12.5 to 25 kg/m
Moderately low	25 to 50 kg/m
Moderate	50 to 100 kg/m
Moderately high	100 to 200 kg/m
High	200 to 400 kg/m
Very high	> 400 kg/m

Grout takes must be recorded systematically for the primary, secondary, and tertiary grouting operations (Fig. 18.7 of *Rock Engineering*). Cross sections showing the elevations of high and low grout take in relation to the geology allow diagnosis of reasons for high losses, and adjustments while the grouting is still in progress.

Slush grouting is a term used for brushing cement mortar grout over the trimmed and inspected rock surface. It is normally employed as a preliminary to dam construction, to fill cracks, even out the foundation surface, and ensure a good bond between rock and concrete or earth.

4.8.3 Plugging major leaks

Conventional grouting is ineffective when groundwater is flowing rapidly through large voids. These must be pretreated to limit grout travel and prevent wastage during subsequent conventional grouting work.

Pretreatments include the filling of relatively narrow channels (10 to 50 mm) with cement-bentonite or cement-sand mixtures. Diesel-fuel and bentonite slurries can be injected rapidly into flowing water; 50-50 slurries are easy to prepare, and contact with water causes rapid development of large clay curds, which can block channels up to 100 mm wide. Cavities in the range 50 to 500 mm are often grouted with sand-gravel–cement-bentonite combinations. Larger cavities require filling with crushed rock or boulders, then further treatment. Other fillers have included sawdust, rice (which swells), cane fibers, processed nut hulls, shredded rubber, plastics or foil, and finely ground organic fiber. Large voids have been sealed by intermittent pumping of neat cement or cement-sand-bentonite grout with calcium chloride accelerator. Sodium silicate and soda ash have been employed successfully as accelerators (Gebhard 1974).

At the Madden dam site in Panama, clay grout (55 percent water, 45 percent clay) was injected through 25- and 178-mm drillholes to seal cavernous limestone beneath the reservoir. Cement grouting was considered impractical and asphalt too costly. The grout traveled 15 m underground and penetrated seams as narrow as 12 mm (Reeves and Ross, 1931).

4.8.4 Aprons and blankets

Many concrete dams have concrete aprons upstream to reduce the hydraulic gradient and increase the length of seepage path. These can be poured or shotcreted. For sealing karst, integral concrete aprons usually perform better than shotcreted carpets (Therond, 1972). Aprons are more durable if the rock joints close to the dam heel are pretreated with a flexible sealant such as soft rubber or asphalt. This helps to prevent cracking when the reservoir is filled.

When the foundations are highly permeable over large areas, the reservoir may be blanketed with clay, concrete, asphalt, or even with rubber or plastic sheeting. At the Senator Wash dam in the United States, a 1-m-thick clay layer was placed and compacted over about 0.5 km^2 as a precautionary measure. On the Molokai water project in Hawaii, a 0.4 km^2 reservoir was lined with butyl rubber on nylon (Gebhard, 1974).

4.8.5 Embankment dam construction

When an earth dam is to be built on a rock foundation, the best way to obtain a tight bond between the earth and the foundation is to make the cleaned rock surface even so the earth embankment can be compacted directly against the rock with heavy rollers. Overhangs are removed and local depressions are filled with concrete or hand-compacted impervious soil (Sherard et al., 1963; Seed, 1972). A cutoff trench is generally needed, carefully blasted to prevent the opening of joints. Another constraint is that close to the cutoff wall, the embarkment is compacted using hand-held mechanical tampers that must be used for compaction in place of heavy rollers.

Phases in construction of a zoned embankment dam are shown in Fig. 4.4. After thorough preparation of the foundation, the embankment is constructed in lifts of limited thickness to achieve thorough compaction and a high density and strength. Clay cores or upstream membranes provide water-tightness, and riprap protects the upstream and downstream faces against erosion.

4.8.6 Quality control during construction

Quality control requires three components: inspection of rock, soil, and groundwater conditions; testing of construction methods and materials; and monitoring of dam behavior.

Inspection, testing, and monitoring continue throughout construction in order to evaluate rock as it becomes exposed, to check the assumptions used in design, to enforce the specifications, and to note any newly discovered fault, seepage, cave, or pocket of weathering.

The work includes monitoring of blast vibration levels to avoid damage to rock and recently poured concrete, and testing of rock reinforcement, grouting, and drainage systems. Foundation quality and treatments are checked by probing, sounding, and testing (Sec. 3.7.8). Detailed documentation is invaluable if problems develop later during construction or during the operational life of the dam.

The information uncovered must be rapidly translated into *design adjustments* if it is to serve a purpose, which is not a problem if the design team is receptive to new information and is well represented on site. Communications between quality control personnel and a design team sitting in a distant head office can present greater problems.

John (1988) remarks that internationally established contractors often prove unable or unwilling to produce at a remote site a product elsewhere taken for granted. Sophisticated designs call for realistic contractual provisions, reliable inspection during construction, and systematic monitoring of performance during and after construction.

Grouting is monitored by comparing grout take amounts and locations during primary, secondary, and tertiary phases, and often by comparing *sonic velocities* before and after treatment. Rodrigues et al. (1983) employed cross-hole and up-hole seismic tests in this manner at the Cabril dam in Portugal. They found an increase in the average longitudinal wave velocity of between 2 and 20 percent, which correlated closely with decrease in permeability and also with grout take. In the Roujanel double curvature arch dam on the river Borne, in jointed mica schists, the speed of sonic compression waves served as the index of quality (Terrassa et al., 1966). Sonic velocity increased by 24 to 58 percent along selected transit directions, sometimes in spite of small grout uptake volumes.

Nearly all medium and large dams include extensive *instrumentation* to measure their behavior (*Rock Engineering,* Chap. 12). Deformation data permit continual updating of the numerical design models as more realistic material parameters are determined. For example, measurements of valley deformation during construction of the 140-m-high Gordon double curvature arch dam in Tasmania allowed the designers to adjust their stiffness assumptions, previously determined by plate load tests on a much smaller scale (Guidici, 1979). The extent of testing and instrumentation for lesser dams depends on the owner's wishes and willingness to bear the cost, weighed against the risks and the insurance policy requirements and costs.

Figure 4.13 shows an example of instrumentation installed in the Itaipu main dam on the Parana River between Brazil and Paraguay. It is a hollow concrete gravity structure 185 m high, flanked by concrete buttress dams, and resting on a horizontally layered basalt foundation (Abrahao et al., 1983). Settlement was measured using three-

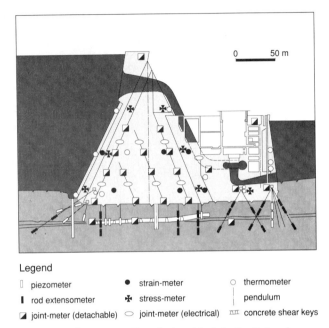

Legend

- ▯ piezometer
- ● strain-meter
- ○ thermometer
- ▮ rod extensometer
- ✲ stress-meter
- pendulum
- ◪ joint-meter (detachable)
- ◌ joint-meter (electrical)
- ⊓⊓ concrete shear keys

Figure 4.13 Intrumentation of a key block in the Itaipu dam on the Parana River, Brazil-Paraguay border (Abrahao et al., 1983)

position rod extensometers and horizontal displacements using inclined rod extensometers and inverted pendulums. Displacements between blocks were monitored using detachable joint meters, and up-lifts were measured using standpipe piezometers installed soon after the grout curtain had been completed.

The U.S. Bureau of Reclamation in their embankment dam instrumentation manual (Bartholomew et al., 1987) state that concern for public welfare makes it imperative that a means be available for gathering information to assess dam performance and safety. In their list of reasons for monitoring they include the need to verify design assumptions and techniques of construction, to diagnose problems and the effectiveness of remedial works, to verify continued satisfactory performance, and to provide data to support or defend against litigation.

Instruments need to be installed early to provide a comprehensive baseline of initial data with which to compare subsequent information. Monitoring should start during the site investigation and continue during dam construction and impounding of the reservoir. Even after this, measurements are usually taken at regular, although extended intervals, and should be available at the time of safety inspections.

4.8.7 Inspection and monitoring during impoundment

Particular vigilance is needed during impoundment, because the dam and reservoir are being tested for the first time under the full head of water. If problems develop during impoundment, the water level should be lowered without delay (but without causing further problems of rapid drawdown), and the situation investigated and rectified.

In most cases, monitoring serves to confirm that all is well. At the Itaipu dam described earlier, the reservoir was filled in five increments stopping at each for interpretation of instrument readings. Settlements of up to 4 mm were recorded during the 2 years of construction, close to the minimum calculated values. Differential settlement between blocks amounted to 2.5 mm at most. No horizontal displacements could be detected. Piezometers indicated good performance of the grout curtain, and the observed uplift pressures were at all times less than assumed in the design. Flows monitored in the drainage tunnels were one-third to one-sixth of the predicted amounts. Further cases of measurements during impounding are given by Kovari and Fritz (1989).

During impoundment, checks should also be made for seepage and piping by monitoring leakage flows and water levels in piezometers. Calibrated weirs with rectangular, trapezoidal, or triangular notches are used to measure flow quantities which are plotted against reservoir level to show the effect of reservoir head on leakage.

The development of *internal erosion* in a foundation or an earth dam is recognized by a sudden increase in stream flow or the appearance of springs or boils downstream, often with muddy water. Leakage waters should be tested for dissolved or suspended solids. Minor leakages can be identified and their sources located by dye tracers such as potassium permanganate or fluorescein, or radioactive isotopes such as rubidium chloride, which can be detected even when diluted to one part in more than ten million of water. Salt ($NaCl$ or $CaCl_2$) tracers can be used as well.

4.9 Long-Term Inspection and Maintenance

4.9.1 Continued monitoring

For reasons of safety, medium- and large-size dams, and also smaller dams with a high hazard rating, are monitored not only during construction, but also at intervals afterward. Readings are continued on a previously established network of geodetic survey monuments, and on extensometer, inclinometer, and piezometer instruments in the dam, foundations, and abutments. Seismometers are installed on or near dams in seismically active regions to record earthquake tremors (*Rock Engineering,* Chap. 12).

4.9.2 Inspection requirements

Dams suffer deterioration and require periodic inspection and repair to maintain them in a safe condition. An informative review is provided by da Silveira (1990) who describes an ICOLD analysis of the deterioration, including failures of 1105 dams in 33 countries.

In many countries, annual, 2- or 5-year inspections are required by statute for dams larger than a certain size. Inspection galleries and shafts are left open to provide access. Records must be kept of leakages through or around the dam, and analyses made to determine the content of suspended and dissolved materials.

An example of legislation in the interests of public safety is provided by the U.S. National Dam Inspection Act of 1972. This called for an inventory of about 49,000 dams, most of them privately owned, of which about 9000 were identified as "high hazard," meaning that in the event of a failure there would be substantial loss of life or property. All dams in this category were to be inspected, along with those in the "intermediate" or "significant" categories believed to represent an immediate danger based on their condition. The Act applied to dams 7.6 m or higher with a maximum water storage capacity of at least 18,500 m^3 and to those with a capacity of at least 61,650 m^3 and higher than 1.82 m.

The U.S. legislation provides various options if a dam is found to be unsafe: remove the dam, increase its height or enhance its stability so that overtopping becomes acceptable, provide a reliable flood warning system, reduce the inflow into the reservoir or increase the size of spillway, require operation at lower reservoir levels to provide more storage for extreme flood events, or even, purchase downstream land in the high hazard zone.

When many years separate construction and the start of routine inspections, an initial comprehensive inspection should establish the condition of the dam and set the pattern for later visits. The aim is to detect hazardous conditions and to recommend any additional investigations that may be needed.

Conditions are evaluated from visual inspection augmented by a review of records, if they exist, giving details of site investigation, design, construction, operation, and maintenance. The inspection should include not just the dam site, but also intakes, spillways, and powerhouse excavations, the reservoir, and even the catchment and downstream areas. At least the following should be checked:

Adequacy of control and warning systems: Hydraulic and hydrological data and their sufficiency, warning system in the event of a threatened failure, adequacy of maintenance and regulating procedures as they pertain to dam safety and operation of the control facilities

Condition of the dam and its foundation and abutments: Adequacy of the design; condition of the earth or concrete; damage caused by trees and other vegetation; settlement, sliding, tilting, cracking, opening of joints; improper functioning of drains and relief wells; development of unacceptably high uplift pressures

Condition of the spillway and control structures: Particularly the operability of gates

Condition of the upstream reservoir and perimeter slopes: Erosion, sliding, or instability; siltation and its effects

Damage caused by water: Uplifting, removal, or deterioration of concrete, rock, or earth materials in the spillway, stilling basin, and at points in the river downstream; signs of leakage and erosion through or around the dam; suspended and dissolved materials in the leakage water; boils or increased seepages of water in the river bed or banks

Immediate reporting is mandatory when the inspectors judge there to be a significant hazard requiring prompt action. Formal written reports follow in all cases, with photographs and drawings to minimize the need for written descriptions. Recommendations are made for additional investigations or analyses if needed.

4.9.3 Maintenance of drains and grout curtains

Old drainage systems must be monitored and maintained if the dam and its foundation are to remain stable. Piezometers are used to detect clogging and to warn when drains require reaming out or replacement. Gravity drainage systems often need little active maintenance, but wells and below-grade galleries must use pumps and filters, usually requiring regular upkeep.

Ley (1974) describes repairs to the drains of a 105-m-high concrete gravity dam in the United States, constructed in the late 1920s on a foundation of metavolcanic rocks, and drained through a porous concrete pipe in the cutoff trench. Seepage had decreased from 34 L/s in the 1930s to 22 L/s in the 1960s. The drains had become partially blocked with hydrated lime, and uplift pressures had increased to dangerous levels. A new foundation drainage system was installed by fan-drilling with NX-sized holes. Piezometers confirmed a substantial reduction in uplift pressures, although little change in seepage quantity could be detected.

The Norris 80-m-high concrete gravity dam was constructed by the Tennessee Valley Authority in 1936. Its spillway apron was provided with an underdrainage system of drillholes into the limestone rock. Twenty years later, before dewatering the apron for a routine inspection, uplift pressures were checked by drilling holes 15 m deep. Pressures were found to have risen to about 3 times the calculated allowable values, and 28 new pressure relief holes were drilled. The old relief holes were found to be filled with sand, mud, and trash. They were cleaned and water momentarily spouted from some of them, but there was no way of fully reinstating the original drainage (Lundin, 1974).

Some grout curtains can also have a limited lifespan. Ley (1974) describes seepage control work at a 36-m-high concrete gravity dam in the United States, constructed in 1922 on weathered and closely jointed granite. Not all the weathered rock had been removed, and leakage and high uplift pressures had persisted since construction. The first 30 years in the life of the dam saw about 12,000 sacks of cement pumped into the foundation. High grout pressures caused breakouts some distance away, and further attempts used more cement and AM9 chemical grout. Drainholes were fan-drilled beneath the dam. At the end of all this treatment, foundation seepage had reduced by about 80 percent and uplift pressures had fallen noticeably.

Arch dams that have shown excessive leakage may be drained and rehabilitated. Typically, the upstream apron is coated with bitumen or concrete. Twenty years after the 1962 completion of the 150-m-high Roselend buttress-arch dam in the French Alps, reservoir improvement was undertaken by complete drainage and sealing of the reservoir near the dam with an extensive concrete heel structure.

4.9.4 Raising and anchoring of old dams

Dams can be anchored to their foundations either to increase the safety factor in cases of suspected underdesign or deterioration, or to allow raising of the dam and the reservoir water level. Long cable anchors passing through concrete and into rock were first used to raise the Cheurfas dam in Algeria (*Rock Engineering*, Sec. 16.2). This 30-m-high gravity masonry dam was built during the 1880s on sandstone bedrock. In 1934 the dam was raised a further 3 m with the help of 37 prestressed anchors to provide additional stability. Each anchor, formed from 630 high-tensile steel wires, carried a working load of 1,000 t. After 20 years, only 3 percent of the prestress had been lost (Mohamed et al., 1969).

In 1978 at the Big Eddy dam near Sudbury, Canada, leaching had affected construction joints in the concrete. Television inspections and permeability testing showed extensive loss of bond in these joints. To increase the safety factor, cable anchors were installed from the crest through the concrete and into the Precambrian bedrock. Also, the tensioned anchors permitted use of grout pressures that otherwise might have been sufficient to cause uplift. Safety factors were augmented to counteract the combined hydraulic and ice thrusts and seismic loadings, which had probably not been considered when the dam was first designed.

References

Abrahao, A. R., Alves Silveira, J. F., and Paes de Barros, F., 1983. "Itaipu Main Dam Foundations: Design and Performance during Construction and Preliminary Filling of the Reservoir." *Proc. 5th Int. Cong. Rock Mech.*, Melbourne, Australia, vol. C, pp. 191–197.

ASCE-USCOLD, 1967. *Current U.S. Practice in the Design and Construction of Arch Dams, Embankment Dams, Concrete Gravity Dams.* Joint ASCE-USCOLD Committee on Large Dams. 131 pp.

———, 1975. *Lessons from Dam Incidents, USA.* ASCE, New York, 387 pp.

ASTM, 1984. *Hydraulic Barriers in Soil and Rock.* American Society for Testing Materials, Technical Publication 874. Four editors, various authors.

Bartholomew, C. L., Murray, B. C., and Goins, D. L., 1987. *Embankment Dam Instrumentation Manual.* U.S. Bureau of Reclamation Engineering & Research Center, Denver, Colo., 250 pp.

Beaujoint, N., and Duffaut, P., 1971. "La Surveillance du Comportement des Fondations de Barrages." *Revue de l'Industrie Minérale - Mines*, July, pp. 1–16.

Bellier, J., 1967. "Le Barrage de Malpasset." Extract from *TRAVAUX*, July, Editions Science et Industrie, Paris, 23 pp.

Best, E. J., 1981. "The Influence of Geology on the Location, Design, and Construction of Water Supply Dams in the Canberra Area." *Bur. Mineral Resources Jour.*, Australian Geology and Geophysics, vol. 6, no. 2, pp. 169–179.

Bonaldi, P., Manfredini, G., Martinetti, S., Ribacchi, R., and Silvestri, T. 1983. "Foundation Rock Behaviour of the Passante Dam (Italy)." *Proc. 5th Int. Congress on Rock Mech.*, Melbourne, Australia, vol. C, pp. C149–C158.

By, T. L., Lund, C., and Korhonen, R., 1988. "Karst Cavity Detection by Means of Seismic Tomography to Help Assess Possible Dam Site in Limestone Terrain." *Proc. Int. Symp. on Rock Mechanics and Power Plants*, Madrid, Spain, vol. 1, pp. 25–31.

Camargo, F. P., Dobereiner, L., Cella, P. R., and Brito, S., 1988. "The Development of Rock Mechanics in Hydroelectric Projects in Brazil." *Proc. Int. Symp. on Rock Mechanics and Power Plants*, Madrid, Spain, vol. 1, pp. 277–288.

Cameron, C. P., Patrick, D. M., Bartholomew, C. O., Hatheway, A. W., and May, J. H., 1988. *Geotechnical Aspects of Rock Erosion in Emergency Spillway Channels—Report 2* (60 pp.), *Analysis of Field and Laboratory Data, and Report 3, Remediation* (63 pp.). U.S. Army Corps of Engineers, Waterways Experiment Station, Tech. Rept. REMR-GT-3

Carlier, M. A., 1974. "Causes of the Failure of the Malpasset Dam." In: *Foundations for Dams. Proc. Eng. Found. Conf.* Pacific Grove, Calif., pp. 5–10.

CFGB (Comité Français des Grandes Barrages) 1964, 1967. "Numerous Reports to the Int. Commission on Large Dams," *Proc., 8th and 9th ICOLD Conferences,* Edinburgh ('64), Istanbul ('67).

Cheng, Y., 1987. "New Development in Seam Treatment of Feitsui Arch Dam Foundation." *Proc. 6th Int. Cong. Rock Mech.,* Montréal, vol. 1, pp. 319–326.

Chitale, V. M., 1983. "Major Dam Foundations in Madhya Pradesh (India)." *Proc. 5th Int. Cong. Rock Mech.,* Melbourne, Australia, vol. C, pp. C159–C163.

Clevenger, W. A., 1974. "When Is Foundation Seepage Unsafe?" In: *Inspection, Maintenance and Rehabilitation of Old Dams. Proc. Eng. Found. Conf.* Pacific Grove, Calif., September 1973, pp. 570–583.

Coyne, A. 1939. *Leçons sur les Grandes Barrages.* Notes quoted in Mary, 1968.

da Silveira, A., 1990. "Some Considerations on the Durability of Dams." *Water Power and Dam Construction,* February 1990, pp. 19–28.

de Andrade, R. M., 1987. "New Techniques for the Determination of the Hydraulic Properties of Fractured Rock Masses." *Proc. 6th Int. Cong. Rock Mech.,* Montréal, vol. 1, pp. 79–85.

Deere, D. U., 1974. "Engineering Geologists' Responsibilities in Dam Foundation Studies." *Proc. Eng. Fdn. Conf. on Foundations for Dams,* Pacific Grove, Calif., ASCE, New York, pp. 417–424.

———, 1976. Dams on Rock Foundations—Some Design Questions." In: *Rock Engineering for Foundations and Slopes, Proc. Speciality Conf.,* ASCE, Boulder, Colo., vol. 2, pp. 55–85.

———, Hendron, A. J., Jr., Patton, F. C., and Cording, E. J., 1967. "Design of Surface and Near-Surface Construction in Rock." Chap. 11 In: *Failure and Breakage of Rock,* Fairhurst, C., (ed.), *Proc. 8th U.S. Symp. on Rock Mech.,* Am. Inst. of Mining Engineers, pp. 237–302.

Dunnicliff, J., 1988. *Geotechnical Instrumentation for Monitoring Field Performance.* John Wiley, New York, 577 pp.

Erban, P.-J., and Gell, K., 1988. "Consideration of the Interaction between Dam and Bedrock in a Coupled Mechanic-Hydraulic FE-Program." *Rock Mechanics and Rock Engineering,* vol. 21, pp. 99–117.

Erguvanli, K., 1979. "Problems on Dam Sites and Reservoirs in Karstic Areas with Some Considerations as to Their Solution." *Bul. Int. Assoc. Eng. Geol.,* no. 20, pp. 173–178.

Fialho Rodrigues, L., Oliveira, R., and Correia de Sousa, A., 1983. "Cabril Dam—Control of the Grouting Effectiveness by Geophysical Seismic Tests." *Proc. 5th Int. Cong. Rock Mech.,* Melbourne, Australia, sec. A, pp. A1–A4.

Fishman, Yu. A., 1979. "Investigations into the Mechanism of the Failure of Concrete Dams Rock Foundations and Their Stability Analysis." *Proc. 4 Int. Cong. Rock Mech.,* Montreux, vol. 2, pp. 147–152.

Fookes, P. G., 1967. "Planning and Stages of Site Investigation." *Engineering Geology,* vol. 6, pp. 63–134.

Freitas de Quadros, E., Correa Filho, D., and Fernandes da Silva, R., 1988. "Methods of Hydrogeotechnical Testing of Dam Foundation." *Proc. Int. Symp. Rock Mech. and Power Plants,* Madrid, Spain, vol. 1, pp. 501–507.

Fujii, T., 1970. "Fault Treatment at Nagawado Dam." *ICOLD. 10th International Congress,* Montréal, Report 037-R59.

Fumagalli, E., 1973. *Statical and Geomechanical Models.* Springer-Verlag, Vienna.

Gebhart, L. R., 1974. "Foundation Seepage Control Options for Existing Dams." In: *Inspection, Maintenance and Rehabilitation of Old Dams." Proc. Eng. Found. Conf.,* Pacific Grove, Calif., September 1973, pp. 660–676.

Gough, D. I., 1978. "Induced Seismicity." Chap. 4 in: *The Assessment and Mitigation of Earthquake Risk,* UNESCO, 341 pp.

———, and Gough, W. I., 1970. "Stress and Deflection in the Lithosphere Near Lake Kariba—I; Load-Induced Earthquakes at Lake Kariba—II." *Geophys. Jour. of the Roy. Astron. Soc.,* 21, pp. 65–78 and 79–101.

Giudici, S., 1979. "Measurement of Rock Deformation in the Abutments of an Arch Dam." *Proc. 4th Int. Cong. Rock Mech.,* Montreux, vol. 2, pp. 167–173.

Haimson, B. C., and Lee, M. Y., 1984. "Development of a Wireline Hydrofracture Technique and Its Use at a Site of Induced Seismicity." *Proc. 25th U.S. Symp. Rock Mech.,* Northwestern University, Evanston, Ill., pp. 194–203.

Harza, R. D., 1974. "Introduction" In: *Eng. Found. Conf. on Foundations for Dams,* Pacific Grove, Calif., ASCE, New York, pp. 1–3.

Hendron, A. J., and Patton, F. D., 1985. "The Vaiont Slide, a Geotechnical Analysis Based on New Geologic Observations of the Failure Surface" (2 vols.), U.S. Army Waterways Experiment Station, Vicksburg, Miss., Tech. Rept. GL-85-5, 104 pp.

———, and ———, 1986. "Civil Engineering Practice," *J. Boston Soc. Civ. Eng.,* vol. 1, no. 2, pp. 65–130.

ICOLD (Int. Commission on Large Dams, and Int. Confs. on Large Dams) various years. Publications of conf. proc., guidelines, and case histories on dams.

Independent Review Panel, 1976. *Failure of Teton Dam; Report to U.S. Dept. of Interior and State of Idaho.* U.S. Government Printing Office, Washington, D.C.

Jaeger, C., 1979. *Rock Mechanics and Engineering.* Cambridge U. Press, New York, 523 pp.

John, K. W., 1988. "Rock Mechanics Input for Design and Construction of Hydro-Electric Power Projects in Developing Countries." *Proc. Int. Symp. Rock Mech. and Power Plants,* Madrid, Spain, vol. 1, pp. 335–346.

Knill, J., 1974. "Engineering Geology Related to Dam Foundations." Panel report, theme VI. vol. 2, *Proc. 2d Int. Cong. Rock Mech.,* Sao Paulo, Brazil.

Kovari, K., and Fritz, P., 1989. "Re-evaluation of the Sliding Stability of Concrete Structures on Rock with Emphasis on European Experience." U.S. Army Corps of Engineers Technical Rept. REMR-GT-12, 67 pp.

Legget, R. F., 1962. *Geology and Engineering.* McGraw Hill, New York, 884 pp.

Ley, J. E., 1974. "Foundations of Existing Dams—Seepage Control." In: *Inspection, Maintenance and Rehabilitation of Old Dams. Proc. Eng. Found. Conf.,* Pacific Grove Calif., September 1973, pp. 584–608.

Link, H., 1969. "The Sliding Stability of Dams." *Water Power,* part I, March, pp. 99–103; part II, April, pp. 135–139; part III, May, pp. 172–179.

Londe, P., and Sabarly, F., 1966. "Permeability Distribution in Arch Dam Foundations." *Proc. 1st Int. Cong. Rock Mech.,* Lisbon, Portugal, vol. 2, p. 517; vol. 3, p. 449.

Lundin, L., 1974. "A Review of TVA's Problems and Treatments of Foundations of Existing Dams." In: *Inspection, Maintenance and Rehabilitation of Old Dams. Proc. Eng. Found. Conf.,* Pacific Grove, Calif., September 1973, pp. 641–659.

Mary, M., 1968. *Arch Dams—History, Accidents and Incidents* (in French), Dunod, Paris, 160 pp.

Mohamed, K., Montel, B., Civard, A., and Luga, R. 1969. "Cheurfas Dam Anchorages: 30 Years of Control and Recent Reinforcement." *Proc. 7th Int. Conf. Soil Mech. Found. Eng.,* Mexico City, Speciality Session no. 15, pp. 167–171.

Müller, L., 1968. "New Considerations on the Vajont Slide." *Rock Mech. Eng. Geol.* vol. 6, pp. 1–91.

———, Bavestrello, F., Rossi, P. P., and Flamigni, F., 1986. "Rock Mechanics Investigations, Design, and Construction of the Ridracoli Dam." *Rock Mech. and Rock Eng.,* vol. 19, pp. 113–142.

Oberti, G., and Fumagalli, E. 1979. "Static-Geomechanical Model of the Ridracoli Arch-Gravity Dam." *Proc. 4th Int. Cong. Rock Mech.,* Montreux, vol. 2, pp. 493–496.

Olivier, H., 1961. "Some Aspects Relating to the Civil Engineering Construction of the Kariba Hydro-Electric Scheme." *Die Siviele Ingenieur in Suid Africa,* April.

Paes de Barros, F., and Colman, J. L., 1982. "ITAIPU Project: The Structural Safety Assessment through Physical Models." *Proc. 14th Int. Cong. on Large Dams (ICOLD),* pp. 1041–1058.

Ransome, F. L., 1928. "Geology of the St. Francis Dam Site." *Economic Geology* (New Haven, Conn.) vol. 23, pp. 553.

Reeves, F., and Ross, C. P., 1931. "A Geologic Study of the Madden Dam Project, Alhajuela, Canal Zone." *U.S. Geol. Survey Bul.* no. 821.

Rocha, D. J. L., and Costa Filho, L. M., 1988. "Uplift Pressures of the Bases and in the Rock Foundations of Gravity Concrete Dams." *Proc. Int. Symp. Rock Mech. and Power Plants,* Madrid, Spain, vol. 1, pp. 509–518.

Rodrigues, F. L., Oliveira, R., and Correia de Sousa, A., 1983. "Cabril Dam—Control of the Grouting Effectiveness by Geophysical Seismic Tests." *Proc. 5th Int. Cong. Rock Mech.*, Melbourne, Australia, sec. A, pp. A1–A4.

Saint Simon, P. G. R., Solymar, Z. V., and Thompson, W. J., 1979. "Damsite Investigation in Soft Rocks of Peace River Valley, Alberta." *Proc. 4th Int. Cong. Rock Mech.*, Montreux, vol. 2, pp. 553–560.

Sapeguine, D. D., Khrapkov, A. A., Chiriaev, R. A., and Goldine, A. L., 1979. "Recherches sur les Fondations Rocheuses pour les Projets des Grands Barrages en Béton." *Proc. 4th Int. Cong. Rock Mech.*, Montreux, vol. 2, pp. 573–579.

Schnitter, N. J., 1967. "A Short History of Dam Engineering." *Water Power*, April 1967, 8 pp.

Scott, G. A., and Dreher, K. J., 1983. "Dynamic Stability of Concrete Dam Foundations." *Proc. 5th Int. Cong. Rock Mech.*, Melbourne, Australia, vol. C, pp. 227–233.

Seed, H. B., 1972. "Foundation and Abutment Treatment for High Embankment Dams on Rock." *J. Soil Mech. Found. Div.*, ASCE, vol. 98, no. SM10, Proc. Paper 9269, pp. 1115–1128.

———, 1979. "Considerations in the Earthquake-Resistant Design of Earth and Rockfill Dams." *Geotechnique*, vol. 29, no. 3, pp. 215–263.

Sherard, J. L., 1974. "Potentially Active Faults in Dam Foundations." In: *Inspection, Maintenance and Rehabilitation of Old Dams. Proc. Eng. Found. Conf.*, Pacific Grove, Calif., September 1973, pp. 768–770.

———, Woodward, R. J., Gizienski, S. F., and Clevenger, W. A., 1963. *Earth and Earth-Rock Dams*. John Wiley, New York, 725 pp.

Stapeldon, D. H., 1976. "Geological Hazards and Water Storage." *Bul. Int. Assoc. Eng. Geol.*, 14, pp. 249–262.

Terrassa, M., Duffaut, P., Garnier, J. C., and Bollo, M. F., 1966. "Auscultation Sismique du Rocher de Fondation du Barrage de Roujanel." *Proc. 1st Int. Cong. Rock Mech.*, Lisbon, Portugal, pp. 597–602.

Therond, R., 1972. "Recherche sur l'étanchéité des Lacs de Barrages en Pays Karstique." Thèse pour le grade de Docteur-Ingénieur, Université de Grenoble, 444 pp. (Available through Electricité de France, Paris).

Ueda, T., and Nakasaki, H., 1987. "Acoustic Emission Monitoring on Foundation Grouting for Rock Masses." *Proc. 6th Int. Cong. Rock Mech.*, Montréal, vol. 1, pp. 569–572.

U.S. Bureau of Reclamation, 1974. *Design of Small Dams*. United States Government Printing Office, Washington D.C., 816 pp.

U.S. Department of the Army, 1977. *Recommended Guidelines for Safety Inspection of Dams: National Program of Inspection of Non-Federal Dams*. DAEN-CWE Circular no. 1110-2-188. Office of the Chief of Engineers, Washington D.C.

U.S. Department of the Interior, 1976. *Failure of Teton Dam*. Report by an independent panel to review cause of Teton Dam failure. U.S. Government Printing Office, Washington D.C.

Van Schalkwyk, A., 1983. "Cost Estimates for Foundation Investigations of Dams." Proc. 5th Int. Congress on Rock Mech., Melbourne, Australia, vol. C, pp. C139–C143.

Wahlstrom, E., 1974. *Dams, Dam Foundations and Reservoir Sites*. Elsevier, New York.

Walters, R. C. S., 1962. *Dam Geology*. Butterworths, London, 335 pp.

Weiyuan, Z., Ruoqiong, Y., and Guangfu, L., 1987a. "An Integral Geomechanical Model Test of Jintan Arch Dam with its Abutment." *Proc. 6th Int. Cong. Rock Mech.*, Montréal, vol. 1, pp. 583–587.

———, ———, and Peng, W., 1987b. "A Three Dimensional Analysis of an Arch Dam Abutment Using a Fracture Damage Model." *Proc. 6th Int. Cong. Rock Mech.*, Montréal, vol. 1, pp. 589–591.

Wilson, D., (ed.), 1981. "Design and Construction of Tailings Dams." *Seminar Proceedings*, Colorado School of Mines Press, Golden, Colo., 280 pp.

Wittke, W., 1990. "Rock Mechanics, Theory and Applications with Case Histories". Springer-Verlag, Berlin (English translation, 1076 pp.)

———and Leonards, G. A., 1985. "Modified Hypothesis for Failure of Malpasset Dam." *Int. Workshop on Dam Failures*, August 6–8, 1985, Purdue Univ., Lafayette, Ind., 74 pp.

Part 2

Rock Engineering Underground

Chapters 5 and 6 are devoted to the civil engineering uses of underground space, tunneling, and the construction of underground caverns for diverse uses. Chapter 7 describes the methods of underground mining, and Chapter 8, the extraction of oil, gas, and geothermal energy.

P2.1 Tunnels and Urbanization

Modern cities are extensively undermined by a network of tunnels for sewers, water mains, roads, and railways. This subterranean "root system" feeds the growth of our urban communities. The underground alternative in cities is often less expensive and less disruptive than construction at the surface.

For transportation routes, sewer and water tunnels, etc., choices are available between twin tunnels and a single larger one; between circular, elliptical, or horseshoe-shaped cross sections; and between boring or blasting. Boring produces a more stable tunnel, and boring machines nowadays can cut economically through most types of ground. Blasting is cyclic, and often slower, although better suited to mixed ground conditions. The blast round is drilled, loaded, and detonated. After waiting for fumes to clear, the crown is scaled, mucking completed, and support extended to the face. Rates of advance using either boring or blasting are limited by how fast the rock can be broken and removed and the ground stabilized. Good ventilation is important when boring and essential when blasting.

Tunnel support cannot be "designed" like the columns and beams of a building. There are too many unknowns. Instead, it is selected empirically, and checked by numerical modeling of idealized conditions.

Support and excavating methods are adjusted during construction, based on the results of monitoring, inspection, and testing. Primary support installed for the contractor's safety is augmented by secondary support at the discretion of the engineer. Stabilization is often provided in several stages.

P2.2 Underground Space

Underground caverns are employed in an increasingly wide range of applications. They accommodate industrial, commercial, and recreational facilities; military bases; and hydroelectric, nuclear, and thermal generating stations. They also provide storage for frozen goods, wines, documents, oil, gas, and drinking water and are used for the processing and disposal of radioactive and chemical wastes and sewage.

By placing ugly noisy structures underground, they are hidden, and the valuable ground surface is preserved for housing, parkland, or agriculture. Risks of damage by fire, earthquake, storm, flood, theft, sabotage, and military attack are greatly reduced. Land acquisition and building costs can be lower than those for construction at the surface, and the operating costs, particularly for heating and cooling, are reduced because of the nearly constant temperature and humidity underground, and the excellent thermal characteristics of rock.

Four alternatives are available: purpose-mined and solution-mined caverns, abandoned mines, and porous aquifers. Purpose-mined caverns can be excavated at nearly any location and to any shape or size. Even larger caverns for fluid storage can remain full of blasted rock, to support the walls. Storage in solution-mined salt caverns and abandoned mines is cheaper, but abandoned mines can be used only if they are stable and suitably shaped and located. Porous aquifers are convenient for storage of air, gas, and oil.

The basic goal is to install permanent linings only sufficient to satisfy the function of the cavern. Blasting is carefully executed to avoid rock damage, and to minimize the support needed. Often the perimeter is presplit. As in any complex project in rock, site data, test results, modeling information, construction control, and monitoring are essential elements for success.

P2.3 Mining and Civil Engineering— Different Objectives

Mining methods differ from those of the civil engineer because of different objectives, and because of constraints imposed by the nature of the ore to be won from the ground.

The miner's main aim is "production," that is, to extract as much high-grade ore as possible, diluted by the least possible quantity of waste rock. The openings are created only as a by-product. For some methods of mining to be successful, the ground is even encouraged to collapse and flow into the excavation. Contrast this with the aim of a tunneler, whose main concerns are the shape, size, and stability of the tunnel or cavern, which must serve a particular long-term purpose. There is little interest in the quality of the excavated product.

The civil engineer can often choose to work underground only where conditions are favorable, and has greater freedom to select the sizes, depths, and directions of tunnels and caverns. The miner has a much more limited choice. Mine excavations often search out seams of sheared, altered, and weakened material. The mining goes where the ore is, even though this may coincide with the worst rock, mechanically speaking. Mines nowadays are reaching depths of several kilometers, where stresses and temperatures are high. Tunnels and caverns, in contrast, usually remain as shallow as the condition of the rock will permit.

Mine openings are continuously migrating through the ore body. As one stope becomes mined out, it is usually backfilled, and another comes into production. In contrast, civil engineering openings, once excavated and lined, are expected to be trouble-free thereafter.

P2.4 Reservoir Engineering

Oil and gas accumulate in a permeable reservoir rock beneath a cap rock that traps the hydrocarbon. Geothermal reservoirs are sites of hot rock, water, or steam that can be harnessed to generate electricity or used for heating. Suitable sites for reservoirs are suggested by geophysical exploration, and by regional modeling of rock stratification, structure, and pressure data from nearby wells. Lithological information is obtained during drilling by examining drill cuttings and rock core, and from downhole logging devices.

Oil companies continue to search for faster and cheaper ways to penetrate rock. Borehole instability at depths of several kilometers is caused both by swelling and spalling. Most borehole overstressing problems are solved by sophisticated formulations of drilling mud and by casing the hole. Rock behavior around boreholes is studied through the use of closed-form equations and numerical modeling.

Completion is the conversion of a borehole to a producing well. Casing is set to keep the drillhole from caving, perforated to give access to the producing formation, and then the formation may be hydraulically fractured to enhance recovery. The highly permeable conduits created by hydraulic fracturing serve to increase the volume of rock drained by the well, and in some cases are used to introduce large amounts of heat to reduce the viscosity of the reservoir fluids. The hydraulic fracture must be induced to penetrate the producing formation but not the caprock.

Production from a conventional oilfield makes use of the natural drive energy available in the reservoir, aided in some cases by injection of gas or water. Tertiary methods using heat or chemicals are required mainly for extraction of oil from tar sands and depleted reservoirs, major fuel sources for the coming centuries. Easily accessible tar sand deposits are being mined from open pits, although various countries are investigating the exploitation of buried deposits by steam injection, fireflood, and other methods of in situ extraction.

Rock mechanics studies play an increasingly important part in reservoir engineering, from the exploratory phases through drilling and production. These studies help to avoid or at least minimize environmental effects such as subsidence, the generation of earth tremors, and the problems associated with disposal of spent fluids.

P2.5 Modes of Rock Behavior

The behavior of rock in a tunnel or a mine depends on the depth and stress conditions and the quality (block size and strength) of the rock. Varying degrees of instability can be broadly classified into the following five categories:

Stable conditions in which the rock, with or without assistance from support or reinforcement, has reached a state of equilibrium. If rock displacements continue, they do so at decreasing rates and without damage to supporting systems.

Gravitational failure conditions in which the roof and sometimes also the walls progressively collapse by raveling or caving, that is, the loosening and falling of blocks. The blocks become detached by sliding along their bounding joints without much distortion or rupture of the rock.

Squeezing conditions, in which the crown, the sidewalls, and sometimes the invert of the excavation converge slowly and continuously by mechanisms of stress-induced viscoplastic flow. There is substantial distortion of intact rock material, often accompanied by creep along joints within the zone of overstressed rock mass.

Swelling conditions in which the rocks exposed near the excavation walls expand by physico-chemical mechanisms associated with the adsorption of water by clay minerals or anhydrite. Note that swelling is the result of mineralogical changes, whereas squeezing is caused by overstressing.

Bursting conditions, in which the rock ruptures explosively by propagation of fractures through previously solid rock. Stored energy is released suddenly and violently. Included in the category are gas outbursts, which occur as the violent ejection of gases and rock into an underground opening.

Which of these five conditions is likely to apply can be forecast, with some considerable degree of approximation, using the diagram shown in Fig. P2.1. The top right quadrant is the size-strength rock mass classification system (*Rock Engineering*, Chap. 3) and shows an example of a tunnel or mine drift to be excavated partly through shale and partly through sandstone. The rock units are located on the diagram according to their block sizes and strengths, as determined by core logging and testing.

The top left quadrant investigates the possibility of gravitational instability (rock falls). If the blocks are much smaller than the diameter of the opening, raveling failures can be anticipated and precautions taken, such as the installation of mesh and shotcrete. Interme-

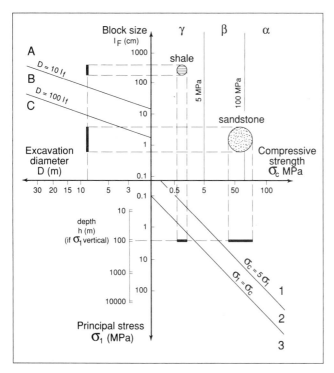

Figure P2.1 Diagram for preliminary evaluation of tunnel stability and potential failure mechanism.

diate values of the ratio of block size to excavation diameter lead to conditions where slab and wedge falls are likely, which can usually be stabilized by rockbolting. As the block size approaches the diameter of the opening, the possibility of gravitational failures becomes increasingly remote.

The lower right quadrant examines the potential for a squeezing or bursting type of failure. Here, the ratio of ground stress to rock strength is critical. The maximum principal stress at a location remote from the opening is amplified by a factor of up to 3 in the walls of the opening. Intact rock failures are unlikely when the ratio of rock strength to maximum far-field stress exceeds a value of about 5. Rupture of intact material becomes increasingly likely as the value of this ratio decreases, and becomes almost inevitable when the ratio approaches a value of 2. The lower the value, the more extensive the zone of intact rock rupture is likely to be. Whether the ruptures, if predicted, will be of a bursting or squeezing type depends on the strength and brittleness of the rock and also on the stiffness of the surrounding loading system.

Chapter 5

Tunnels

5.1 History and Applications of Tunneling

5.1.1 Evolution of tunneling methods

The earliest utility tunnels were constructed more than 3000 years ago by the Babylonians, Aztecs, and Incas in the search for precious metals, and by the peoples of India, Persia, and Egypt. Until about the nineteenth century, tunnels in hard rock were excavated by building a fire at the face, then spraying water on the hot surface. Progress amounted to about 1 m per week.

In the early days, the "drill steel" was driven by sledgehammer, and penetration rates of about 1 m/h were the best that could be achieved. Improved drill bits led to much more efficient drilling and blasting, and opened the door for invention of the rockbolt. Pneumatic-percussive rock drills were first employed in the latter part of the nineteenth century for blasthole drilling in the Frejus tunnel between France and Italy, and the Hoosac tunnel in Massachusetts (Ottosson and Cameron, 1976). Tungsten carbide drill bits, introduced in 1940, gave much improved penetration rates and increased bit life.

Technology improved dramatically from the mid-1900s (Fig. 5.1). Jacklegs were replaced by jumbos with as many as six booms. Heavier and more powerful equipment allowed drilling of larger (35- and 45-mm) blastholes. Pneumatic AN/FO loading equipment was mounted on the jumbo, reducing by about 20 percent the time required to charge the blastholes of a tunnel round, and cutting the cost of explosives to about one-half. Hydraulic drills introduced in the mid-1970s nearly doubled the drilling rate and reduced the noise and mist levels dramatically (Johnsen, 1988).

5.1.2 Tunneling today

Demand for underground construction follows the growth of urban communities. Modern cities are extensively undermined by tunnels, a

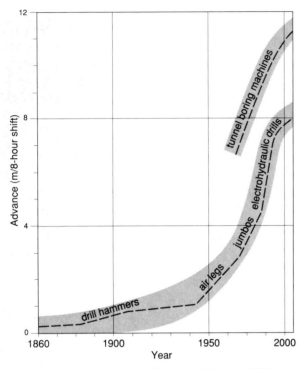

Figure 5.1 Progress in tunneling (after Johnsen, 1988).

less expensive and less disruptive alternative to construction at the surface. Long-distance transportation networks, often partially underground, connect cities to each other.

Tunnels to carry water or sewage account for about 77 percent of the total, railway and subway tunnels for about 15 percent and highway tunnels for the balance of about 8 percent. About 1530 km of tunnels were completed in the United States between 1968 and 1975 at a cost of $5.2 billion. Japan alone had 1424 km of tunnel under construction by early 1980, of which 862 km were for waterways, 269 km for railways, and 212 km for roads (Tanimoto, 1980). Each year, about 150 km of hydropower tunnels are being added to the total of 3000 km in Norway alone (Selmer-Olsen and Broch, 1982; Broch, 1989). Because of islands and deep fjords extending many kilometers inland, the Norwegians have more than 750 road tunnels, and are working to replace the remaining 250 ferry crossings.

The length and diameter of a tunnel depends on its use. Road and railway tunnels are usually shorter than 3 km, but have large diameters to accommodate vehicles. Water tunnels are typically 24 to 40 km long, but of smaller diameter, being dimensioned for the anticipated rates of flow. The longest continuous rock tunnel in the world, the 120-km Päijänne tunnel, supplies water to Helsinki in Finland. Perhaps two-thirds of all tunnels are excavated at diameters of between 4 and 5 m, but the size can vary from as large as 11 m (a water intake bored for the Mangla hydroelectric project in Pakistan) to as small as 1.2 m, bored by special minituneling machines.

5.2 Site Investigation

5.2.1 Objectives and constraints

The goal of a site investigation for a tunnel is to evaluate ground conditions along the route, and to assess their impact on the following aspects of planning, design, and construction:

Tunnel route and grade, size and shape, portal and shaft locations

Ground stabilization requirements before, during, and after construction

Excavating methods and rates of advance

Safety, stability, groundwater inflows, risk of encountering corrosive or unusually hot groundwater, explosive or toxic gases

Environmental effects of tunneling on overlying structures and on the regional groundwater regime

Because of their length, tunnels have low investigation budgets judged on a per kilometer basis. Predictions can be accurate for shallow tunnels in well-known, uniform ground, but are unreliable in unfamiliar, steep, or variable terrain. Shallow tunnels also *require* a more accurate investigation, because of the risk of collapse of thin cover, and the possibility of encountering costly mixed-face conditions. Investigations become more expensive and less informative as they go deeper. More often than not, geotechnical information remains fragmentary even up to the awarding of the contract, and is never as detailed as for more localized projects such as foundations.

To counteract this problem, the tunnel itself is often used as a means of further exploration. The common *observational method* of design allows flexibility for adjustments to support systems and excavating methods while construction proceeds. However, the range of possible adjustments may in practice be limited because of the need to make certain key decisions in advance, such as to choose between blasting and boring and between shotcrete and precast concrete lining methods.

Investigation techniques and objectives vary according to the depth of the proposed tunnel, and whether it is to pass beneath land or water (West et al., 1981, and *Rock Engineering,* Chap. 6).

5.2.2 Methods of investigation

5.2.2.1 Shallow and deep exploration

For shallow tunnels, exploratory drilling is the main method of investigation, and accurate delineation of bedrock topography is the most important objective. Many hundreds of meters of core are recovered in a typical investigation (Fig. 5.2). The tunnel elevation should be selected to pass through either soil or rock, not along the interface where conditions are likely to vary greatly from place to place. Zones of poor ground or mixed-face tunneling need to be identified, and avoided if possible. Ground can be pretreated by grouting or drainage where the problems are most severe.

For deeper tunnels, drillholes are expensive and can wander off alignment, sometimes by tens of meters. They may have to be drilled through many meters of irrelevant ground to reach the last few meters of interest. Because of budget limitations, they are spaced far apart. They must be used to their full potential, not just as a source of core, but also for downhole observations and tests. Outcrops, when available, are fully utilized. Geophysical methods and geological interpretation are employed to extrapolate downward and interpolate between points at which observations are made. Extrapolation is difficult in mountainous regions because of topography and the folding and faulting associated with mountain building.

Figure 5.2 Core storage racks for a South African water supply tunnel, which links with the Lesotho Highlands water supply system. Note the convenience of extending the sides of the core boxes to form handles, omitting lids, and storing the boxes on angle-iron slides.

5.2.2.2 Offshore exploration. Drilling over water is expensive because of the need for a platform and attendant barges, and the high cost of delays caused by rough water. Drillholes have to be staggered to either side of the alignment and grouted on completion to reduce the risk of hazardous inflows when driving the tunnel.

Divers with geotechnical experience can explore the extent and nature of underwater rock exposures and the thickness and types of sediment. Outcrops on shore may be close enough for extrapolation. To reduce the need for drilling, geophysical methods, particularly seismic profiling and sidescan sonar, are employed to determine the depth of water and thickness of soil and rock above the tunnel.

5.2.2.3 Horizontal drillholes. Long horizontal drillholes (*Rock Engineering,* Sec. 6.5.2) avoid the wasted drilling associated with vertical holes. Information can be obtained also by monitoring the penetration rate as a function of thrust and rotation speed, by monitoring groundwater pressures and discharges to detect high-pressure water, and by inspecting the drillhole walls with television cameras or ultrasonic scanners. The drillhole gives access for in situ testing along the alignment, and for installation of monitoring instruments if the hole passes close to the tunnel perimeter.

5.2.2.4 Pilot tunnels. An exploratory (pilot) tunnel offers a number of important advantages over an investigation by drillholes alone. Pilot tunnels and exploratory shafts give a better view of the rock than that obtained by core or borehole camera evaluations, allowing joint orientations, persistences, roughnesses, and filling materials to be assessed. They provide access for large-scale rock testing, water inflow evaluation, and trials of blasting and support. With the help of monitoring instruments, the pilot tunnel serves as a model of the full-scale prototype. The same instrumentation can continue to be employed during full-scale excavation. A small pilot tunnel tends to be more stable than the prototype, and allows the engineer to evaluate full-scale stabilization requirements in comparative safety.

Pilot tunnels also provide access for ground treatment including grouting, drainage, and bolting. A pilot bore can intercept and relieve zones of excessive water pressure. Pretreated ground can be excavated much more rapidly without risk of instability or sudden in-rushes of water, and without stopping to grout ahead of the full-bore tunnel face.

Pilot tunnels can be situated in the crown, invert, or center of the full-scale tunnel section. Those close to the crown must avoid cutting into or damaging the crown rock, and must be well supported to prevent loosening of the future tunnel's perimeter. A pilot bore within the final profile is expanded by slashing or reaming its walls, which reduces damage to the final tunnel profile.

Alternatively, a pilot tunnel can be bored to one side or above or below the main tunnel. If below the invert of the main tunnel, it can provide drainage during construction, and also can carry mucking vehicles or conveyors. This leaves the full-bore tunnel clear for lining, and for early installation of mechanical and electrical equipment. A completed pilot bore can play a long-term role as a service tunnel or as part of the ventilation system.

5.2.3 Site investigation reports

Site investigation reports describe the work done, methods used, factual results (including borehole logs and test data), inferred ground characteristics, and predicted tunneling conditions. Plans and cross sections, the key to a useful report, show graphically how conditions vary from place to place. Interpolation between points where conditions are known, although often difficult, must be attempted.

All parts of the report should be made available to prospective bidders. The report should explain the reliability and limitations of data and interpretations. General disclaimers of responsibility are best avoided (Sec. 5.3.7).

5.3 Planning Considerations

5.3.1 Tunneling or cut-and-cover construction

The engineer often has to choose among construction at surface, cut-and-cover construction in which a box or tube is inserted in a trench which is then backfilled (Fig. 5.3), or fully underground tunneling by blasting or boring machine. A sunken tube is a further alternative for underwater work. Many shallower tunnels start in open cut, and proceed into cut-and-cover and fully below-ground construction. Cut and cover is usually cost-effective down to a depth of about 10 m, but is less attractive in urban areas where it disrupts traffic, creates noise and disturbance, and requires expensive shoring to protect the foundations of nearby buildings.

5.3.2 Portal and shaft locations

Tunnels start and end in portals if they penetrate horizontally through a hill or mountain, or in shafts if they go deep below water or relatively flat terrain. Shafts are acceptable when a tunnel is to carry water, oil, or gas, but not usually for entering or leaving a road or rail

Figure 5.3 Cut-and-cover tunnel through horizontally bedded dolomitic limestone, Hamilton, Ontario.

tunnel. Temporary shafts, which may later be sealed, divide longer tunnels into sections to allow simultaneous construction at more than two working faces; the separate tunnel sections are often let under separate contracts. Permanent shafts at intervals are needed for ventilation, to give access to subway stations, and for reasons of safety.

At portals, adverse factors can combine to give the worst conditions for both excavation and support. The portal is the location of least cover, where any loss of ground quickly migrates upward to cause settlements at the surface, where the rock is most weathered and the joints most open, and where problems of slope instability combine with those of tunneling alone. It is also the place where the contractor starts, and therefore has the least experience of the ground conditions, and where blasting and support are still being adjusted and may not yet be adequate. Pre-reinforcement of the portal area, intensive monitoring of ground movements, and careful blasting of the initial sections are needed in most cases (Craig and Brockman, 1971).

The ideal portal location, other factors being equal, is at an outcrop where the quality of rock can be confirmed as sound, and where the rock above the tunnel crown is at least several meters thick.

Shafts tend to be more stable than portals and tunnels because gravity acts in a favorable direction, along the shaft axis, and because horizontal stress components are often similar in magnitude. The worst problems of shaft sinking usually coincide with the soil-rock interface, where the soils are often bouldery and the rock loose, weathered, and permeable. Pregrouting or even freezing may be needed at this level to contain water inflows and to stabilize the ground.

5.3.3 Alignment and grade

Alignment refers to the route followed by the tunnel on plan, and *grade* (gradient) refers to the depth and inclination of the tunnel *invert* (floor) and *crown* (roof). Alignment and grade are selected according to the function of the tunnel, the ground conditions, and the excavating methods.

A tunnel in uniform ground usually follows the shortest route between portals or shafts to minimize costs for excavation and support. However, the invert elevations of some types of tunnel are predetermined with little scope for adjustment. Tunnels for gravity drainage are aligned and graded according to hydraulic requirements, and those for vehicles have to satisfy visibility, climbing, and braking requirements for roads and railways.

When the ground is variable, alignment and grade should be adjusted to avoid tunneling through unstable ground or formations that are difficult to excavate or to seal against water inflows. Weak zones should be crossed in the shortest practical distance by orienting the tunnel in the dip direction of the discontinuities (Palmstrom and Berthelsen, 1988). Shallow tunnels should be located to pass entirely through either rock or soil, avoiding the soil-rock interface that frequently coincides with incompetent ground, difficult excavation, and high water inflows. *Mixed face* tunneling, passing frequently from soil into rock and back again, is usually costly and slow, and is best avoided.

Further constraints are imposed by the excavating techniques and equipment. Tunnel boring machines can negotiate only certain limits of grade and curvature (*Rock Engineering,* Chap. 15). For example, the minimum practical radius of curvature for a 6-m-diameter tunnel excavated by full-face TBM is usually between 120 and 140 m (O'Rourke, 1984). No such restrictions apply for tunnels excavated by roadheader or by blasting, although those excavated by blasting may need to be deeper to avoid damage to thin or fragile rock cover or to foundations close to the tunnel crown.

5.3.4 Cross-section shape

Tunnel cross sections are commonly circular, elliptical, or horseshoe-shaped. The traditional horseshoe shape evolved mainly from the need to construct vertical walls supporting a stable arch of brick or stone resting on a flat invert or roadway.

A circular cross section is ideal for full-flowing water tunnels because it provides the greatest area for flow, and permits use of a high-strength tubular steel liner. An egg-shaped cross section works well for sewers and storm water drainage tunnels. The small-radius invert acts as a narrow self-scouring channel during periods of low flow, and the upper and larger parts of the cross section provide a high flow capacity for surge discharge during storms and floods.

A transportation tunnel requires a more or less rectangular area for traffic. Only part of a circular bore can be used by traffic, although some of the remaining space in the invert and crown can be put to good use for ventilation, drainage, and other services. In a circular two-lane vehicular or rail tunnel, traffic occupies 50 percent of the total area, whereas in a three-lane tunnel the occupation factor is reduced to only 40 percent (Muir-Wood, 1972). An elliptical three-lane tunnel by comparison raises the occupation factor to about 52 percent.

When choosing a tunnel shape, rock quality and ground stress conditions should be considered. A circular cross section is theoretically the most stable (gives the least stress concentration) in isotropic ground with equal all-around stress, whereas an elliptical section with the long axis aligned with the maximum principal stress may be best when the stresses are unequal. This leads to large spans and small heights for the common condition in which the horizontal component of stress in rock is greater than the vertical. The design of a cavern must be checked to exclude the possibility of slabbing, buckling, and rockbursting in the extensive, highly stressed floor and roof.

Egger (1983) analyzes the roof stability of shallow tunnels for three cases of isotropic, horizontally stratified, and orthogonally jointed rock. His method assumes that failure will occur by downward movement of a vertical-walled slab of rock as wide as the tunnel, which is observed in many cases where the overburden is thinner than two to three tunnel diameters. The best shape of tunnel roof is then defined by a major principal stress trajectory.

Choice of cross section is related to choice of excavating method. A circular tunnel lends itself to excavation by full-face boring machine, requires the least volume of excavation for a given cross-sectional area, and is often the least expensive alternative even though part of the cross section may be wasted or backfilled during construction. An elliptical cross section can be mined by joining two closely spaced circular bores. A twin-bore subway tunnel, for example, is often joined into a single ellipse at each station.

Rectangular tunnels are often more appropriate when blasting through horizontally bedded sedimentary formations. Blasting tends to give a flat roof, whether required or not, and there is little point in trying to preserve a curved arch unless one is strictly needed.

5.3.5 Single or twin tunnels

Often the engineer has a choice between twin tunnels and a single larger one. The single tunnel tends to be less expensive when the quality of rock is high, and when rock cover is ample above the crown. Twin tunnels become more attractive if the rock cover is thin, weak, or broken, because smaller tunnels are easier and cheaper to support. The cost of excavating additional rock is more than offset by savings in support costs, and by fewer delays caused by ground instability. At shallow depth beneath cities, instability has consequences not only for the tunnel, but also for overlying buildings and services that could be damaged by subsidence.

The distance separating twin tunnels should be sufficient for the intervening pillar to be self-supporting. The pillar dimension depends on stress levels and rock strength, and should be checked by a stress analysis. In South African mining practice, long parallel service excavations are usually separated by at least twice the sum of the diameters of the two excavations. An effective diameter is assumed to encompass the zone of fracturing; hence, greater separations are needed for tunnels excavated by blasting.

A much greater separation, of a hundred or more meters, may be needed if a tunnel is to be excavated alongside another that has previously been lined. An existing liner can easily be damaged by the bedding plane sliding that accompanies release of high horizontal ground stresses (*Rock Engineering,* Sec. 5.1.3).

5.3.6 Estimating construction costs

The typical North American tunnel excavated during the 1970s, with a diameter of 5 to 7 m, cost $3400 per lineal meter to excavate, stabilize, and line. However, costs vary from place to place and country to country because of differences in the geology and in the costs of labor, equipment, and materials, from as little as $800 per lineal meter for a tunnel excavated through shale by a boring machine to as much as $20,000 per meter for a difficult and large underwater tunnel. Costs typically increase in proportion to the cross-sectional area of the tunnel, but escalate substantially if delays occur, such as when excavation must stop to allow grouting from the face.

In moderately good ground, for a typical road tunnel, excavation usually accounts for about 55 percent, primary support for about 15 percent, and lining for about 30 percent of the cost. However, in difficult ground, the combined primary support and lining can cost as much as 70 percent of the total, with excavation becoming a less significant part of the total.

Computer-aided models are used increasingly to predict tunneling costs. For example, the program COSTUN provides estimates for large-scale shaft, tunnel, and cut-and-cover construction work in soil and rock (Wheby and Cikanek, 1973; Bennett, 1981). It divides construction into excavation, mucking, and support. Costs are assessed according to the size of the tunnel and rock quality, and adjusted for the contractor's overhead and profit. The total estimate is obtained as the sum of labor, materials, and equipment for each phase of construction, and is adjusted using indexes that compensate for changes in costs between Chicago in 1969, where COSTUN was calibrated, and the time and location of the current project.

Kaiser and Gale (1985) evaluated COSTUN for Canadian tunnels through limestone, siltstone, and sandstone. They modified the input to accept not only the original RQD parameter of rock mass quality, but also three alternatives, RSR, RMR, and Q (Sec. 5.4.2). The index Q was found to give a better prediction of costs. A cost model for tunnel boring has been developed at the University of Trondheim in Norway, based on indexes of drilling rate, bit wear, and intensity of jointing (Blindheim et al., 1983).

5.3.7 Specifications

There are two basic classes of specification: prescriptive and performance. A prescriptive specification stipulates the materials, components, and procedures of construction, whereas a performance one leaves the methods of construction up to the contractor, and states only the required product. A performance specification is often preferred for tunneling work, because an experienced contractor can be relied on to select the best methods for completing the work quickly and safely.

The contractor usually retains full responsibility for safety during the tunneling work, whereas the engineer is responsible for controlling costs and for long-term safety within, above, and around the tunnel. The contractor usually determines the initial (primary) support. However, to the extent that this plays an important role in achieving long-term stability in modern rock tunnels, the engineer should ensure that primary support is sufficient for both short- and long-term needs. The secondary liner is often added only after complete stabilization has been achieved, to give an extra measure of safety and for

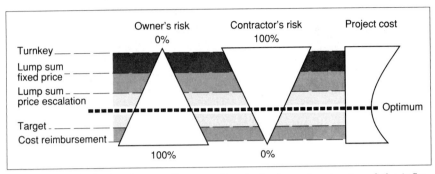

Figure 5.4 Risk sharing in tunneling according to the type of contract, and the influence on the cost of the work (Kleivan, 1988).

hydraulic and aesthetic reasons. Poor primary support can increase secondary support requirements and costs, and at worst, can result in subsidence, immediate or delayed.

Risks should be shared equitably among the parties in a tunneling project. Various types of contracts are available, each with a different balance of risk between owner and contractor (Fig. 5.4). National and international committees have published guidelines on contractual sharing of risk (USNCTT, 1974; ITA, 1988); some of the geotechnical recommendations are summarized as follows:

Changed conditions: The contract should provide for an equitable adjustment in the contract price when conditions underground differ materially from those that a competent contractor would have expected from the site investigation. This tends to reduce overall cost by eliminating the contractor's need to include a large contingency amount in the bid price.

Disclosure of information: All subsurface information, not only facts, but also geotechnical interpretations, should be disclosed to prospective bidders. Full disclosure obliges the geotechnical engineer to make clear the basis for the interpretations and the amount of doubt or uncertainty to be placed on them.

Disclaimers: There should be no disclaimer that attempts to protect the engineer and owner from the consequences of errors or omissions in the site investigation. Adequate resources should be employed in the investigation, and the results should stand on their merits.

Prequalification: Tunneling is specialized work, and owners should seek bids only from contractors who are both technically and financially capable of completing the work to the required standards.

Disputes: Arrangements should be made for early settlement of disputes, if possible on site, and without delaying the work. A conciliation procedure should be established, such as by a review board with two members appointed by the owner and the contractor, and a third selected by these two appointed members. If mediation proves ineffective, arbitration should be considered before litigation. During the work, a countersigned diary of events should be maintained.

Ground characterization and support: The contract documents should classify the ground along the tunnel and give primary support provisions for a broad range of ground classes. Procedures should be specified whereby the owner and contractor can quickly agree on changes to the work and payment as a result of encountering site conditions different from those expected. Bills of quantities should be structured so that the price of any changes to the work can be easily established.

Water problems: The site investigation should attempt to establish the pressure, temperature, and chemical composition of groundwater along the route and the permeability of the rock formations. The report should state the expected inflows in the tunnel, and whether adverse effects might result from any lowering of the water table as a result of tunneling. The contract should specify any precautions required of the contractor to limit inflows, and should make provision for monitoring of water inflows.

Tolerances should not be too stringent, or they will increase the cost of tunneling. For example, a deviation in alignment of ±300 mm over a distance of 30 m is common for tunnels excavated by full-face boring machines. This is some 4 to 5 times the tolerance normally associated with concrete placement at surface.

5.4 Excavating Methods

5.4.1 Tunnel blasting

Rock in the path of a tunnel has to be removed quickly and cheaply, whereas rock remaining in the walls must be left undamaged. Good blast design and control ensure a smooth and undamaged perimeter.

The typical *cycle of tunnel operations* is shown in Fig. 5.5. First the round is drilled and loaded; holes may be drilled for rockbolting at the same time, also using the jumbo. The round is detonated. After waiting for fumes to clear, the crown is scaled, making use of the muckpile both for access and to provide support to the face. Mucking is then completed, and shotcrete is extended to the face.

The trend in blasthole drilling for tunnels is toward use of jumbos

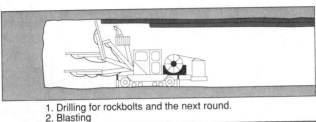

1. Drilling for rockbolts and the next round.
2. Blasting

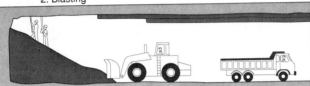

3. Scaling and mucking out the last round.

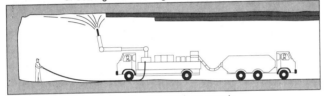

4. Shotcreting the two previous rounds.

Figure 5.5 Modern mechanized tunneling cycle (from Grimstad and Barton, 1988).

with computer-controlled, hydraulically operated drills (Fig. 5.6). The first holes to be detonated (those with the shortest delays) create a "cut," an opening toward which the rest of the rock is successively blasted. To create a smooth wall, the perimeter holes are often smaller in diameter and loaded with reduced charge density (*Rock Engineering,* Sec. 13.5).

Tunnels of diameter smaller than about 8 m and larger headings in good quality rock are excavated *full face* by drilling and blasting the entire face in a single round. As rock conditions deteriorate or the heading becomes larger, a *top heading and bench* method is employed in which an upper section (heading) is removed first, followed by installation of crown support and then removal of the remaining bench (Fig. 5.7a). This facilitates both mucking and the installation of support. In really bad ground, a smaller pilot drift is advanced and stabilized in the crown, and the remaining rock is removed in several stages.

Typically, a tunnel is advanced by blasting one to three rounds per day. The *length of advance* ("pull") per round is limited by the quality of the rock and by the diameter of the excavation. Advance varies

Figure 5.6 Jumbo drilling a blast round in quartzite at the CELS tunnel in Italy. (*Photo courtesy Prof. Sebastiano Pelizza, Politecnico di Torino*)

from 50 to 95 percent of the heading width, and from about 0.5 m in very broken ground that requires immediate support to as much as 3 m in massive and self-supporting rock in a large-diameter excavation.

Most tunneling problems arise not in average conditions, but at a few locations of extremely poor rock. Palmstrom and Berthelsen (1988) describe a 400-m-thick weak zone in an undersea tunnel that was penetrated with a combination of short blasting rounds and rapid application of fiber-reinforced shotcrete with concrete lining after each round.

5.4.2 Boring or blasting?

Boring is less disruptive than blasting and produces a more stable tunnel requiring less support (*Rock Engineering,* Chap. 15). *Full-face*

(a)

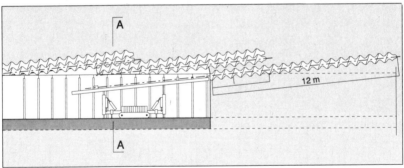

Section A-A

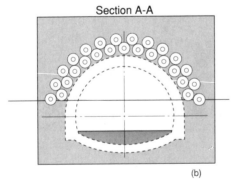

(b)

Figure 5.7 Torino-Frejus motorway, Giaglione Tunnel, Italy. (*a*) Photo showing bench excavation and completion of support to the final invert level. Note the overlapping umbrellas of grout tubes at changes of section; (*b*) diagram of umbrella grouting technique. (*Courtesy Prof. Sebastiano Pelizza, Politecnico di Torino*).

boring machines nowadays can cut economically through most types of ground. However, they are expensive to buy and to transport and assemble on site. Mobilization of a TBM becomes worthwhile only if the tunnel is longer than about a kilometer.

Blasting is usually cheaper than boring if ground conditions change often along the route. The cutting heads and tools, thrust capacities, and shielding provisions that are suitable for boring through soft ground are usually unsuitable for hard rock. Unexpectedly hard or soft rock, or higher than expected levels of ground stress, can delay the work and damage the machine.

Progress is not just a function of the rate of penetration of the cutter head when continuously boring, but of the overall rate of advance taking into account the time needed to service the machine, replace damaged cutters, install emergency support, grout to overcome water problems, and blast if the machine is brought to a standstill. When problems occur, a full-face TBM can seldom be removed from the tunnel without complete dismantling. An open cutter head allows access for work in front of the retracted head, for example, to blast hard rock lenses, but a machine with a closed head can be freed only by mining around the side and continuing to blast until reaching ground that is once again borable.

Roadheader machines that excavate only part of the face at any one time are much more maneuverable. Although not capable of penetrating massive or high-strength rock (uniaxial compressive strength should be less than about 60 MPa, and the abrasivity should be low), they can be introduced easily into the tunnel for limited lengths of driveage at locations where the rocks are suitable, and can be withdrawn to a safe distance to allow blasting.

Shafts and inclines may be benched with the two halves of the shaft bottom blasted alternately, or may be blasted full face. *Raise boring* is an attractive alternative if access is available from below (Fig. 5.8). A pilot hole is drilled at a diameter of about 28 cm. The reamer, typically 2 m in diameter, is then mounted on the drive shaft and pulled and rotated upward. In hard rock, the pilot hole normally advances at a rate of 1 to 3 m/h, and the reamer at 0.5 to 2.0 m/h. A 350-m-long shaft can be completed in 3 months instead of 7 months by drilling and blasting (Roald and Oiseth, 1988). Shaft lengths of more than 600 m are feasible under favorable conditions.

Tunnels can be excavated in a similar manner, by reaming. In Japan, a 3.4-km-long pilot tunnel 3.6 m in diameter was reamed to a final diameter of 6.1 m with outstanding results (Tanimoto et al., 1988). Advance rates reached a peak of 420 m per month for the pilot bore, and 390 m per month for the full bore excavation, compared with 300 m per month achieved by longhole drilling and blasting. The concrete

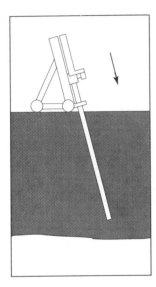 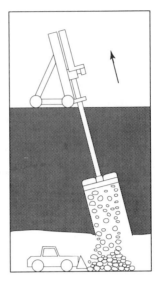

Figure 5.8 Raise boring by downward drilling of pilot hole, followed by upward reaming (Roald and Oiseth, 1988).

liner was placed simultaneously with reaming work at a speed of 400 m per month.

5.4.3 Muck handling

Rates of advance are limited by the rate at which the excavated materials ("muck") can be removed from a tunnel or shaft. In a tunnel, the choice between a rail or a wheeled vehicle system depends mainly on tunnel diameter and length. The longer the tunnel, the greater the time and cost of haulage. Rail systems are more often used in long, small tunnels, whereas dump trucks provide a more efficient system in tunnel excavations that have a larger cross section. Trucks can negotiate a steeper grade than rail systems.

Conveyor belt systems have particular merit in large tunnels through soft rocks where muck volumes are great. Conveyor mucking, although more common in mines than in tunnels, is attractive for ultra-high-speed removal of tunnel spoil. A 1.2-m-wide belt can easily carry a continuous load of 1000 t/h (Franklin and Matich, 1990).

Shafts are slower than tunnels to construct because excavation must be interrupted repeatedly to remove broken rock from the base of the shaft. Raise boring removes this problem, because the broken rock falls clear, and the cutter operates more efficiently when unimpeded by debris (Sec. 5.4.2).

5.4.4 Ventilation

Good ventilation is important when boring and essential when blasting. A ventilation duct is hung from the crown, and extended to keep pace with advance of the heading. Vehicle tunnels also need a permanent ventilation system, usually carried in the invert beneath the road pavement. When the tunnel is long, ventilation can take up a substantial portion of the cross section and require a much greater excavated volume than would be needed for traffic alone.

Long vehicular tunnels connect islands in the Japanese archipelago and link Great Britain with France. Because of ventilation difficulties in these and other long under-water tunnels, exhaust-emitting wheeled vehicles are usually transported from portal to portal by train.

5.5 Support and Stabilization

5.5.1 Support types and strategies

5.5.1.1 Primary and secondary support. There are three major reasons for ground support: to maintain stable conditions above and around the tunnel, to provide safe working conditions and steady progress during construction, and to provide a liner with long-term performance characteristics of water-tightness and stability.

The terms *temporary* and *permanent* support, now largely obsolete, relate to the days when timber was used for shoring, later to be removed and replaced by masonry. Modern supports do not rot away, but the concept of two-stage support persists in the terms *primary* (or initial) and *secondary* support. The purpose of primary support, installed by the contractor as required, is to ensure safety during construction and stability above the tunnel. Secondary support, installed at the discretion of the tunnel engineer, consists of extra materials to augment safety or to improve hydraulic characteristics. The dividing line is arbitrary, often defined only for convenience of the contract, because tunnels are usually stable with primary support alone. Tunnel stabilization is now often provided in not two but several stages.

5.5.1.2 Stability during construction. Ground support that lasts only for the period of construction is provided by the *shield of a boring machine* or by an independently mounted *shield* or *face-breasting* system or by *freezing* or *compressed air*. The choice depends on the method of excavation and on the expected stresses and groundwater pressures, and particularly on the *stand time* of the unsupported excavation. Unsupported stand times can vary from minutes in raveling or running ground to days or even months or years in competent rock.

The most "temporary" form of support, used only in soil or loose rock, is the tunneling shield that protects the heading and moves with the advancing face. The shields employed in rock tunneling nearly always form a part of a boring machine, and are hardly ever used when a tunnel is blasted. Shield support is followed immediately either by rockbolting or by liner installation.

Rigid shields can be troublesome because of unpredicted convergence and swelling. In rock, they are often replaced by flexible *finger shields,* or removed entirely. The 12.8-km-long 3-m-diameter Stillwater tunnel in Utah was excavated at a depth of 600 m through a hard shale. Boring began in 1977 using a full shield and a precast concrete segmental liner (Cording, 1984). The rate of advance reached 600 m per month, but slowed to 210 m in 1.5 months as the shield became stuck several times in shear zones. The shield was freed by mining around its upper perimeter and installing steel supports. After several such episodes, tunneling was stopped and the contract was terminated for "convenience of the government."

The nearby Hades and Rhodes tunnel was excavated with a TBM that had only a partial shield in the crown. Easier access to the face allowed steel *spiling* (an umbrella of steel channels or tubes filled with concrete) to be extended over the top of the cutter head to support rock that tended to ravel. The tunnel was advanced through flowing sand and squeezing and raveling shales, when necessary by hand mining in a series of small drifts in front of the machine. The Stillwater completion contract, let to a different contractor, made use of a machine with a shield composed of 12 blade-grippers which supported the ground and also served as reaction against the thrust of the cutter head. The blade-grippers could be "walked" forward one at a time or in groups, and could accommodate radial displacement of the ground (the next generation of machine used four blade-grippers and worked even better). The same four-piece unbolted concrete liner segments used in the previous Stillwater contract were installed in the tail of the shield, and grouted in place with a mixture of sand, cement, and fly ash.

Pelizza et al. (1989) describe *forepoling* to protect a 12-m-wide tunnel section in Italy (Fig. 5.7). A thin cover of sand, gravel, and boulders at the portal contained neolithic remains and medieval burial grounds that were not to be disturbed. The tunnel was advanced 500 m beneath a grouted umbrella arch. Every 9 m, holes 12 m long were drilled in an overlapping fan pattern, into which perforated steel pipes were installed and grouted. The liner was completed with wire mesh, shotcrete, and ribs at 0.75-m centers.

Timber supports are slow to erect, obstruct the tunnel, impair ventilation, and are less durable than rockbolts and shotcrete. Nevertheless, timber cribbing continues to serve a useful purpose in highly

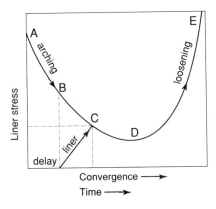

Figure 5.9 Conceptual relationship between tunnel liner pressure, convergence, and time.

stressed ground. By yielding and crushing, it absorbs energy and gives warning of highly stressed conditions. Many tunneling specifications call for standby timber for use in emergencies.

Steel sets with an H-shaped cross section are heavier and more expensive than *steel ribs* with a U-shaped cross section. Assembled from curved segments to form a free-standing arch, they fit imperfectly in an irregularly blasted tunnel, and have to be *blocked* with pieces of timber to hold them in position and bring them into contact with the rock at as many points as possible. *Lagging boards* of timber or, for durability, precast concrete or steel are inserted between the flanges of the H-sections to span the gaps between adjacent sets. The gaps that inevitably remain between rock and lagging allow the rock to move and loosen, and to develop greater pressures on the support members than if movement were more confined as with shotcreted and bolted support systems (Fig. 5.9, discussed in Sec. 5.5.4.3).

Ground freezing was employed in Oslo, Norway, to stabilize a section of railway tunnel (Josang, 1983). A gravel-filled canyon 20 m wide at the top reached to within half a meter of the tunnel crown, making normal blasting impossible. Cut-and-cover excavation in downtown Oslo would have created almost impossible traffic problems. As an alternative, 56 freeze pipes were installed at 1-m spacing by drilling from a nearby tunnel. A two-stage calcium chloride brine freezing system created a frozen arch spanning the rock canyon.

5.5.1.3 Rockbolts, shotcrete, and the NATM. Rockbolts are used alone when the rock consists of large, durable blocks, but are better used in combination with shotcrete, wire mesh and steel ribs, straps, or crown plates when the rock is weak or broken (Fig. 5.10). For this type of support, and the ground control philosophy that goes with it, Professor

Figure 5.10 Shotcrete and steel rib reinforcement, Rubira road tunnel, Barcelona, Spain. (*Photo courtesy TABASA, Spain, and BRGM, France*)

L. V. Rabcewicz coined the name *New Austrian Tunneling Method* (NATM), in contrast to the "old Austrian method" which employed mainly timber.

The NATM uses reinforcement and coating materials in proportions that are adjusted to suit the ground. Close monitoring of convergences and liner pressures, an essential part of the NATM, allows support to be added progressively when and where needed. The end result is an economical (although slow to install) system that minimizes both ground displacements and support quantities, and gives excellent control in badly crushed, squeezing, and swelling rock, or even soil.

The best rock (strength more than 4 times the ground stress) needs either no support, or *rockbolts* in the crown only. Progressively deteriorating conditions require an initial "first aid" coating of *shotcrete,* 20 to 40 mm thick, that helps the rock arch and support itself, followed if necessary by bolting and additional layers of shotcrete with *wire mesh* between each layer, (now often replaced by *steel fiber* reinforcement). Near faults and zones of potential ground squeeze, the shotcrete layers are further reinforced by flexible lightweight *steel ribs* with a U-shaped cross section. These may be constructed from overlapping telescopic segments to accommodate rock squeeze without becoming overstressed, and are pulled tight against the rock surface by rockbolts. They are widely spaced when only nominal support is

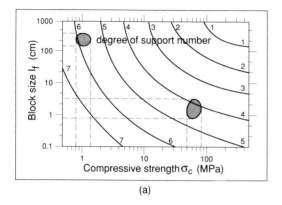

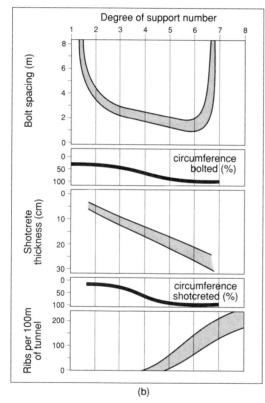

Figure 5.11 Relationship between degree of support number and requirements for NATM support. (a) Size-strength classification contoured to show degree of support number; (b) primary support requirements as a function of rock mass quality (Franklin, 1976).

needed, but are as close as 0.5 to 1.0 m where extremes of ground squeeze are expected (Fig. 5.11). Steel ribs are nowadays often replaced by *lattice girders* fabricated from reinforcing steel. These are lighter in weight and easier to attach to the rock. When shotcreted, they form a reinforced concrete beam (Fig. 5.12).

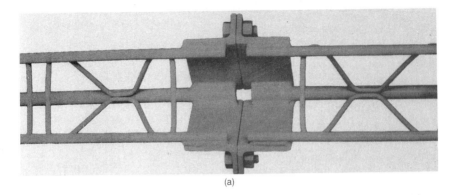

Figure 5.12 Lattice girder reinforcement for a shotcreted tunnel. (a) Reinforcing girder; (b) partially shotcreted girders and mesh. (*Photos courtesy P. Kaiser*)

The NATM was introduced in Austria, France, Germany, Switzerland, and Italy and its use rapidly spread worldwide. One of the earliest applications was in the Frankfurt underground railway tunnel, which was driven in 1969 through interlaminated clay marl, chalk, and sandstone. Cover amounted to only to 6.2 m where the line passed beneath the historic Roemer Building. Settlements were limited to between 36 and 44 mm and the building remained undamaged (Louis, 1972). The method has been used successfully in the Seikan tunnel in

Japan (Nakahara, 1976), and also in the Emisor Central sewer tunnel in Mexico City, where it was necessary to counteract heavy inflows of water and sand issuing from closely jointed volcanic rock. To stabilize the tunnel, 100 mm of shotcrete was applied immediately, and the arch was then supported with 3-m-long Perfobolts and a second 100-mm layer of shotcrete (Maples, 1973).

The support used in the NATM is flexible in two senses. It is mechanically flexible and can deform to accommodate rock squeeze while continuing to act as a coherent membrane. It is versatile in that it can be adjusted to match support requirements that vary from place to place. The more traditional alternative of installing a single support system throughout a tunnel requires a design to resist the worst, not just the average, anticipated ground conditions. The single system alternative is simple and requires less skill, control, inspection, monitoring, and testing. It is usually best for short tunnels of small diameter, when ground conditions are well known and uniform. The NATM is nearly always best for longer and larger tunnels in conditions of lesser known and variable ground. Precast liners are, however, much quicker to install, and are ideal for rapid tunneling through uniform ground.

Crown plates can be a very effective means of support in bored tunnels. In Toronto, contractors have made rapid progress through highly stressed shale using crown plates and four rock dowels per plate, two to the left and two to the right of the tunnel crown. The resin-bonded rock dowels are installed by an operator working from a platform near the TBM cutter head. Installation is completed in just a few minutes while the TBM gripper pads are being relocated for the next cutting cycle.

Reinforced shotcrete is strong, resilient, and durable. Thorpe and Heuze (1986) describe tests on various tunnel liner systems subjected to *intense blasting*. One chamber was heavily reinforced with fully grouted rockbolts, 2.4 m long on 1.2-m centers, and with ribs covered with wire mesh and at least 50 mm of fiber-reinforced shotcrete. High-explosive charges were detonated and the response of the rock monitored. Upon reentering the chamber after the test, the reinforcement was found completely intact with no spalling or cracking of the shotcrete lining. In contrast, concrete and reinforced concrete did not perform well, nor did steel liners unless backpacked with cellular concrete. Selmer-Olsen (1983) found that 25-year-old shotcrete linings remained for the most part intact, with only minor local spalling caused by oil and clay coatings on the rock. Bond problems can occur, however, when thin layers of shotcrete are applied to soft shale.

During recent years, additives such as *microsilica* and *steel fiber reinforcement* have made possible a high strength of shotcrete to replace the older, more labor-intensive method of shotcrete reinforced by welded wire mesh. Wet mix shotcrete now appears superior to the dry mix process.

5.5.1.4 Steel liners. Many tunnels of small diameter are supported by steel *liner plate* welded in place to form a tube, and back-grouted with portland cement grout, or back-packed with pea gravel to fill the void between steel and rock. Material costs are high, but the method requires only simple construction techniques, so high-quality support can be achieved quite quickly and with little supervision, monitoring, or testing.

Steel liners are often used for high-pressure water tunnels, such as the penstocks of hydroelectric power plants, particularly in sections where internal water pressures approach the minimum in situ rock stress. They give guaranteed strength and freedom from leaks. Water tunnels can remain unlined if rock quality is excellent or if weaker sections are stabilized by shotcrete and bolts (Sec. 5.5.2.4).

5.5.1.5 Concrete liners. *Cast-in-place* liners are generally best when the tunnel has an irregular blasted cross section. Concrete is pumped behind a traveling form, by direct-acting or squeeze-type pumps, with air slugging near the end of the discharge pipe. Concrete slump is normally between 100 and 125 mm to ensure adequate flow down the sides of the form. Fly ash can replace part of the cement. It tends to improve the flow characteristics of the concrete and to reduce cracking by reducing the heat of hydration. Vibrators clamped to the form or immersed in the freshly pumped mix can improve the density and strength of the liner. Reinforcing steel often is not required, although it helps to prevent the development of shrinkage cracks. Forms are collapsed and moved to the next lining location when the concrete has reached a compressive strength of between 4.2 and 5.5 MPa, often 8 to 10 h after placing.

Groundwater containing sulfates is especially destructive for linings of concrete that is made from ordinary portland cement, and *sulfate-resistant cements* are often specified for tunneling applications.

The patented *Bernold System* makes use of pressed and perforated steel sheets which act as both form and reinforcement. The curved segments are coupled together by pins. Concrete pumped behind this steel mesh flows around the annular gap and partially emerges from the mesh openings to form a high-strength structural liner (*Rock Engineering,* Sec. 17.2.3.4).

Precast segments are usually best in tunnels that have a circular cross section as excavated by full-face TBM, but only if there is little

Figure 5.13 Concrete segmental lining placed inside a shotcrete and steel rib structural lining, Dupont Circle subway station, Washington D.C. (*Photo courtesy U.S. National Committee on Tunneling Technology*)

or no collapse or overbreak. Concrete quality tends to be higher because of the improved control possible in a manufacturing plant at the ground surface (Fig. 5.13). The segments can be reinforced or not, depending on the application. The joints interlock and contain a caulking groove so that they can be made watertight after erection. The earliest "precast" units were brick and stone masonry. Later designs of cast-iron segment, as used in the London underground railway system (in clay), were flanged and bolted together. Precast segments may also be bolted, but more often are self-supporting. They are easily damaged, and must be handled with care when transported to the tunnel face; they are best installed by special erecting arms built into the trailing gear of the boring machine.

5.5.2 Support strategies

5.5.2.1 Selection or design? Tunnel support systems cannot be "designed" in the same sense as the columns and beams of a building. There are too many unknowns. Ground conditions are variable. The loads that develop when installing segments, thrusting the TBM forward, and grouting the liner are difficult to estimate and often more damaging than pressure from the ground. Instead, tunnel support is selected empirically. It may be checked using analytical solutions and a range of assumed ground conditions to ensure that stresses are acceptable.

5.5.2.2 Merits of stiff and flexible liners. A common error is to assume that the higher the ground pressure, the thicker and more rigid the liner must be. The opposite is true. Rigid liners attract stress without appreciably reducing distortion; they support unequal horizontal and vertical stresses at the expense of high bending moments. In contrast, systems such as reinforced shotcrete and unbolted segmental liners, by flexing, tend to equalize contact pressures, and develop only small bending moments.

Kuesel (1987) remarks that a lining is like a balloon; if you poke it in at one point, it will try to bulge out at surrounding points. Outward deformation is resisted by passive pressure of the ground. Thus it follows that "nothing improves the performance of a lining like a good grout job," and that axial stiffness is much more important to a lining than flexural strength. Increasing a lining's flexural stiffness cannot significantly restrain ground movement but will increase parasitic lining stresses resulting from imposed ground movement. The stiffness of a liner is proportional to the cube of its thickness, so a small reduction in thickness can greatly increase flexibility.

5.5.2.3 Choice and liner type and thickness. The first step is to select either a flexible system such as the NATM, or a segmental or cast-in-place system that relies on a liner to carry the rock loads. Excavation and support methods are interdependent, and one cannot be selected without regard to the other. If the tunnel is to be excavated by full-face boring, shotcrete will be difficult to apply close to the face. On the other hand, the smooth, circular tunnel produced by boring is ideal for receiving a precast segmental or tubular lining.

Cost is a further consideration. Selmer-Olsen (1983) reports that in Norway a 20-cm-thick reinforced shotcrete liner is more expensive and slower to produce than an unreinforced cast concrete lining 30 cm thick. However, thinner coatings of shotcrete, in the many cases where these are sufficient, are less expensive than cast or precast concrete, and allow the tunnel to advance more rapidly.

In the case of NATM, support measures are specified for each class of ground within the range of variation expected. Empirical design criteria, discussed below, are ideal for this purpose. Decisions to use one or another class of support are made during construction, when actual rock quality and performance can be observed.

Liner thickness is then selected. A secondary concrete lining placed after the ground with its primary support has deformed and stabilized is subjected only to subsequent loads—construction stresses, groundwater pressures, long-term ground creep, and the effects of any future construction such as excavation of parallel or intersecting tunnels. If a lining must resist hydrostatic pressure (external or internal), this probably governs the design. Groundwater pressures are eliminated, if possible, either by grouting leaking seams or by draining the lining exterior. The concrete liner for the Peachtree Center Station in Atlanta was designed to support the pressure of contact grout between the lining and the rock (Kuesel, 1987).

The thickness of *cast-in-place concrete liners* has to be sufficient to permit flow of concrete from the slick line in the crown to all locations between the form and the rock. Typically a minimum clearance of 200 mm is specified for tunnels up to about 9 m in diameter, and 300 mm for larger tunnels. The total concrete thickness includes this clearance together with the thickness of any primary support left in place. When steel sets are used, the total thickness is often much greater than needed to support the ground.

The thickness of *segmental liners* is controlled by the need to prevent damage during handling, and by requirements of the joints. Bolted segmental linings usually have a minimum thickness of 200 mm for concrete and 150 mm for metallic segments. Unbolted, knuckle-jointed linings may be as thin as 125 mm.

5.5.2.4 Unlined tunnels. *Unlined pressure tunnels* and shafts were introduced in Norwegian hydropower construction in response to a shortage of steel after the first world war; more than 100 km of unlined pressure tunnels exist today, and one unlined shaft carries a water head of almost 1000 m (Palmstrom, 1988; Broch, 1989). The rock must be impermeable, and the cover above the tunnel must be sufficiently thick to ensure that the minor principal stress, computed by finite element analysis, is not exceeded by the water pressure.

The shaft or tunnel is filled in steps with continuous monitoring to determine leakage, which usually is in the range 0.5 to 5.0 L/s/km. As much as 5 percent of construction costs are saved by eliminating concrete and steel, and by earlier completion of the work and startup of the power plant. According to Vasilescu et al. (1971) the maximum velocities of water in unlined pressure tunnels should be kept below about 2 m/s.

As much as two-thirds of the cost of *road tunnels* in hard rock can be saved if expensive concrete linings are avoided (Martin, 1982). Alpine tunnels are lined with shotcrete, then a waterproof membrane, then further concrete to keep the inside of the tunnel completely dry. In contrast, about 98 percent of Norwegian road tunnels are lined only at portals and in sections of poor rock. Smooth blasting is used to enhance stability. Infiltration is controlled by local canopies and by insulation as discussed in Sec. 5.6.3. Lighting is also avoided to reduce costs, and the tunnels are generally curved slightly near entrances and exits to make it safer for drivers inside using full headlights. Ventilation is by natural draft.

5.5.2.5 Swelling and squeezing ground. *Squeezing,* defined as ductile yielding of overstressed ground, is often confused with swelling, although the latter is a quite different mechanism of expansion accompanied by an increase in water content (*Rock Engineering,* Sec. 10.4). Often seams of swelling clay are subexcavated and replaced by a pad of rock wool or other flexible material before shotcreting, to allow room for expansion.

At the other end of the scale from the mostly stable hard rock conditions of Scandinavia are squeezing and swelling conditions found in areas such as the Himalayas. Development of hydropower in the Himalayan foothills requires tunnels to be driven through intensely folded and thrusted rocks where the Asian plate overrides the Indian plate (Saini et al., 1989).

Convergences of 20 percent of the tunnel diameter have been experienced, and support pressures may reach 2 MPa when rigid supports are used. Near water-bearing faults, rock fragments flow into the tunnel as a slurry. Rock temperatures as high as 40°C were reported at Bhabaha Hydroelectric Project. The tunnel atmosphere exceeded 32°C, requiring an extensive system to refrigerate the tunnel air. Various combustible and toxic gases have been encountered such as in the Loktaka Hydel project where 15 people were killed and many had burn injuries because of explosion and fire.

At the Chibro Khodri tunnel, a thrust zone about 300 m wide in red shales delayed completion by nearly 6 years. The tunnel was divided into three of smaller diameter. A flexible support system with steel arches and loose backfill was adopted in preference to a stiff support system that attracted higher loads.

In one water tunnel in the Himalayas, the ground squeezed during construction to the extent that the tunnel diameter was reduced by more than 1 m, and the steel ribs provided for primary support were twisted. However, when the tunnel was remined to recover the cross section and replace the ribs, the engineers found that the ground had stabilized and required no structural support at all (Kuesel, 1987).

Nakano (quoted in Tanimoto, 1986) defined rock squeezing behavior in terms of a *competence factor* C_f, determined as the ratio of rock strength divided by overburden pressure:

$$C_f = \frac{\sigma_c}{\gamma D}$$

where σ_c = uniaxial compressive strength
γ = density of the overburden
D = depth of cover

Based on observations in more than 50 tunnels in Japan, the squeezing behavior of tunnels was related to the competence factor in the following manner:

$C_f > 10$, slight or no rock pressure

C_f between 4 and 10, loosening rock pressure

C_f between 2 and 4, light to moderate squeezing

$C_f < 2$, heavy to very heavy squeezing

Cording (personal communication) notes that stress slabbing may be expected in brittle rocks if $C_f < 6$.

Squeezing and swelling conditions require a completely circular liner including an invert arch. The support should be flexible enough to deform without significant damage. Extremes of ground squeeze can be counteracted by excavating a diameter larger than the one required or by installing a deformable interface between rock and liner or by prestressing the liner against the rock to reduce the amount of swelling (Kovari et al., 1981; Wittke, 1981). Swelling also can be minimized by preserving the natural water content of the rock, applying a sealing coat (e.g., shotcrete) to prevent wetting and drying.

The 9.1-km Nabetachiyama Tunnel in the Hokuriku region of Japan, described by Tanimoto (1980), was driven through mudstone with competence factors in the range 0.26 to 3.22 and liquid limits in the range 60 to 168 percent, indicative of both squeezing and swelling conditions. The middle section of the tunnel encountered large faults, intense jointing, and inflammable gas. Plastic zones extended for more than 10 m around the tunnel and ahead of the face. Pre-reinforcement was tried, including spiling and face reinforcement using steel and glass fiber bolts. However, drilling deeper than 3 m proved impossible because of the squeezing of liquified debris. Supports installed included 216-mm-diameter pipe struts at 0.75-m-centers, a 17-cm shotcrete liner, and rockbolts at one for every square meter of surface.

5.5.3 Empirical design

5.5.3.1 Definitions. Judgment and experience are major ingredients in rock engineering design, and are the sole ingredients in the absence of other more formal methods (*Rock Engineering,* Sec. 7.1.5). Our ability to judge and forecast is always limited by the extent and quality of our experience. For example, to estimate the support needed in a tunnel we relate our impression of the ground conditions to our knowledge of the performance of support systems on previous projects in similar ground. Unfortunately we may be at a loss if either the particular ground or the tunneling procedure itself is unfamiliar.

Empirical design solves the problem of our own limited experience by making available the accumulated experience of others. It requires three steps beyond simple judgment: description of *ground quality* by a quantitative classification system, to provide a universal language whereby the global experience gained working in ground of many different qualities can be related to future projects; description of *ground performance* by a formalized quantitative system, which defines such parameters as unsupported stand time and support requirements; and *correlation* of ground quality to performance by comparison of results from a variety of projects over a full spectrum of ground conditions (case histories).

Purely empirical methods are those that predict tunnel support requirements from a knowledge of ground conditions, using a preestablished correlation. Semiempirical methods predict rock loads empirically as an intermediate step, and then go from rock loads to support requirements via a simple theoretical model of support behavior.

The many empirical design alternatives give no more than a rough estimate of average support requirements, but often are sufficient for practical purposes when tempered by engineering judgment. Empirical predictions often prove closer to the "truth" than the apparently more precise predictions of numerical modeling. Being based on real data, they provide a standard against which theoretical predictions are judged.

5.5.3.2 Deere's method. Deere et al. (1969) gave criteria for support evaluation in rock tunnels based on the rock quality designation (RQD) (*Rock Engineering,* Sec. 3.1.2.2). For RQD values greater than 60 they recommended support consisting of rockbolts, mesh, and strapping, whereas for RQD values less than 40, steel sets or ribs were specified. RQD values of between 40 and 60 called for linear interpolation of support requirements.

5.5.3.3 RSR system. The rock structure rating (RSR) system was developed in the United States by Wickham et al. (1972 and 1974). The initial version was based on an evaluation of 134 sections in 33 tunnels, and the final one on 187 sections in 53 tunnels. The tunnels were mainly supported by steel sets and designed using the Terzaghi tunnel support classification. RSR is determined by adding three weighted parameters; *A,* which represents the geological conditions (rock type, rock quality, degree of weathering, and geological structure); *B,* which depends on the joint spacings and orientations with respect to the tunnel axis; and C, which rates the groundwater inflow and condition of the joints. Tables and charts are used to determine these parameters.

5.5.3.4 RMR system. The rock mass rating (RMR) system, developed in South Africa in 1973, has been described in various publications (Bieniawski 1973, 1974, 1976, 1979, 1988, and 1989). RMR is the sum of six properties: uniaxial compressive strength, RQD, joint spacing, quality of the joints, groundwater conditions, and joint orientation. Tables allow determination of the parameters as a guide to the selection of excavation and support procedures for openings from 4 to 12 m in diameter.

5.5.3.5 Size-strength system. The size-strength system, although used as a general purpose classification, was developed in 1973 mainly to predict NATM support requirements in tunnels (Franklin, 1976). Figure 5.11 gives the predicted requirements for shotcrete, rockbolts, and ribs as a function of the degree of support number obtained from the size-strength diagram, Fig. 3.11 in *Rock Engineering,* Chap. 3.

5.5.3.6 Q System. The Q system was developed at the Norwegian Geotechnical Institute (NGI) by Barton et al. (1974, 1975, 1976, 1988) and considers six parameters: RQD (rock quality designation), J_n (joint set number), J_r (joint roughness number), J_a (joint alteration number), J_w (joint water reduction factor), and SRF (stress reduction factor). The Tunneling quality Q is expressed as the product of ratios of pairs of these parameters as follows:

$$Q = \frac{\text{RQD}}{J_n} \frac{J_r}{J_a} \frac{J_w}{\text{SRF}}$$

Numerical values of Q range from 0.001 for exceptionally poor-quality squeezing ground up to 1000 for exceptionally good-quality rock, which is practically unjointed.

Support recommendations are based on 212 case histories of underground openings, mostly in igneous rocks. Figure 5.14 presents a simplified version (Grimstad and Barton, 1988). The equivalent span is found by dividing the actual dimension by a factor representing the safety requirement. For underground power stations the factor is 1.0, whereas it is 1.0 to 1.3 for road tunnels, and 1.6 for water tunnels.

5.5.3.7 Comparisons and evaluations. Various authors give a relationship between the RMR and Q classifications in the form: RMR = A log Q + B, where A is typically in the range of 9 to 14, and B is in the

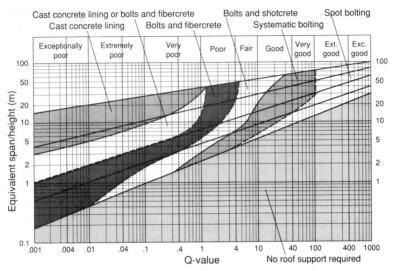

Figure 5.14 Simplified diagram for design of rock support based on the Q system (Grimstad and Barton, 1988).

range of 35 to 55. Kaiser and Gale (1985), for example, suggested the relationship RMR = 8.7 ln$Q$ + 38. They found that both RMR and Q systems accurately predicted the no-support limit, but tended to overestimate the support required for the tunnel walls. The Q system gave a better forecast of support quantities.

Butler and Franklin (1990) describe a computerized, knowledge-based expert system called Classex that combines Q, RMR, and RQD ground classification systems. An expert system can contain a much larger store of rules and facts than can be remembered by any single person. Data are checked, and judgmental errors are minimized. Carefully worded questions guide the user in obtaining the most applicable predictions of rock mass behavior and support requirements.

The literature abounds with examples in which predicted opening sizes, support measures, or stand-up times were considerably different from those found to be necessary. One reason is that users of empirical methods are often aware of only part of the extensive literature on the subject. The RMR system has been modified over the years and is spread among many publications. The Q system has been clarified and updated by Stacey and Page (1986) and by Kirsten (1988), among others. Einstein et al. (1979, 1983) presented a bibliography of empirical tunnel design methods, and an evaluation of the relative merits of five of these. They found many ambiguities. They recommended studying the base cases for each system, or at least understanding its development, assumptions, and limitations. All empirical design systems are evolving, and generally speaking, only the most recent version of a system should be used.

5.5.4 Semiempirical methods

5.5.4.1 Rock load methods.
Rock load methods make use of rock mass classifications to predict loads acting on tunnel support, allowing design of supports to resist these loads. Separate predictions are often made for the vertical and horizontal rock load components.

The first attempts at estimating rock loads were made by Ritter in 1879 and by Kommerell and Bierbaumer in the early twentieth century (Steiner et al., 1980). Better known are the predictions of Karl Terzaghi (Terzaghi, 1946), who, as well as being a pioneer of soil mechanics, was a major contributor to rock engineering. All of the early approaches ignore important load-determining factors such as flexibility of supports, gaps between supports and rock, groundwater and ground stress conditions, and extent of blast-induced damage. The introduction of rockbolts and shotcrete to replace passive supports of heavy steel means that rock loads are much lighter than they used to be and probably much less than predicted by the early empirical formulas. Modern tunneling methods that use a TBM or carefully controlled blasting, followed by support with rockbolts and shotcrete, preserve the rock in a sounder condition, encourage arching, and reduce the burden substantially.

5.5.4.2 Terzaghi's method.
A liner has to support the entire weight of overlying rock and soil only in the extreme case of a shallow tunnel where the rock contains smooth vertical joints and where little or no horizontal stress acts to enhance friction. Tunnel linings in mountainous terrain obviously do not support the weight of thousands of meters of material. Instead, stresses are redistributed around the opening by dilation and mobilization of shear strength along the joints, in a mechanism known as *arching*. The lining has to support only those stresses not carried by the rock arch.

The rock burden can be visualized as the weight of a potential roof fall bounded by the arched rock above and the tunnel crown below. Dimensions and hence weight of this burden can be estimated, and are related to the type of rock, the jointing, and the width and diameter of the opening. Terzaghi expressed loads acting on the liner in terms of the opening width B, height H, and rock mass characteristics (Fig. 5.15). His method has been revised by Cording to better suit modern support systems and for use in larger caverns (Chap. 6).

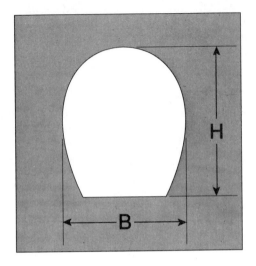

Rock Condition	Rock Load	Remarks
1. Hard and intact	Zero	Light lining, required only if spalling or popping occurs
2. Hard, stratified, or schistose	0 to 0.50 B	Light support
3. Massive, moderately jointed	0 to 0.25 B	Load may change critically from point to point
4. Moderately blocky and seamy	0.25 B to 0.35 $(B + H)$	No side pressure
5. Very blocky and seamy	0.35 to 0.10 $(B + H)$	Little or no side pressure
6. Completely crushed but chemically intact	0.10 $(B + H)$	Considerable side pressure—required support for lower end of ribs, or circular ribs
7. Squeezing rock, moderate depth	1.10 to 2.10 $(B + H)$	Heavy side pressure—invert struts required, circular ribs recommended
8. Squeezing rock, great depth	2.10 to 4.50 $(B + H)$	Heavy side pressure—invert struts required, circular ribs recommended
9. Swelling rock	Up to 76 m irrespective of $(B + H)$	Circular ribs required—in extreme cases use yielding support

Figure 5.15 Terzaghi's rock mass classification and prediction of rock loads on steel sets and lagging. The table relates to saturated rock; load values for cases 4 through 6 can be halved if the tunnel is permanently above the water table. (Terzaghi, 1946).

5.5.4.3 Ground response curves. Ground response curves show the relationship between support load and tunnel convergence. They demonstrate how the load-carrying requirement of a liner varies according to the delay between excavating the heading and installing the support. As shown in Fig. 5.9, a very stiff liner installed at the tunnel face permits no convergence, and attracts a high stress (close to point A). If a delay is permitted and also as the liner deforms, stability is reached at the lesser stress C. Arching has been allowed to develop, and the rock contributes greatly to supporting itself. If support is delayed beyond point D, arching is lost and the liner must carry a burden of loosened rock corresponding to the much greater stress E.

The required support capacity at first decreases because of the arching that accompanies initial, small rock movements. Then, if the rock is allowed to loosen further, the trend is reversed and the required capacity increases dramatically as arching is destroyed. It is common NATM practice to defer the installation of the support, and even to construct initial lining courses with missing sections ("blockouts") to permit axial yielding, in order to assist the development of ground resistance through deformation (Kuesel, 1987).

Quantitative ground response curves have been developed for various models of material behavior (e.g., Brown et al., 1983, and the convergence-confinement method, Wong and Kaiser, 1987). The extent of convergence before supports begin to carry load depends on the delay between excavation and installation, on the gap between the support and the rock, and on the flexibility of the support system. The curves intersect at a point where further convergence is prevented; at this point, loads imposed by the rock match reactions from the support system.

Arching is three dimensional and includes a contribution from the unsupported face of the tunnel. As shown in Fig. 5.16, the stress level rises to a maximum ahead of the advancing face and drops to zero where the tunnel is unsupported. Pressure on the liner rises again to a peak at point E, a short distance back from the leading edge of the liner, before stabilizing at point F. The magnitude of the peak lining pressure depends on the distance behind the advancing face at which the liner is installed.

5.5.5 Analytical methods

Methods to compute liner stresses include elastic closed-form solutions, beam-spring models, and beam-continuum models such as those based on the finite element method (*Rock Engineering*, Sec. 7.2). Their merits and limitations have been reviewed by the ASCE Technical Committee on Tunnel Lining Design (O'Rourke, 1984). They con-

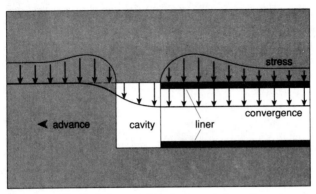

Figure 5.16 Vertical stress and displacement along the crown of a tunnel (Eisenstein et al., 1984).

cluded that because of the considerable ground variations, analytical studies are useful only for investigating the sensitivity of lining sections to variations of ground and liner characteristics and for placing bounds on possible lining behavior. Kuesel (1987) remarks that the proper application of stress analysis to tunnel linings is in parametric studies of lining behavior. By the time a relatively thick and rigid secondary liner is installed, most rock stress redistribution has occurred; subsequent bending moments that develop are usually inconsequential.

Charts that assist in predicting liner stresses have been published by Detournay and St. John (1988). An advanced theoretical model to predict ground stresses, strains, and displacements around a circular tunnel is provided by Hisatake et al. (1989); it includes realistic peak and residual strength criteria and nonlinear stress-strain relationships. Grimstad and Barton (1988) describe use of the Universal Distinct Element Code (UDEC) with input determined on the basis of the parameters of the Q system. They claim that the numerical procedures provide a wealth of information concerning stresses, deformations, and joint displacements, which assist the designer in assessing the effect of the chosen rock support.

The predictions of numerical modeling are no more reliable than the data on which they are based; ultimately, the most reliable data are obtained from *back-analysis* of large-scale excavations in identical ground. This method has been demonstrated by Sakurai (1983) at a project in Washuzan, Japan, where twin highway tunnels with a combined span of 24 m were to be excavated directly on top of twin railway tunnels with the same dimensions (Fig. 5.17). His "field measurement aided design technique" made use of sliding micrometer and inclinometer measurements from the ground surface, as well as

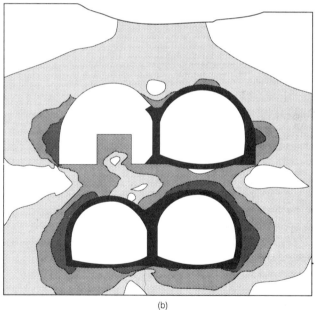

Figure 5.17 Washuzan project, Japan. Twin highway tunnels were to be excavated above twin railway tunnels. (a) The four tunnels in weathered granite (*Photo courtesy S. Sakurai*); (b) an example of the maximum shear strain distribution around the four tunnels (Sakurai, 1983).

extensometer and convergence measurements from the interior of the tunnel during excavation of the first of the four tunnels. Back-calculation using the finite element method, and comparison of predicted and measured displacements gave an equivalent modulus of deformability for the rock mass, which was then applied to design of the complete four-tunnel prototype.

5.6 Control of Groundwater and Gas

5.6.1 Hazards of water and gas

5.6.1.1 Infiltration. Inflow into tunnels below the water table depends on the spacing and tightness of joints and the head of water above the crown. Wet conditions make tunneling more difficult, slower, and more expensive (Fig. 5.18). Accidents are more frequent because wet machines and materials are difficult to handle. Drilling and blasting can be hazardous, and the methods and types of explosive may have to be modified to avoid misfires (Jones, 1983).

During the service life of a pedestrian or vehicular tunnel, even small amounts of seepage can lead to unsafe conditions, icing of highway tunnels, and wet walkways in subway stations and pedestrian tunnels. Water infiltration also leads to deterioration of architectural finishes and corrosion of steel supports.

One way to avoid these problems in the completed tunnel is to specify in the contract documents a permissible rate of infiltration or leakage. Rates permitted for rapid transit tunnels in the United States

Figure 5.18 Wet conditions in the Wolverine tunnel. Flow from grout packers in the tunnel face amounted to 50 L/s at a pressure of 2 MPa. (*Photo courtesy P. Kaiser*)

have typically ranged from 0.2 to 1.7 L/day/m^2 of inside surface area. Those specified for the Stockholm subway were between 1.9 and 9.5 L/day/m^2, and those for U.S. waste water conveyance tunnels have ranged from 2.1 to 14.8 L/day/m^2 (O'Rourke, 1984). Specified rates may be compared with actual inflows by monitoring the water produced by the tunnel pumping system, using an overflow weir with a calibrated notch.

5.6.1.2 Leakage. Exfiltration (leakage) is of concern when tunnels carry high-pressure water or toxic fluids. Underground excavations containing fluids such as petroleum products at near-atmospheric pressures can be left unlined if the rock quality is high, and if the excavation is below the water table so that the fluids are contained by inward seepage of groundwater (Chap. 6). Fluids of greater toxicity or at higher pressures must be contained by an impervious liner.

Leakage of water can be hazardous, particularly when under pressure. High-velocity water is erosive, and water at low velocity but moderate to high pressure can trigger liner instability or a landslide. If the soils or joint fillings around the tunnel are susceptible to swelling, any cracking of the liner will soon lead to swelling pressures and further damage.

5.6.1.3 Pumping and water handling problems. High-capacity pumping equipment is needed to prevent flooding when inflows are substantial, such as in karstic limestone. Recently in Ontario, a tunnel to carry cooling water to the Bruce Generating Station power plant had to be diverted to bypass a limestone cavern connected with the lake above. Even this bypass tunnel had to be extensively grouted.

Hot water at depth can be more of a problem than cold water. In 1950 to 1955 the Tecolate tunnel through mountains northwest of Santa Barbara, California, was driven for 10.3 km through sedimentary strata (Trefzger, 1966). Groundwater, heated by passage through a zone of geologically recent faulting, issued from the face at temperatures of up to 41°C and at rates of up to 580 L/s. One inflow of 180 L/s held up construction for 16 months and resisted all grouting attempts. To cope with the almost unbearable conditions, workers traveled to and from the heading immersed up to their necks in cold water in mine cars.

5.6.1.4 Instability and blowouts. High-velocity water passing through loose or low-durability rocks can erode the rock material and enlarge the joints. In 1958, a tunnel for the Stockholm subway encountered a sudden and extreme inflow that washed out a silt-filled fault. Groundwater from an overlying gravel esker entered at an estimated 100 to

170 L/s and flooded much of the tunnel system. The fault had to be sealed off by divers (Morfeldt, 1969).

Frequent and severe problems are associated with sudden blowouts when a pocket of high-pressure water is unexpectedly penetrated by the tunnel heading. The San Jacinto tunnel near Banning, California, was driven through massive granite. At a point only 50 m from the shaft, the heading broke through a clay gouge barrier into the heavily jointed and water-bearing hanging wall of a fault. An estimated 480 L/s of water surged into the tunnel, accompanied by over 760 m^3 of rock debris. Subsequent mapping located 21 similar faults along the tunnel line (Thompson, 1966).

5.6.1.5 Depression of the water table. The underdrainage effect of tunneling and consequent lowering of the water table can affect regional water supplies or generate subsidence as a result of consolidation of overlying soils (*Rock Engineering,* Sec. 4.1.4.6). Environmental risks should always be evaluated. Drainage can be permitted if the inflows will be slight, when subsidence will not be an issue, and when permanent lowering of the water table will not be objectionable. Otherwise, inflows during construction must be reduced by pregrouting, or the water table can be recharged through wells during construction until such time as the installed liner provides a more permanent barrier to infiltration.

5.6.1.6 Gas hazards. Natural gases are found in pockets in some rock formations (Chap. 8). Gases such as methane, when they seep into a tunnel, present a serious hazard because of their toxicity and particularly their explosion potential. Carbon monoxide and hydrogen sulphide are toxic but not explosive, and even an inert gas like carbon dioxide can cause suffocation if it is allowed to displace air. When tunneling through formations known or suspected to be gassy, exploratory drillholes should be sampled to estimate the amount and chemical content of the emitted gases. Air samples should be taken and analyzed at the face during tunnel driving. Smoking and naked flames are to be avoided, and good ventilation is essential.

5.6.2 Predicting inflows

Groundwater enters mainly through joints and faults. The longer the tunnel, the greater the inflows and possibility of intersecting zones of fractured and water-bearing ground. More important than precise prediction of inflow amounts is a reliable forecast of the nature and magnitude of potential problems. This requires a thorough investigation and an interpretation of the geology and the groundwater regime, with measurements of hydraulic conductivity of the rock mass, and good geological mapping. When sudden inflows appear possible, the specification should call for probing ahead of the face to relieve the pressures in a controlled manner through small drillholes, before they are released at full force by the heading itself.

Long-term inflows need to be evaluated considering both present and future policies for regional groundwater extraction. Some New York City subway tunnels were operated for years with little water infiltration, but changes in land use and water management policy caused the groundwater levels to rise, and the seepage to greatly increase (O'Rourke, 1984).

Closed-form solutions for predicting groundwater inflows given by Goodman et al. (1965) include models for inflow through tunnel walls without drawdown, inflow through tunnel walls with a declining water table, advance of the tunnel through a water-bearing zone with drawdown, and transient flow of water through the face of a tunnel. The equation for inflow in situations where drawdown cannot occur, for example, beneath a lake, river, or prolific aquifer, is as follows:

$$q = \frac{2K(H + h)}{2.3 \log (r/2h)}$$

where q is the steady state inflow per unit length through the walls of a tunnel completely surrounded by a water-bearing formation having hydraulic conductivity K, H is the depth of the overlying lake or river, h is the distance from the tunnel crown to the ground surface, and r is the tunnel radius.

Bello-Maldonado (1983) provides a model for inflow prediction for a tunnel driven at constant rate through free and recharged aquifers with both constant and variable head alternatives. Solutions like these can be used for approximate forecasting of pumping requirements in uniform and rather simple rock conditions. When conditions are not so simple, such as when major inflows are likely at faults, the closed-form solutions must be replaced by numerical modeling.

5.6.3 Control of infiltration and leakage

Methods for control of infiltration and leakage include drainage and grouting, often used in combination. The techniques are described in *Rock Engineering,* Chap. 18.

5.6.3.1 Short-term control of groundwater.
During construction, the tunnel invert must be kept as dry as possible to maintain safe conditions for traffic, particularly in shaley ground that can turn to mud. The invert is provided with a lateral ditch and a series of sumps for

water collection and pumping. Conditions for construction traffic can be improved if the invert is paved with durable broken rock or lean concrete.

Moderate inflows can be controlled by *pumping from within the tunnel*. Groundwater pressures high enough to cause instability of the rock walls are relieved by drilling drainholes if the increased inflows can be handled by pumping.

More substantial inflows and higher pressures, when anticipated from the site investigation or by probing ahead of the face, are reduced by *grouting* through holes drilled from surface, from a pilot bore, or from the heading. In extreme cases when other methods prove ineffective, water inflows can be contained within acceptable limits by *ground freezing methods* (Rock Engineering, Sec. 18.3) or by working in *compressed air* if the water pressures are not too great or by *pumping to lower the water table* using wellpoints drilled from the surface. The costs are high, and the environmental consequences must be evaluated. These alternatives are a last resort and are more commonly needed in soil than in rock.

Tunnels through the shale and limestone bedrock of Oslo, Norway, have to be driven under conditions of practically zero infiltration to prevent lowering of the groundwater table, and consolidation and settlement of the overlying marine clays (Asting, 1983). As an alternative to permanent recharge wells, recent tunnels have been pregrouted by fan-drilling ahead of the face. The treatment is repeated at 20-m intervals, consolidating a 3-m annulus where tunnels are to be blasted, and 1.5 m where they are to be bored. Cement and chemical grouts are used, depending on whether water acceptance in grout holes is above or below about 2 lugeon units. Pore pressure measurements monitor success of the treatment. Pregrouting has been found far more efficient than postgrouting.

5.6.3.2 Long-term watertightness.

When a tunnel remains unlined, water cascading onto a road surface or rail track is evidently unacceptable, particularly if exposed to winter freezing. Local seepages are treated inexpensively in Norwegian road tunnels by installing canopies of corrugated aluminum. Water is diverted to the sides, collected in drains and transported out of the tunnel. An insulating sandwich of rock wool or polyurethane foam between aluminum or fiberglass prevents water from freezing (Martin, 1982).

For waterproofing of lined tunnels, grouting is more reliable and less expensive than drainage. The permanence of pumping systems depends on the composition of the groundwater; high concentrations of calcium and iron hydroxides lead to clogged pumps and drainage channels. Combining a grout curtain with shallow drains to mop up residual seepage has sometimes been useful (Kuesel, 1987).

Grouting behind a liner (*back-grouting*), from underground in the completed tunnel, is a more localized and less costly procedure than grouting from surface before the tunnel is driven, and is the preferred method whenever inflows are sufficiently small to be tolerated during construction. A further reason for back-grouting is to secure full, continuous contact between the lining and the ground—the most efficient tunnel lining is one that mobilizes the strength of the ground by permitting controlled ground deformation.

Joints in segmental liners are sealed by water stops, gaskets, and caulking sealants. Cast-in-place liners can be sealed by fixing an impervious membrane against the rock or shotcrete before erecting the concrete form, attached to pins in the rock using a heat welding method that avoids puncturing (Fig. 5.19).

Tunnels to carry high-pressure water or toxic fluids are often lined with a steel tube back-grouted with concrete. At the North Fork Stanislaus River hydroelectric project in California (Schleiss, 1988) a penstock tunnel 2.16 km long and 4.27 m in diameter was excavated by a TBM at an average rate of 26 m/day through quartz-mica schist and quartzite. Primary support was required only through a 50-m

Figure 5.19 Waterproof membrane and concrete formwork, Torino-Frejus motorway, Giaglione Tunnel, Italy. (*Photo courtesy Prof. S. Pelizza, Politecnico di Torino*)

sheared zone. The tunnel was designed to carry an internal pressure of 7.5 MPa (a head of about 770 m of water). External water pressures were also high, and even though drains were installed, a steel liner was considered necessary for when the tunnel was empty. The design calculations required measurements of convergence and groundwater infiltration, plate load tests, and use of the TBM gripper pads to load the rock mass while measuring deformations.

5.7 Construction Control

5.7.1 Construction control program

Tunneling projects require a carefully planned program of *construction control*. The program is threefold: inspection of changes in ground conditions, testing of support materials, and monitoring of ground and support behavior. Excavation and support provisions are confirmed or adjusted depending on the results.

5.7.2 Recording changes in ground conditions

Variations in ground conditions are recorded by inspecting the tunnel headings and often by probing ahead of the face. Blasted headings are inspected after the fumes from the round have cleared, during scaling, and before mucking. TBM headings are inspected at least daily, before the rock is obscured by shotcrete or other forms of support (in fully shielded machines, rock inspection can be difficult or impossible). A systematic record should be kept of rock quality parameters (block size, strength, and the nature of jointing), of information on faults and water inflows, and of any underbreak, overbreak, or instability (*Rock Engineering,* Chaps. 2 and 3). The ground is classified using the system selected as standard for the project, and the actual ground quality is compared with predictions made in the site investigation report.

When drillholes from surface are few or conditions variable, the tunnel engineer relies greatly on exploratory drilling ahead of the face. To avoid interrupting tunnel excavation, special lateral galleries can be dug or blasted to accommodate probe-drilling equipment. Rotary drills with downhole mandrel thrust systems can achieve penetration distances of between 200 and 500 m in horizontal boring (*Rock Engineering,* Secs. 6.5.1 and 6.5.2). Diamond coring is best, because it provides not only a drillhole but also a core. Percussive open hole rock drilling is quicker, although less informative. Ground conditions must be inferred from rates of penetration of the drill bit and from inflows and pressures of water.

5.7.3 Inspecting and testing of materials

At the start of a tunneling project, rockbolts, mesh, and liner segments are tested to ensure compliance with specifications. Each batch of materials usually comes with a factory certificate of steel properties. Tests should be conducted on site to confirm the holding capacity of rockbolts in a representative range of rock conditions (*Rock Engineering,* Sec. 16.1.7.5). Testing equipment and instruments should be calibrated at this time. Storage and handling should be arranged to ensure that materials will arrive undamaged at the installation site.

In the course of tunneling, support practices need to be inspected by qualified persons. In a well-managed construction project, contractors and engineers work together to remedy deficiencies in materials and methods before they become problems. Variable materials like shotcrete require daily testing of samples (*Rock Engineering,* Sec. 17.2.5.8).

5.7.4 Monitoring of ground and support behavior

Usually a few *representative cross sections* of the tunnel are selected and fully instrumented to measure movements of the liner and surrounding rock, settlements at surface (shallow tunnels only), loads and pressures on support systems, and groundwater pressures. In addition, simple *convergence measurements* are made at a much greater number of cross sections for day-to-day control of rock behavior and support performance. A full range of options for instrumentation is shown in Fig. 5.20. Actual arrangements depend on the size and diameter of tunnel, variability of the ground, ease of access, the method of lining, and on whether the tunnel is blasted or bored (*Rock Engineering,* Chap. 12).

To highlight the trends in ground and support behavior, the results are clearer if presented as graphs rather than as tables of readings. When the graphs and inspections of rock quality at the face show improving conditions, support can be reduced and advance rates increased. When conditions are deteriorating, advance must be slower and the support augmented. Advance rates should be limited to allow installation of the required support close to the working face; unsafe conditions develop if support lags behind.

5.8 Tunnel Maintenance

Tunnel linings have a limited life, and those constructed many years ago using different and often inferior methods are starting to show their age. The railway tunnels of Europe, built in the rail boom of the

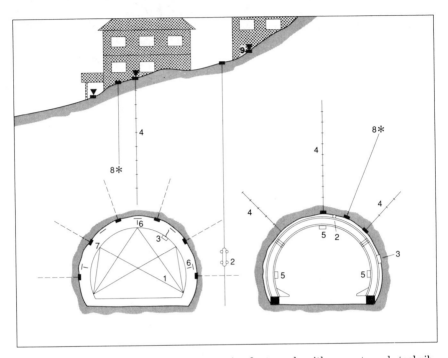

Figure 5.20 Typical patterns of instrumentation for tunnels with concrete and steel rib liners. (1) Convergence measurements; (2) inclinometers; (3) crack gauges; (4) multiple extensometers; (5) vibrating wire strain gauges; (6) pressure cells; (7) load cells; (8) piezometers; (9) settlement gauges and leveling monuments.

nineteenth century, were lined mostly with brick. In the 1960s in France, two express trains encountered roof fall debris, killing many of the passengers. The spotlight was suddenly focused on detection and avoidance of such problems.

Special track cars can detect changes in rail alignment such as those caused by floor heave. The Swiss Matisa track recording car, for example, measures horizontal and vertical curvature and the distance between the rails. A British track recorder ejects paint between the rails wherever it detects a serious fault, while moving at speeds of up to 200 km/h. Similar but slower vehicles measure clearances between the train and the liner. The "porcupine" method used in the Federal Republic of Germany has a measuring accuracy of 20 to 30 mm. The "Castan" type used by the French SNCF records photographically.

In old water tunnels, a roof fall will obstruct the flow of water or sewage. Falls are more difficult to detect because inspection requires dewatering, which itself can cause problems. An old intake tunnel, the sole source of water for the town of Sault Saint Marie in Ontario, was placed at risk not long ago by the need to blast power plant foundations in the river bed above the tunnel. A search through old records showed that divers' inspections every few years had always been accompanied by falls of sandstone. Had these falls continued between dives, they would have long ago filled the tunnel with broken rock. The divers had triggered the falls by lowering the water pressure to make diving easier. Under normal conditions, without transient seepage pressures, the unlined tunnel appeared quite stable. Controls were placed on levels of blast vibration, and the construction work above was completed without incident.

Lining deterioration often occurs because of inadequate contact grouting, inadequate cover of concrete over steel, or rotting of embedded timber.

Water leakages that develop over the years can be a source of problems. The 2.4-m-diameter Carter Lake pressure tunnel in Colorado was excavated in 1954 through conglomerates, sandstones, and siltstones. In spite of injecting over 2000 bags of cement in 1969, by late 1971, hillside leakage had increased from about 75 to 750 L/min and longitudinal cracks had appeared in the liner. In 1972, 3500 L of AM9 chemical grout reduced the escape of water to about 5 L/min. Chemical grouting was chosen as an alternative to a steel liner (Gebhart, 1974).

References

AFTES (Association Française des Traveaux en Souterrain), 1984. Updated recommendations from working groups on various topics including safety, blasting, convergence-confinement design, maintenance and repairs to tunnels, etc. *Tunnels et Ouvrages Souterrains,* Numéro Spécial, November, pp. 2–169.

———, 1989. *Recommendations Concerning Grouting of Underground Works.* Working Group Report 8 (translated into English), 48 p.

Asting, G., 1983. "Experiences with Pregrouting in Sewage Tunnels in the Oslo Area." In: *Norwegian Tunnelling Technology,* Norwegian Soil and Rock Eng. Assoc., Pub. 2, pp. 57–63.

Barton, N. R., 1976. "Recent Experiences with the Q-System of Tunnel Support Design." *Proc. Symp. Exploration for Rock Eng.,* Johannesburg, Balkema, vol. 1, pp. 107–117; session report, vol. 2, pp. 167–172.

———, 1983. "Application of Q-System and Index Tests to Estimate Shear Strength and Deformability of Rock Masses." *Proc. Int. Symp. Eng. Geol. and Underground Construction,* IAEG, Lisbon, vol. 2, pp. II 51–II 70.

———, 1988. "Rock Mass Classification and Tunnel Reinforcement Selection Using the Q-System." *Rock Classification Systems for Engineering Purposes,* ASTM STP 984, Louis Kirkaldie (ed.), ASTM, Philadelphia, pp. 59–88.

———, Lien, R., and Lunde, J., 1974. "Engineering Classification of Rock Masses for the Design of Tunnel Supports." *Rock Mech.,* vol. 6., pp. 189–236.

———, ———, and ———, 1975. "Estimation of Support Requirements for Underground Excavations." *Proc. 16th U.S. Symp. Rock Mech.,* ASCE, New York, pp. 163–177; discussion pp. 234–241.

Bello-Maldonado, A. A., 1983. "Post-Construction Seepage Towards Tunnels in Variable Head Aquifers." *Proc. 5th Int. Cong. Rock Mech.,* Melbourne, Australia, vol. B, pp. 111–117.

Bennett, R. D., 1981. *Tunnel Cost Estimating Methods.* U.S. Army Eng. Waterways Experiment Station, Technical Report GL-81-10, 238 pp.

Bieniawski, Z. T., 1973. "Engineering Classification of Jointed Rock Masses." *Trans. South African Inst. Civ. Eng.*, vol. 15, pp. 335–344.

———, 1974. "Geomechanics Classification of Rock Masses and Its Application in Tunneling." *Proc. 3d Int. Congr. Rock Mech.*, Denver, vol. IIA, pp. 27–32.

———, 1976. "Rock Mass Classifications in Rock Engineering." *Proc. Symp. Exploration for Rock Eng.*, Johannesburg, Balkema, vol. 1, pp. 97–106; session report, vol. 2, pp. 167–172.

———, 1979. "The Geomechanics Classification in Rock Engineering Applications." *Proc. 4th Int Congr. Rock Mech.*, Montreaux, ISRM, vol. 2, pp. 41–48.

———, 1984. *Rock Mechanics Design in Mining and Tunneling.* A. A. Balkema Publ., Amsterdam.

———, 1988. "The Rock Mass Rating (RMR) System (Geomechanics classification) in Engineering Practice." *Rock Classification Systems for Engineering Purposes*, ASTM STP 984, Louis Kirkaldie (ed.), ASTM, Philadelphia, pp. 17–34.

———, 1989. *Engineering Rock Mass Classifications.*, John Wiley, New York, 251 pp.

Blindheim, O. T., Dahl Johnsen, E., and Johannessen, O., 1983. "Criteria for the Selection of Fullface Tunnel Boring or Conventional Tunneling." In: *Norwegian Tunnelling Technology,* Norwegian Soil and Rock Eng. Assoc., Pub. 2, pp. 33–38.

Brekke, T. L., and Ripley, B. D., 1987. *Design Guidelines for Pressure Tunnels and Shafts.* Report AP5273, Research Project 1745-17, prepared for Electric Power Research Insititute, Palo Alto, Calif., U. of Calif., Berkeley.

Broch, E., 1988. "Site Investigations." In: *Norwegian Tunnelling Today,* Norwegian Soil and Rock Eng. Assoc., Pub. 5, pp. 49–52.

———, 1989. *The Technical, Economical, Environmental Disclosures of the Underground Employment in the Future.* General Rept., Suolosottosuolo; Int. Cong. GeoEng., Turin, Italy, 12 pp..

Brown, E. T., Bray, J. W., Ladanyi, B., and Hoek, E., 1983. "Ground Response Curves for Rock Tunnels." *J. of Geotechnical Eng.*, ASCE, vol. 109, no. 1, pp. 15–39.

Butler, A. G., and Franklin, J. A., 1990. "Classex: An Expert System for Rock Mass Classification. *Proc. Int. Symp. Rock Mech.*, Mbabane, Swaziland, pp. 73–80.

Cording, E., 1984. "State of the Art: Rock Tunneling." In: *Tunneling in Soil and Rock,* K. Y. Lo (ed.), ASCE, New York, pp. 77–106.

Craig, C. L., and Brockman, L. R., 1971. "Survey of Tunnel Portal Construction at U.S. Army Corps of Engineers Projects." *Proc. Symp. Underground Rock Chambers,* Phoenix, Ariz., pp. 167–201, Amer. Soc. Civ. Eng., New York.

Deere, D. U., 1968. "Geological Considerations." In: *Rock Mech. in Eng. Practice,* R. G. Stagg and D. C. Zienkiewicz (eds.), John Wiley, New York, pp. 1–20.

———, and Deere, D. W., 1988. "The Rock Quality Designation (RQD) in Practice." In: *Rock Classification Systems for Eng. Purposes,* ASTM STP 984, Louis Kirkaldie (ed.), ASTM, Philadelphia, pp. 91–101.

———and ———, 1989. *Rock Quality Designation (RQD) after Twenty Years.* U.S. Army Corps of Eng. Contract Report GL-89-1, Waterways Experiment Station, Vicksburg, MS, 67 pp.

———, Hendron, A. J., Jr., Patton, F. D. and Cording, E. J., 1967. "Design of Surface and Near-Surface Construction in Rock." In: *Failure and Breakage in Rock,* Charles Fairhurst (ed.), SME, AIME, New York, pp. 237–302.

———, Peck, R. B., Monsees, J. E., and Schmidt, B., 1969. *Design of Tunnel Liners and Support Systems.* U.S. Department of Transportation, Washington D.C., MTIS No. PB-183799, 404 pp.

———, ———, Parker, H. W., Monsees, J. E. and Schmidt, B., 1970. *Design of Tunnel Support Systems.* Highway Research Record, No. 339, Highway Research Board, pp. 26–33.

Detournay, E., and St. John, C. M., 1988. "Design Charts for a Deep Circular Tunnel under Non-Uniform Loading." *Rock Mech. and Rock Eng.*, vol. 21, pp. 119–137.

Duffaut, P., and Piraud, J., 1975. "Soutènement des Tunnels Profonds Autrefois et Aujourd'hui." *Industrie Minérale,* N° Spécial, Cahier 7 du Comité Française de Mécanique des Roches, pp 1–17.

Egger, P., 1983. "Roof Stability of Shallow Tunnels in Isotropic and Jointed Rock." *Proc. 5th Int. Congress Rock Mech.*, Melbourne, Australia, vol. C, pp. 295–301.

Einstein, H. H., Steiner, W., and Baecher, G. B., 1979. "Assessment of Empirical Design Methods for Tunnels in Rock." *Proc. 4th RETC,* AIME, New York, vol. 1, pp. 683–706.

———, Thompson, D. E., Azzouz, A. S., O'Reilly, K. P., Schultz, M. S., and Ordun, S., 1983. "Comparison of Five Empirical Tunnel Classification Methods—Accuracy, Effect of Subjectivity and Available Information." *Proc. 5th Int. Cong. Rock Mech.,* Melbourne, Australia, vol. C, pp. 303–313.

Eisenstein, Z., Heinz, H., and Negro, A., 1984. "On Three Dimensional Ground Response to Tunnelling." In: *Tunnelling in Soil and Rock,* K. Y. Lo (ed.), ASCE, New York, pp. 107–127.

Fairhurst, C., and Lin, D., 1985. "Fuzzy Methodology in Tunnel Support Design." *Proc. 26th U.S. Symp. Rock Mech.,* Balkema, Rotterdam, vol. 1, pp. 269–278.

Franklin, J. A., 1976. "An Observation Approach to the Selection and Control of Rock Tunnel Linings." *Proc. Conf. Shotcrete for Ground Support,* ASCE, Easton, Md., October 3–8, pp. 556–596.

———, and Matich, F., 1990. "Conceptual Design of a Tunnel Beneath the Northumberland Strait, Canada." *Proc. 2d Symp. Strait Crossings,* Trondheim, Norway, pp. 181–184.

Gebhart, L. R., 1974. "Foundation Seepage Control Options for Existing Dams." In: *Inspection, Maintenance and Rehabilitation of Old Dams. Proc. Eng. Foundation Conference,* Pacific Grove, Calif., September 1973, pp. 660–676.

Goodman, R. E., Moye, D. G., van Schalkwyk, A., and Javadel I., 1965. "Groundwater Inflows during Tunnel Driving." *Eng. Geol.,* vol. 2, pp. 39–56.

Grimstad, E., and Barton, N., 1988. "Design and Methods of Rock Support." In: *Norwegian Tunnelling Today,* Norwegian Soil and Rock Eng. Assoc., Pub. 5, pp. 59–64.

Gysel, M., 1987. "Design of Tunnels in Swelling Rock." *Rock Mech. and Rock Eng.,* vol. 20, pp. 219–242.

Hisatake, M., Cording, E. J., Ito, T., Sakurai, S., and Phien-Weja, N., 1989. "Effects of Non-Linearity in Strength Reduction of Rocks on Tunnel Movements." *Proc. ISRM Symp. Rock at Great Depth,* Pau, France, pp. 553–560.

Hochmuth, W., Krischke, A., and Weber, J., 1987. "Subway Construction in Munich, Developments in Tunnelling with Shotcrete Support." *Rock Mech. and Rock Eng.,* vol. 20, pp. 1–38.

Hoek, E., and Brown, E. T., 1980. *Underground Excavations in Rock.* Institution of Mining and Metallurgy, London, 527 pp.

ITA (International Tunnelling Association), 1988. "Recommendations on the Contractual Sharing of Risks." *Tunneling and Underground Space Technology,* vol. 3, no. 2, pp 103–140.

John, K. W., 1988. "Rock Mechanics Input for Design and Construction of Hydro-Electric Power Projects in Developing Countries," *Proc. Int. Symp. of Rock Mechanics and Power Plants, Madrid,* Spain, vol. 1, pp. 335–346.

Johnsen, M. G., 1988. "History and Development." In: *Norwegian Tunnelling Today,* Norwegian Soil and Rock Eng. Assoc., Pub. 5, pp. 11–16.

Jones, M., 1983. "Pumps and Pipes for Underground Water Services." *Tunnels and Tunnelling,* December, pp. 35–40.

Josang, T., 1983. "Ground Freezing Techniques Used for Tunnelling in Oslo City Center." In: *Norwegian Tunnelling Technology,* Norwegian Soil and Rock Eng. Assoc., Pub. 2, pp. 39–44.

Kaiser, T. K., and Gale, A. D., 1985. "Evaluation of Cost and Empirical Support Design at B.C. Rail Tumbler Ridge Tunnels." In: *Canadian Tunneling,* Tunneling Assoc. of Canada, pp. 77–106.

Kuesel, T. R., 1987. "Principles of Tunnel Lining Design." *Tunnels and Tunnelling,* April, pp. 25–28.

Kirsten, H. A. D., 1988. "Discussion on Q-System." In: *Rock Classification Systems for Eng. Purposes,* ASTM STP 984, Louis Kirkaldie (ed.), ASTM, Philadelphia, pp. 85–88.

Kleivan, E., 1988. "NoTCoS—Norwegian Tunnelling Contract System." In: *Norwegian Tunnelling Today,* Norwegian Soil and Rock Eng. Assoc., Pub. no. 5, pp. 67–72.

Kovari, K., Madsen, F. T., and Amstad, C., 1981. "Tunneling with Yielding Support in Swelling Rocks." *Proc. International Symp. Weak Rock,* Tokyo, pp. 1019–1026.

Laubscher, D. H., 1977. "Geomechanics Classification of Jointed Rock Masses—Mining Applications." *Trans. Inst. Min. Metall.* (UK), vol. 86, sec. A, pp. A1–A8.

Louis, C., 1972. "Construction of Tunnels by the New Austrian Method." *Revue de L'Industrie Minérale,* April 15 (in French).

Maples, W. A., 1973. "Shotcrete Applications in the Americas." *Proc. Eng. Found. Conference on Use of Shotcrete for Underground Structural Support,* South Berwick, Maine, pp. 58–64.

Martin, D. (ed.), 1982. "Norway Drives for Low Cost Road Tunnels." In: *Norwegian Hardrock Tunnelling,* Norwegian Soil and Rock Eng. Assoc., Pub. 1, pp. 95–99.

Morfeldt, C. O., 1969. "Significance of Groundwater at Rock Constructions of Different Types." *Proc. Int. Symp. Large Permanent Underground Openings,* Oslo, pp. 305–317.

Muir-Wood, A. M., 1972. "Tunnels for Roads and Motorways." *J. Eng. Geol.,* vol. 5, pp. 111–126.

———, 1975. "The Circular Tunnel in Elastic Ground." *Géotechnique,* vol. 25, no. 1, pp. 115–127.

Nakahara, A., 1976. "Shotcrete Application for the Seikan Tunnel." *Proc. Eng. Foundation Conference,* Easton, Md, October 4–8, A.C.I Pub. SP-54, pp. 460–474.

Nguyen, V. U., and Ashworth, E., 1985. "Rock Mass Classification by Fuzzy Sets." *Proc. 26th U.S. Symp. Rock Mech.,* Balkema, Rotterdam, vol. 2, pp. 937–946.

O'Rourke, T. D. (ed.), 1984. *Guidelines for Tunnel Lining Design.* Tech. Committee on Tunnel Lining Design, Underground Technology Research Council, ASCE. Amer. Soc. Civ. Eng., New York, 82 pp.

Ottosson, L., and Cameron, T. L., 1976. "Hydraulic Percussive Rock Drills—A Proved Concept in Tunneling," in M. J. Jones (ed.), *Tunneling '76* Inst. Min. Metall., London, pp. 277–285.

Palmstrom, A., 1988. "Unlined High Pressure Tunnels and Shafts." In: *Norwegian Tunnelling Today,* Norwegian Soil and Rock Eng. Assoc., Pub. 5, pp. 73–79.

———, and Berthelsen, O., 1988. "The Significance of Weakness Zones in Rock Tunnelling." *Proc. Int. Symp. Rock Mech. and Power Plants,* Madrid, Spain, vol. 1, pp. 381–388.

Panet, M., 1986. "Calcul du Soutènement des Tunnels à Section Circulaire par la Méthode Convergence-Confinement avec un Champ de Contraintes Initiales Anisotrope." *Tunnels et Ouvrages Souterrains,* Numéro Spécial, November.

Pelizza, S., Barisone, G., Campo, F., and Corona, E., 1989. "Rapid Umbrella-Arch Excavation of a Tunnel in Cohesionless Material under an Archeological Site." *Proc. Int. Cong. Progress and Innovation in Tunnelling,* Toronto, vol. 2, pp. 885–891.

Pera, J. (for a working group of 16) 1984. "Entretien et Répartion" (Tunnel Maintenance and Rehabilitation). *Tunnels et Ouvrages Souterrains,* no. 77, September–October, pp. 228–232.

Presley, C. K., 1981. "The Drilled Shaft Approach to Development of a New Uranium Mine." *Proc. 38th Ann. Mtg. & Conv.,* Canad. Diamond Drill. Assn., Winnipeg, paper no. 4, 17 pp.

Reith, J. L. (for a working group of 13), 1984. L'Étanchéité des Ouvrages en Souterrain." *Tunnels et Ouvrages Souterrains,* Numéro Spécial, November, pp. 69–130.

Roald, S., and Oiseth, T., 1988. "Reaming of Shafts and Tunnels." In: *Norwegian Tunnelling Today,* Norwegian Soil and Rock Eng. Assoc., Pub. 5, pp. 97–102.

Rogers, G. K., and Haycocks, C., 1989. "Rock Classification for Portal Design." *Proc. 30th U.S. Symp. Rock Mech.,* Morgantown, W. Va., Balkema, Rotterdam, pp. 23–30.

Rutledge, J. C., and Preston, R. L., 1978. "Experience with Engineering Classifications of Rock." *Proc. Int. Tunneling Symp.,* Tokyo, pp. A3.1–A3.7

Saini, G. S., Dubee, A. K., Singh, B., 1989. "Severe Tunnelling Problems in Young Himalayan Rocks for Deep Underground Opening." *Proc. Int. Symp. Rock at Great Depth,* Pau, France, pp. 677–685.

Sakurai, S., 1983. "Field Measurements for the Design of the Washuzan Tunnel in Japan." *Proc. 5th Int. Cong. Rock Mech.,* Melbourne, Australia, vol. A, pp. 215–218.

Sandström, G. E., 1963. *History of Tunnelling.* Barrie & Rockcliff, London, 427 pp.

———, 1963. *Tunnels*. Holt Reinhart, New York, 427 pp.
Schleiss, A., 1988. "Design Criteria Applied for the Lower Pressure Tunnel of the North Fork Stanislaus River Hydroelectric Project in California." *Rock Mech. Rock Eng.*, vol. 21, pp. 161–181.
Selmer-Olsen, R., and Broch, E., 1982. "General Design Procedure for Underground Openings in Norway." In: *Norwegian Hardrock Tunnelling*, Norwegian Soil and Rock Eng. Assoc., Pub. 1, pp. 11–18.
———, 1983. "Examples of the Behaviour of Shotcrete Linings Underground." In: *Norwegian Tunnelling Technology*, Norwegian Soil and Rock Eng. Assoc., Pub. 2, pp. 71–75.
Serafim, J. L., and Periera, 1983. "Considerations of the Geomechanics Classification of Bieniawski." *Proc. Int. Symp. Eng. Geol. Underground Construction*, Balkema, Rotterdam, vol. 1, pp. II.33–II.42.
Singh, B., 1988. "Analysis of Squeezing Pressure Phenomenon in Tunnels." *Proc. Int. Symp. Underground Eng.*, New Delhi, vol. 1, pp. 3–14.
Stacey, T. S., and Page, C. H., 1986. *Practical Handbook for Underground Rock Mechanics*, Trans. Tech. Publications, Germany, 144 pp.
Steiner, W., Einstein, H. H., and Azzouz, A. S., 1980. *Improved Design of Tunnel Supports—Tunneling Practices in Austria and Germany*. Report no. UMTA-MA-0100-800-7, U.S. Dept. of Transportation, Washington, D.C.
Szechy, K., 1966. *The Art of Tunnelling*. Akademiai Kiado, Budapest, 891 pp.
Tanimoto, C., 1980. "Tunnelling in Rock with Rock Bolts and Shotcrete." Lecture notes, University of Kyoto, Japan (in English), 243 pp.
———, Yoshikawa, T., and Hohjo, A., 1988. "Rapid Excavation of a 5.7 km Headrace Tunnel in Shin'aimo Hydroelectric Power Plant Project—Geostatistical Assessment and TBM Application." *Proc. ISRM Symp. Rock Mech. and Power Plants*, Madrid, Spain, vol. 1, pp. 397–405.
Terzaghi, K., 1946. "Introduction to Tunnel Geology." In: *Rock Tunneling with Steel Supports*, Proctor & White (eds.), Commercial Shearing and Stamping Co., Youngstown, Ohio.
Thompson, T. F., 1966. "San Jacinto Tunnel." In: *Engineering Geology in Southern California*, Assoc. Eng. Geol., Spec. Publ., pp. 104–107.
Thorpe, R. K., and Heuze, F. E., 1986. "Preliminary Studies of Reinforcement Dynamics for a Re-usable Underground Test Chamber." *Proc. 27th U.S. Symp. Rock Mech.*, Tuscaloosa, Ala., pp. 801–807.
Trefzger, R. E., 1966. "Tecolote Tunnel." *Engineering Geology in Southern California*, Assoc. Eng. Geol. Special Publ., pp. 108–113.
USNCTT, 1974. *Better Contracting for Underground Construction*. U.S. National Committee on Tunneling Technology, Subcommittee on Contracting Practices, National Academy of Sciences, Washington, D.C.
Vasilescu, M. S., Benziger, C. P., and Kwiatkowski, R. W., 1971. "Design of Rock Caverns for Hydraulic Projects." *Proc. Symp. Underground Rock Chambers*, ASCE, New York, pp. 21–50.
Wahlstrom, E. E., 1973. *Tunnelling in Rock*. Elsevier, New York.
West, G., Carter, P. G., Dumbleton, M. J., and Lake, L. M., 1981. "Site Investigation for Tunnels." *Int. J. Rock Mech. Min. Sci. & Geomech. Abstr.*, vol. 18, pp. 345–367.
Wheby, F. T., and Cikanek, E. M., 1973. *A Computer Program for Estimating Costs of Tunneling* (COSTUN). Report to Federal Railroad Administration—U.S. Department of Transportation, NTIS Report No. PB228 740, 180 pp.
Wickham, G. E., Tiedemann, H. R., and Skinner, E. H., 1972. "Support Determinations Based on Geologic Predictions." *Proc. 1st Rapid Excavation and Tunneling Conf.*, Chicago, vol. 1, pp. 43–64.
———, ———, and ———, 1974. "Ground Support Prediction Model—RSR Concept." *Proc. 2d Rapid Excavation and Tunneling Conf.*, San Francisco, vol. 1, pp. 691–707.
Wittke, W., 1981. "Some Aspects of the Design and Construction of Tunnels in Swelling Rock." *Proc. Int. Symp. Weak Rock*, Tokyo, pp. 1333–1334.
Wong, R. C. K., and Kaiser, P. K., 1987. "Performance of a Small Tunnel in Clay Shale." *Can. Geotech. J.*, vol. 24, no. 2, pp. 297–307.

Chapter

6

Caverns and Underground Space

6.1 Natural and Artificial Caverns

6.1.1 Geological caves

Solution of calcareous rocks has created natural caverns with unsupported arched roofs far more extensive than those of any excavation. The largest known single opening, with a span of more than 270 m, is found in the Sarawak Caverns in the island of Borneo. Famous European limestone grottos include the Verna cavern in the French Pyrenees, with a span of 250 m, and the Torca del Carlista across the border in Spain, which is 500 m long, 230 m wide, and 125 m high. In Italy, the Grotta Gigante (giant cave) measures 240 by 180 by 138 m high (Duffaut, 1982).

In North America, rooms in the Carlsbad Caverns in New Mexico have spans of 200 m with sections partially supported by thin stalactites and stalagmites. The Mammoth Caves in Kentucky and the Underground Sea in Tennessee are similar but smaller limestone caves.

Some large natural caverns in volcanic rocks are formed when a molten lava chamber is suddenly drained, leaving the already solidified outer layer as the walls of the chamber. Examples in Hawaii contain stalagmites, stalactites, and curtains, all formed as the dripping lava solidified.

Natural caves have long been used as dwellings. Solution caves in China carry traces of episodic roof falls and human artifacts dating back hundreds of thousands of years (Kerisel, 1987). Caves in the Pyrenees of Spain and France contain some of the most important artifacts of Neanderthal and Cro-Magnon people.

6.1.2 Artificial caverns

It was natural for people to assist nature by extending natural caverns and by mining new dwellings in soft rock formations. At Cappadocia in Turkey, volcanic tuff afforded easy excavation and excellent stability; many of the ancient troglodytic caves remain inhabited to this day, and provide excellent protection against heat in summer and cold in winter. The Pharaohs of Egypt, when they abandoned pyramid construction, excavated large burial chambers in limestone rocks of the Valley of Kings (Curtis and Rutherford, 1981).

The Catacombs of Paris and Rome, dug out of soft limestone around the fourth century, provided refuge for persecuted Christians. In our times, the opal-seekers in the hot dry climate of central Australia excavate their homes in the very mudstone deposits from which they mine opals, yet service these dwellings with electricity and modern appliances.

Natural caverns spanning 270 m may be compared with the widest *civil engineering chamber,* a hydroelectric powerhouse with a 35-m span, built near Grenoble, France, in the 1930s. This cavern was lined with concrete for stability.

Large unsupported *mine openings* include the 120-m span inclined (30° and 50°) stopes in an iron mine in France (Maury, 1977*a*), and rooms 42 by 100 by 10 m high in a potash mine in New Brunswick, Canada, at a depth of 950 m. Spans exceeding 50 m have been mined by solution in uniform domal salt. In Scandinavian mines, unsupported spans of 80 to 100 m have been achieved in good rock conditions. Stopes at the Oka mine near Montreal, Canada, approach 150 by 200 m for heights of up to 200 m, at depths of 500 m.

6.2 Uses and Benefits of Underground Space

6.2.1 Variety of uses

Rock caverns accommodate an increasing variety of goods and services; underground space provides important advantages when compared with facilities at the surface (McCreath and Mitchell, 1978; Duffaut, 1980; Boivin, 1985). Some of the main uses are:

> Industrial, commercial, residential, and recreational facilities including hospitals, laboratories, factories, schools, offices, churches, theatres, ice hockey arenas, concert halls, and restaurants

Subway systems including stations and tunnels

Military installations including naval bases with underground dry docks, munitions stores, and civil defense shelters

Warehousing facilities for frozen goods, wines, and archival documents (Fig. 6.1)

Underground hydroelectric, nuclear, and thermal generating stations, with their machine halls, penstocks, tailrace and diversion tunnels, and surge chambers

Peak energy storage facilities including pumped water, compressed air, and superconductive storage

Storage facilities for bulk oil, gas, liquified (cryogenic) gas, and drinking water

Processing and disposal of radioactive waste, chemical waste, and sewage

In just one small Scandinavian country, Finland, the total volume of civil engineering rock excavation is about 5,000,000 m^3 per year. Uses include hydroelectric power stations, substations, and fluid storage, as well as civil defense facilities and bulk storage of goods. About 300 underground projects were completed in 1986, of which one-fifth were occupied by public buildings.

6.2.2 Environmental and other benefits

Underground construction has some important environmental and technical advantages:

Ugly and obtrusive structures are hidden, and valuable surface space is preserved for housing, parkland, or agriculture.

Risks of damage by fire, explosion, earthquake, storm, flood, theft, sabotage, and military attack are greatly reduced.

For example, if the Chernobyl plant had been underground and designed with filters, the consequences of the nuclear accident would have been far less severe. Control of acid rain is a further reason for putting power stations underground. In addition, underground space can make good economic sense:

Land acquisition and building costs can be lower than those for construction at the surface.

The rock floors and walls of caverns provide, at no further cost, a high-strength medium for foundations and containment of fluids

Figure 6.1 Storage of foodstuffs in Stockholm, area 5000 m^2 and temperature $-32°C$ (Construction Skanska, Sweden).

under pressure. There is no need for load-bearing walls and floors, roofs that must take water or wind loads, or insulated exterior walls with windows.

Costs can be further reduced if caverns are converted from previous mining uses, or mined by solution of soluble rock.

Excavated materials can sometimes be sold as aggregate or chemicals.

Operating costs, particularly for heating and cooling, are reduced because of the nearly constant temperature and humidity underground, and because of the excellent insulating and thermal storage characteristics of rock.

6.3 Underground Warehouses, Offices, and Factories

6.3.1 Utilization of underground space

Kansas City is often cited as an example where underground space has been developed for commercial and industrial uses. Over 3000 people work in the Kansas City facilities, which house cold storage chambers, sensitive manufacturing operations, offices, parking areas, and many other amenities (McCreath and Mitchell, 1978; Woodard, 1980). The site was expanded from a converted room and pillar mine in high-quality limestone that continues to be mined for use as aggregate. Access is from the base of river cliffs, so trucks and cars can drive directly into the space. The competent, flat-lying strata give excellent conditions for 10- to 12-m room spans with minimal support, and groundwater problems are negligible. A uniform annual temperature of 14°C results in minimal heating requirements. Other underground storage facilities in the United States are located in Pittsburgh, Philadephia, New York City, and Salt Lake City.

Rock caverns have for some time been used as *cold stores* for fruit and vegetables at normal refrigerator temperatures (+2°C to +5°C), and for frozen foods like fish, meat, and ice cream at deep-freeze temperatures (−25°C to −30°C). Energy consumption for deep freezer storage is 75 percent and refrigerator storage only 25 percent of similar surface stores (Broch, 1989). Insurance premiums are reduced because in the event of a breakdown in the cooling system, the large cold reservoir provided by the rock will maintain the temperatures at operating levels for at least several weeks.

In Bordeaux, Burgundy, and particularly in Champagne, thousands of kilometers of caves have been excavated to provide *wine storage* under ideal aging conditions. Cognac and Armagnac brandies are also left to age in large casks underground for many years.

Archives, in particular for the banking industry, must provide many years of storage for paper and microfilm records; the underground environment reduces deterioration and enhances security from theft or fire. In the Kansas City storage facility, concrete-lined buildings house computer tapes, microfilm, television tapes, master prints of movie films, and other valuable records (Bennett-Smith, 1977).

Underground *civil defense installations* in the Nordic countries are used in peacetime to accommodate *sports arenas* including sprint tracks, shooting galleries, handball courts, and swimming pools. More than 10 such dual-purpose facilities are now in daily use in Norway (Broch, 1989). Energy consumption for running the Gjøvik underground swimming pool (Fig. 6.2) is about half of what would be required for a similar building on the surface. To build the facility, 11,000 m^3 of rock were excavated and transported to a nearby marina that was under construction.

The Holmlia facility in suburban Oslo provides a blast-resistant and gas-tight underground shelter for about 7000 people. The floor area of 7550 m^2 includes a 25- by 45-m sports hall and a swimming pool with six 25-m lanes. Excavated rock amounting to 53,000 m^3 was used in nearby construction. When completed in 1983, the facility cost \$7.5 million, of which 8 percent was for design, supervision, and administration, 67 percent for civil engineering works, and 25 percent for heating, sanitary, ventilation, and electrical equipment (Broch, 1989).

Liljestrand (1986) reports that underground sports and leisure facilities in Finland cost about 20 to 40 percent more to construct than corresponding facilities at the surface. However, savings on the cost of land range from \$50,000 to \$500,000 per facility, and energy consumption is about 15 percent less. A further important factor favoring underground structures is the avoidance of "poor" architecture, almost synonymous with large, low-cost structures at the surface.

6.3.2 Benefits and costs

The main economic attraction of underground space for human use is the low *heating and cooling costs.* In 1978, 29 percent of the total U.S. energy consumption was for space heating, cooling, and humidity control. By housing people and facilities underground, energy savings of 50 percent are common, and savings of up to 90 percent have sometimes been achieved (for example in the Kansas City facility). Energy savings in Finland range from 20 percent for a swimming pool at +26°C to as much as 74 percent for ice hockey arenas at a temperature of +7°C.

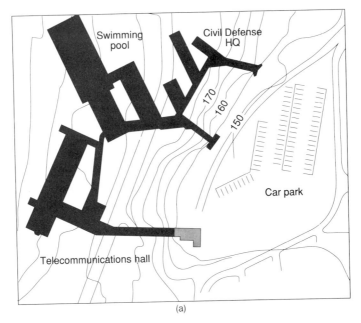

(a)

(b)

Figure 6.2 Subsurface swimming pool, civil defense, and telecommunications installations at Gjøvik, Norway (Broch, 1989). (a) Layout of underground facilities; (b) swimming pool. (*Photos courtesy Fortifikasjon A.S. and Noteby A.S.*).

The steady ambient temperature allows precise design of heating and cooling equipment. At a depth of only 4 m, annual and daily temperature fluctuations are smoothed to less than 2°C. In Toronto, for example, ground temperatures are a steady 10°C within a few meters of the surface, and from there downward, the geothermal gradient accounts for an increase of about 1°C for every 80 meters of depth.

The costs of excavating and stabilizing an entire facility can be competitive with surface construction costs. 1985 estimates for the city of Sudbury, Ontario, were in the range $250 (Canadian) to $380/m^2 of floor space including cavern excavation and stabilization, compared with a range of $210 to $540/m^2 charged locally for unfinished warehouse space at surface. Sudbury is a mining town, where the necessary rock engineering skills are readily available. The rocks, however, are less than ideal, being brittle metasediments without the convenient horizontal bedding that reduces blasting and support costs.

One disadvantage of underground living space is the lack of natural light. This can be remedied by careful architecture, taking advantage of expanded portals where the facility is excavated in from a hillside or natural cliff, and installing skylights where a shaft or opening can be excavated to meet the surface. Artificial lighting can be used with imagination (Fig. 6.3), or daylight can be "piped" underground using fiber optics.

Figure 6.3 Lighting from beneath a walkway at Science North museum, Sudbury, Canada, reveals the natural rock structure. The exposed rock forms an important component of the building's architecture.

6.4 Power Plants and Energy Storage

6.4.1 Hydroelectricity

An underground hydroelectric facility consists typically of a powerhouse cavern (Figs. 6.4 and 6.5) with upstream high-pressure penstock pipes or tunnels and with downstream low-pressure tailrace channels or tunnels. To prevent hydraulic shock during closure of the turbine valves, surge chambers or compressed air cushions (Sec. 6.4.4) are connected to the penstock and tailrace. Further interconnecting systems of tunnels are required to divert the river, and for access during construction and operation. Illustrated in Fig. 6.6, the Churchill Falls underground hydroelectric complex in Labrador, Canada, includes a powerhouse 296 m long, 25 m wide, and 50 m high, and 11 inclined penstocks each 366 m long.

Underground siting of a hydroelectric power station is preferred if the rock is competent and durable enough to require only minimal support. Underground turbines can be located at elevations that take full advantage of the available head of water. Penstock tunnels, lined with steel or concrete, or unlined in good rock conditions (Sec. 5.5.2.4), confine high-pressure water at a cost less than that of steel penstock tubes at surface.

6.4.2 Pumped storage

Pumped storage systems are used to meet short-term electrical power demands and to smooth out daily demand fluctuations. They work well with power generation systems, generally thermal stations, that cannot easily be regulated to meet short-term peaks. A pumped storage system consists of upper and lower storage reservoirs connected to each other and to an intermediate pumping and generating station by a high-pressure tunnel or penstock system. Water from the upper reservoir is used to generate electricity during the day when demand is high. By night, the turbines become pumps operated by the excess electrical output, and recharge the storage reservoir for the next day.

Various configurations are possible depending on topography, depth and thickness of soil, and the quality of rock on site. One approach is to use a vertical shaft linking a lower storage reservoir to one at the surface, with the power station housed underground. The upper and lower reservoirs may be underground, but more often they take advantage of existing lakes and rivers, disused quarries, or excavated open cuts. Underground reservoirs of a million cubic meters capacity have been constructed.

Energy storage schemes operate at maximum efficiency with heads in the order of 300 m and relatively short hydraulic systems. A natu-

272　Rock Engineering Underground

(a)

(b)

Figure 6.4 Manicouagan-5 hydropower machine hall. (*a*) Top heading; (*b*) anchored concrete beams to support 425-t cranes. (*Photos courtesy D. Nguyen, Hydro Québec*)

Figure 6.5 LG2 powerhouse, La Grande hydroelectric project, James Bay, Canada. (a) During assembly of turbines. (*Photo courtesy Hydro Québec*).

Figure 6.5 *(Continued)* LG2 powerhouse, La Grande hydroelectric project, James Bay, Canada. *(b)* Completed powerhouse chamber. *(Photo courtesy Hydro Québec)*

ral catchment is unnecessary if the upper reservoir is watertight. At the lower reservoir site, unless there is considerable natural inflow, leakage must be reduced to a minimum.

Dolezalová et al. (1982) discuss a typical pumped storage project in Czechoslovakia with an open-air reservoir and an underground power station. The cavern 120 m long, 33 m wide, and 52 m high was mined in faulted schist. Planning and construction for pumped storage in weak shales and chalk are discussed by Hodgson (1986).

6.4.3 Compressed air energy storage

Rock caverns make ideal containment vessels for compressed air, which can be employed as an alternative to water in a pumped storage scheme (Morfeldt, 1972). Air is compressed during off-peak hours, and then released for combustion in a two-stage gas turbine to generate electricity when demand is at its peak. Exhaust gases reheat the air emerging from the cavern, giving a further boost to generation efficiency.

Alternatives include storing the air over a water bed under a moderate *constant pressure* from a water column, or compressing it to 10 MPa in a *constant volume* system, allowing the pressure to drop during use. Leakage through rock joints must be minimized. Low-

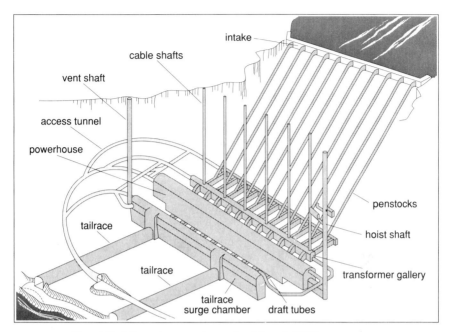

Figure 6.6 Churchill falls underground complex (Gagné, 1972).

pressure caverns, usually shallow, can be left unlined if surrounding rock can be kept saturated; the pressure of joint water should exceed that of the stored gas. High-pressure caverns must be lined or at least grouted. Compressed air energy storage systems have been constructed in salt rock, taking advantage of its easy solution mining and self-sealing qualities.

As an alternative to caverns, compressed air systems can use *porous rock strata*. Design must take into account the effects of pressure and thermal cycling on the competence, permeability, and porosity of the storage rock, of the overlying cap rock, and of the charge-discharge wellbores. Air pressures fluctuate and may reach 2 MPa or more. The air is likely to be hot (100 to 300°C) because of compression. The best reservoir rocks appear to be uniform-grained high-silica sandstones with slight to moderate cementation. These have the right combination of porosity, permeability, and mechanical and chemical stability; some carbonate rocks may also be suitable (Pincus, 1978).

6.4.4 Compressed air cushions

Air cushions replace surge shafts at several Norwegian underground hydroelectric plants. Caverns with volumes of up to 100,000 m^3 and pressures of about 5 MPa are successfully operating without any air loss through the rock (Broch, 1989).

To be safe, the water pressure in the rock mass should be greater than the gas pressure in the cavern. This means that under normal conditions, for a gas pressure of 10 MPa, the depth of the cavern should be more than 1000 m. However, high-pressure gases can be accommodated at shallower depths with the precaution of a *water umbrella,* achieved by injecting water around the cavern at a pressure higher than the contained gas.

6.4.5 Superconductive energy storage

Superconductive magnets maintained at a low temperature may be used in future energy storage schemes. The magnets are charged with electric current during periods of low consumption, and discharged during demand peaks. The method is more efficient than pumped storage, and free from topographic constraints.

A system proposed for Wisconsin consisted of a 1000 to 10,000 MWh storage unit with three solenoids about a common axis, each in a donut-shaped cavern with a radius up to 100 m and a cross section 5 by 10 m (Haimson et al., 1978). Sites in granite and quartzite were investigated. Design had to take into account the high radial and axial loads applied to the tunnel walls by magnetic forces. The rock must be stiff and dry for the operations to be safe, and properties must be adequate to withstand many thousands of load cycles.

6.4.6 Hot water

Energy can be stored in limited quantities by more conventional means. In Sweden, two heat storage facilities are operating in water-filled rock caverns with capacities of 15,000 and 100,000 m^3. The larger cavern at Lyckebo is used to heat 550 dwellings. The water is warmed by solar collectors during the long Scandinavian summer days and then is used for heating during winter (Pilebro et al., 1987; Broch, 1989).

A cube of rock with 10-m side length (1000 m^3) can store 24 MWh, which is the annual heat demand of a single family house in northern Europe. In a pilot project at the University of Luleå in Sweden, 115,000 m^3 of crystalline rock is perforated by 120 drillholes to a depth of 65 m. During the summer season, waste heat from a gas-fired generation plant is used to heat water, which when circulated, raises the rock temperature from 30 to 60°C. About half of the 2 GWh of stored heat is recovered during the winter by circulating cold water, and is used to heat one of the university buildings (Nordell, 1988).

6.5 Storage of Bulk Fluids

6.5.1 Hydrocarbon storage requirements and costs

Hydrocarbons (crude or refined oil, liquified or unliquified gas) can be stored in surface tanks, shallow buried tanks, or in underground caverns or reservoirs. Four arrangements are possible for underground storage in rock formations: solution caverns in soluble rocks, purpose-mined caverns, abandoned mines, and aquifers (Fig. 6.7). Bulk storage (more than 500,000 m^3) is typically needed for fuels, whereas smaller chambers (50,000 to 100,000 m^3) are employed for kerosene, butane, propane, propylene, and other chemical products.

Underground oil storage originated with the desire of industrial nations to maintain supplies of crude oil for bridging possible shortages. Sweden began storing oil in caverns in the 1940s (Daerga et al., 1986).

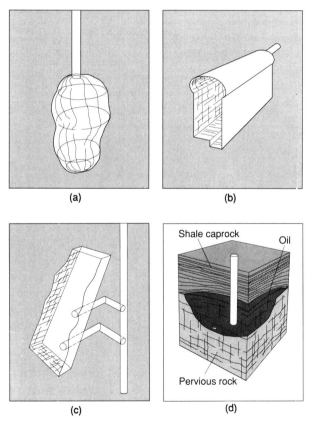

Figure 6.7 Alternatives for underground storage. (a) Solution-mined cavern; (b) excavated cavern, (c) abandoned mine; (d) porous aquifer.

Finland followed in the early 1960s (Johanssen and Lahtinen, 1976). The Scandinavian countries have some of the best rocks for unlined caverns, strong and massive granites with a high water table. Elsewhere, the technology has been extended to less favorable soft rock formations (Maury, 1985).

Advantages of underground hydrocarbon storage include the ability to maintain stocks close to densely populated areas, protection against fire and sabotage, virtually unlimited storage volumes, and often, reduced costs.

When the volume of bulk storage exceeds about 5000 to 10,000 m^3, underground storage becomes competitive with steel or concrete tanks at the surface. In Scandinavia, costs have ranged from \$25 to \$50/m^3 for volumes of between 100,000 and 1,000,000 m^3 (Broch, 1989). The cost of bulk storage of unpressurized liquids in mined caverns, according to Maury (1985), is about 75 percent of that for storage at surface. Storage in worked-out mines is even less expensive, about 30 percent of the cost of surface storage. A cost of \$28.60/m^3 (\$4.55 per barrel) was estimated in 1976 for oil storage caverns in the United States. Mechanical and electrical equipment brought this to an all-inclusive cost of \$33.20/m^3 (\$5.28 per barrel) (Jansson, 1974).

6.5.2 Storage alternatives

6.5.2.1 Solution-mined caverns.

Storage of fluids in solution-mined salt caverns offers advantages of cost and security over the equivalent surface storage (Dreyer, 1982). Mining by solution is of course restricted to soluble formations, usually salt domes or thick-bedded salt deposits. Solution mining can rapidly and cheaply create cavities of several hundred meters in height with volumes of a million cubic meters. The caverns provide storage for liquids such as petroleum and liquified natural gas (LNG), and for gaseous products such as natural gas and compressed air.

Large salt cavities tend to lose part of their available volume by creep or collapse (Bérest, 1989). According to sonar measurements, a pear-shaped cavern of 90,000-m^3 capacity at a depth of 1450 m in Tersanne, France, had lost about 35 percent of its volume in 10 years. The Eminence cavern in the United States, situated in a salt dome at a depth of 1860 m, with an initial height of 230 m and a diameter of 20 m, had lost 40 percent of its volume in a 2-year period, and eventually lost more than 60 percent in 6 years. Later, it was enlarged easily by further solution mining. A cavern of 68,000 m^3 capacity in Kiel, West Germany, leached into a salt dome at a depth of 1350 m, suffered a volume loss of 7500 m^3 after 45 days, and a further loss of 1900 m^3 5 months later (Bérest et al., 1986). Some of this was probably the result of spalling of the walls, and some was no doubt caused by viscoplastic deformation.

Salt cavern closure rates can be kept to a minimum by dissolving a more nearly spherical initial shape, cooling stored fluids, minimizing pressure cycling, and maintaining the cavern working pressure as close to overburden stress as possible.

6.5.2.2 Excavated caverns. Purpose-mined caverns for liquid storage can be excavated at nearly any location, provided that the water table is not too deep; water provides containment and makes lining of the cavern unnecessary (Figs. 6.8 and 6.9). Caverns can in principle be mined to any desired shape or size, although rock grouting and reinforcing costs may be high if site conditions are difficult.

An alternative to stabilizing the walls is to leave the cavern filled and supported by blasted rock (Bogdanoff, 1986). These *rubble-filled caverns* can be mined by shrinkage stoping or block caving (Sec. 7.2) with long blastholes and substantial charges because damage to the rock walls is unimportant. Sufficient rock is removed from lower drawpoints to account for bulking, leaving the stope nearly full of rubble at a bulk porosity of 40 percent. The caverns can be up to 4 times the normal size, are cheap to excavate, and are very stable.

Rubble-filled caverns may be particularly suitable for storage of *thermal energy* (Sec. 6.4.6). Not only the cavern walls, but also the blocks themselves contribute to the thermal capacity. Quality of the rock need not be high; on the contrary, a moderately weak rock contributes to the caving process and is more easily mined. The rock should not be so weak that it breaks into a fine sand, which plugs the void space and makes heat recovery inefficient.

6.5.2.3 Abandoned mines. Abandoned mines are cheaper than purpose-mined facilities; the cavity already exists. However, they can be used only if they are stable, suitably shaped and located, and if they can be sealed to contain the product. The main problem is support, because mining operations place little emphasis on long-term support.

Solids or liquids can be introduced into the workings either in containers via shafts or access tunnels, or through pipes as a slurry using pneumatic pumping methods or in a gravity feed system (Schneider, 1988). The irregular geometry of the mine makes drawdown and filling schemes complicated. Control of groundwater inflows presents a further problem.

6.5.2.4 Aquifers. Aquifers are porous, permeable reservoirs. They offer no cavity as such, but instead offer the possibility of storing fluids

Figure 6.8 Storage cavern for LPG (butane and propane) at Killingholme, U.K.. (*Photo courtesy GEOSTOCK, France*)

Figure 6.9 Storage cavern for LPG (butane) at Lavera, south of France. (*Photo courtesy GEOSTOCK, France*)

in pores and fissures. Porous aquifers are used for storage of light and immiscible air, gas, and oil (Goodall et al., 1988). Water in the pores is displaced downward by the stored phase which is under a pressure higher than the ambient fluid pressure. Capillary effects prevent channeling and assure high recovery of the stored product. Peripheral recharge wells may be used to control the pressure in the water beneath and beside the storage volume to ensure a hydraulic barrier.

6.5.3 Oil storage

Typical configurations of storage include parallel caverns, room and pillar arrangements, or interconnecting tunnels. Large caverns are possible only in competent rock. The water table should be high, and the rock impermeable and durable in the environment of temperature, pressure, and chemistry of the stored liquids. For example, viscous crude oil may have to be maintained at 80°C for ease of pumping (de Laguérie, 1985).

Lining can be avoided because of *hydrodynamic sealing;* most oils are immiscible (they float on water), and are kept in the cavern by the inward flow of surrounding groundwater (Fig. 6.10).

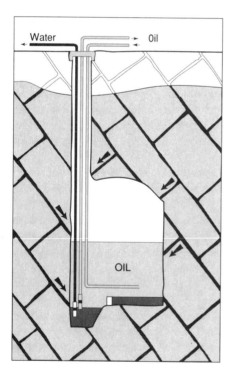

Figure 6.10 Oil storage caverns—containment by groundwater (Hagconsult, Sweden).

Seepage water, termed *bedwater*, accumulates at the bottom of each chamber and can be drawn off separately. In the *moving waterbed* method, the upper surface of the hydrocarbon is kept at a constant level by automatic pumping. When filling the store, water is displaced as hydrocarbon is pumped in, and when emptying it, a corresponding quantity of water is introduced. In the alternative *fixed waterbed* method, infiltrating water is pumped out to maintain a waterbed thickness of about 0.5 m. In *dry bottom storage,* a concrete floor is laid, and seepage water is collected from sumps.

A groundwater barrier must be maintained above unlined caverns used to store volatile petroleum products, to prevent escape of the gas that accumulates in the crown. A groundwater barrier is also required between adjacent caverns carrying different products. The water pressure around the cavern must exceed the vapor pressure at storage temperature; otherwise the groundwater blows out of the joints and gas seeps to surface. In Sweden, a minimum 5 m of groundwater head above the cavern crown appears to be a widely accepted value. When not available naturally, it must be provided by injection. Caverns must be lined or thoroughly grouted if the rock is either very permeable or unstable.

Daerga et al. (1986) describe a problem of *bacteria and fungus* carried into the cavern by groundwater and collecting in the waterbed. Jet fuel stored in underground galleries degraded so badly that it had to be re-refined; microbial action at the oil-waterbed interface was identified as the cause.

6.5.4 Cryogenic storage of liquified gas

To store natural gas economically, it can either be compressed or cooled. Many gaseous hydrocarbons can be stored in liquified form. Required temperatures at atmospheric pressure are $-6°C$ for butane, $-40°C$ for propane and propylene, $-100°C$ for ethylene, $-161°C$ for liquified natural gas (LNG or methane), and $-196°C$ for liquid nitrogen.

Temperatures down to $-40°C$ can be attained easily underground, whereas lower temperatures pose serious problems. Steep temperature gradients and thermal stresses develop around the openings. Contraction upon cooling opens and extends joints and can create new fissures, with further fracture propagation caused by pressures within the cavern, so the rock is no longer impervious (Maury, 1985; Inada and Taniguchi, 1986; Broch, 1989).

Water influx can fill a cavern with ice. This reduces the volume available for storage, and heats the liquified product, which then requires further refrigeration or bleeding to remove gas. Cooling can generate ice lenses in the rock. Liquified gases seeping through the jointed rock boil when they encounter warmer regions; they displace groundwater and cause elevated pore pressures far from the cavern.

The Vexin liquid propane storage site in France was constructed in chalk at a depth of 150 m (Maury, 1977b). Rooms were designed at 8.5 m height and 6.5 m width in this relatively weak rock. No serious instabilities were observed although the stresses measured in the walls of the horseshoe-shaped rooms were slightly greater than the strength of the chalk. No support was needed, although the contractor installed rockbolts as a safety measure.

Very cold caverns (below about $-100°C$) require an impermeable lining to prevent the product from infiltrating the rock. The lining is usually made of plastic foam, which also serves as insulation.

6.5.5 Compressed gas storage

Gas is usually best stored in porous aquifers or depleted oil and gas reservoirs, preferably reservoirs in pinnacle limestone reefs, which tend to be better sealed laterally. The requirements are similar to those for compressed air storage (Sec. 6.4.3), except that low temperatures may be considered if the natural gas is stored at relatively low pressures.

Gas caverns on a large scale have been mined in salt beds in the regions of Tersanne and Etrez in France. At Tersanne, 14 caverns are operational, and at Etrez, 28 caverns ultimately will be developed. In southern Ontario, depleted reefal limestones at depths of 800 m store natural gas from Alberta during the summer for the high demands of the Canadian winter.

6.5.6 Drinking water

Drinking water stored in caverns is easier to keep clean, and its temperature remains constant. Reservoirs at the surface are more susceptible to pollution from the atmosphere and from shallow, contaminated groundwater.

Two water storage caverns were excavated in Precambrian gneiss to supply the town of Kristiansund in Norway, each with a volume of 8000 m^3 and lined with 100 m^3 of shotcrete (Broch and Odegaard, 1983). Cost was estimated at $80/$m^3$ of storage volume.

In Trondheim, also in Norway, two caverns 12 m wide and 10 m high were combined to give a water storage capacity of 20,000 m^3 (Broch, 1989). The water is kept cool, and maintenance costs are low. Storage in caverns larger than 10,000 m^3 appears to be less expensive than in surface tanks of concrete. The Trondheim underground chamber was 20 to 25 percent less expensive than a surface facility, even without taking into account the cost of the land.

6.6 Geological Repositories for Radioactive Waste

6.6.1 The nature of wastes

Radioactive waste includes *high-level* spent reactor fuel, *intermediate-level* wastes from weapons manufacture and research reactors, and *low-level* waste such as rags, papers, filters, and discarded protective clothing. Other types of waste include *transuranic* waste, containing artificial elements that are heavier than uranium and long lived but often low level in terms of total radioactivity, and *tailings,* which are the radioactive rock and soil by-products of uranium mining and milling.

6.6.2 A growing problem

Nuclear power is generated in more than 25 countries, and worldwide production was projected to reach 2000 billion killowatt hours by 1990. By 1989, there were more than 100 operable nuclear reactors in the United States and another 30 with construction permits (IAEG, 1989). Current energy production creates about 2000 metric tons of spent fuel per year, and the existing stockpile of U.S. spent fuel is about 15,000 metric tons of uranium. The U.S. Department of Environment predicted that about 130,000 metric tons of uranium high-level waste would be generated by the year 2020 (Eriksson, 1989).

Radioactive wastes from nuclear generating plants continue to be stored in temporary surface facilities while a permanent solution is sought. The largest producer of nuclear power in Canada, Ontario Hydro, stores spent fuel rods in "swimming pools" on the generating sites.

The United States plans to store weapons-related radioactive wastes, less dangerous than commercial reactor wastes, in underground repositories in New Mexico and Nevada. Salt repositories at the Asse Mine in West Germany and the Bartensleben mine in East Germany are used for low- and intermediate-level waste storage, and also house experimental facilities for study of high-level waste disposal in salt.

Existing wastes present a problem of some urgency whether or not any further wastes are generated. The U.S. Nuclear Waste Policy Act of 1983 aims to provide safe and permanent disposal no later than 1998.

6.6.3 Selection of the host rock

Geological solutions that have been largely rejected to date include dropping "secure" containers into continental drift subduction zone trenches, embedding containers in deep-sea oozes, and injecting cement grouts "laced" with high-level wastes into deep shale formations. The alternative of disposal in deep underground geological "repositories" appears to be the safest long-term solution.

Investigations in 16 countries are summarized in a booklet published by the Organization for Economic Cooperation and Development (OECD-NEA, 1982). Canada plans to store wastes in granites at a depth of 1 km; Sweden and Switzerland also favor granitic sites. The German repository program focuses on salt strata in the Zechstein Formations. France is exploring granite, schist, salt, and clay sites, the latter in conjunction with Belgium. Clay sites remain of interest to Italy, Japan, and Britain.

Having investigated salt, basalt, and tuff, the United States in 1987 selected the tuff option at the Yucca Mountain site in Nevada. Its most favorable attributes are the location of the candidate horizon (a devitrified welded ash-flow tuff) 200 to 400 m above the groundwater table, the aridity of the region, the sorptive qualities of underlying strata, the isolation of the site, and the close proximity of the Nevada test site (Eriksson, 1989). In addition, the waste isolation pilot project (WIPP) facility in bedded salt in New Mexico at a depth of 660 m is intended to receive transuranic waste generated by U.S. Defense facilities.

6.6.4 Design of a hard rock repository

6.6.4.1 Conceptual design. Repository life must be very long, and design assumptions have to be valid for thousands of years in the case of high-level waste (St. John, 1982), and also for transuranic wastes which are less dangerous but very long lived. Much can happen during such a time period. For example, an ice age might develop, subsidence and uplift can occur, and erosion can be considerable. Therefore, repositories must be placed at such depths that changes in the geological and climatic conditions do not allow the waste to escape to the surface.

The repository geometry favored by most countries is similar to a large mining complex: waste canisters will be lowered down shafts to emplacement levels, moved to their final rock cavern locations by a shielded transport vehicle, and the cavern backfilled.

The intent of the U.S. program, for example, is to permit retrieval of waste for up to 50 years after emplacement. The tunnels are then to be backfilled and the shafts plugged and sealed with engineered "buffer" material. "Release limits" are specified for various radionuclides, in total, equal to about 10 percent of natural background radiation in the United States. Environmental protection regulations say that for the first 10,000 years after repository closure, there must be no more than a 10 percent probability of releases exceeding the release limits, and no more than a 0.1 percent probability of releases exceeding 10 times the release limits. Groundwater travel time between the repository and the "accessible environment" must be greater than 1000 years.

Repositories are sited in the best possible rock, and the adits and chambers are sized conservatively so that rock mass competence is not an issue. Designers must, however, try to envisage and model a broad range of physical and mechanical aspects of rock, waste, and buffer behavior including thermal, chemical, hydrogeological, and radiation effects (Côme 1984; Rometsch, 1985). For example, high-level radioactive waste emits heat and, depending on insulation, may raise temperatures in the adjacent rock to about 235°C. Thermal expansion superimposed on initially high stresses, and combined with the swelling pressure of clay in the backfill, could open existing joints and perhaps generate new fissures.

6.6.4.2 Preventing contaminant migration.
Transport by groundwater is acknowledged to be the most probable mechanism for moving the toxic particles from the repository to the biosphere. Canisters damaged by radiation might corrode in the presence of aggressive waters, and the groundwater might then convey radionuclides from the repository. Hence the search for a deep host rock with little or no groundwater circulation.

The rock surrounding the repository should be practically impermeable, or made so by grouting. Individual joints can conduct water and contaminants rapidly over great distances, and must be sealed. Contaminant migration is studied using discrete element models for flow in jointed rock; continuum solutions are invalid in this application (*Rock Engineering,* Sec. 4.4.3).

Geological barriers are expected to be able to contain wastes for very long periods of time. Some corroborative evidence for this is provided by the existence of many natural deposits of radioactive minerals similar to waste materials.

6.6.4.3 Buffers and vault sealing.
As a first line of defense, the radionuclides may be vitrified in a dense glassy mass of synthetic rock-like material. Vitrified wastes are then further protected by an *engineered barrier system* which includes canisters, materials placed over and around the canisters, and barriers used to seal the underground facility. This supplements the natural barrier provided by the rock.

The vault must eventually be backfilled and sealed to make it inaccessible. Vault sealing entails grouting, borehole sealing, buffer packing, and backfilling (Gyenge, 1980). Grouting may be required around the service and ventilation shafts, disposal rooms, tunnel junctions, and service headings. Exploratory drillholes have to be carefully plugged. Waste canisters are surrounded by buffer packing for mechanical support and to provide an ion-exchange chemical barrier. For the first period of high thermal flux, the repository may be air cooled, and a pumped hydraulic barrier installed (Svemar and Sagefors, 1987).

Radionuclides are adsorbed by clay minerals; clay cations can be easily replaced by radionuclides, which have a higher valence and a lower mobility in solution. Repositories in clays and shales are attractive for this reason, and waste facilities in other rocks propose to use clay buffers of high cation exchange capacity to isolate canisters from the jointed rock mass (Pusch, 1985).

6.6.5 Repositories in salt

Rock salt is practically impermeable to gas and liquids because of the tightness of the structure and the absence of open natural joints and fissures. Moreover, the high plastic deformability of rock salt hinders the development and maintenance of open artificial fissures through which liquids and gases could leak out. In addition, solution mining provides a means for creating large storage capacities at economic costs (Langer, 1989). When the solution mining phase is finished, the brine-filled caverns are emptied. Emplacement of wastes in these empty caverns is via the access well.

Repository design in salt must account in great detail for the thermomechanical behavior of salt, which is extremely sensitive to temperature (*Rock Engineering,* Chap. 10; Hardy and Langer, 1982; 1988). Water is a secondary concern, except in shaft sealing, because creep will seal up fissures created during construction. Once mine closure is complete, the potential for migration of large brine volumes must be considered as negligible.

Salt repositories are designed to close totally. Granular *salt fill* (halite) is packed around the canisters, and is stowed to fill completed rooms. As slow closure continues, aided by the temperature, the backfill compacts and the full lithostatic stress will eventually act on the containers (Bérest et al., 1989; Fordham, 1989; Fordham et al., 1989).

Difficult questions remain to be answered. For example, might a dense, hot canister slowly sink through the salt? Could the heated repository zone rise, being less dense and more mobile than the surrounding cooler salt? Might the salt fracture under these conditions, opening up solution channels? Could future dome growth bring canisters to surface? There also remains some debate on the nature of the constitutive laws for these materials (Dusseault, 1989). It is difficult to accept the use of behavioral laws from short-term laboratory tests at creep strain rates of 10^{-10} to $10^{-7} s^{-1}$, to apply in the field case, where strain will persist for many centuries at rates slower than $10^{-11} s^{-1}$

6.6.6 Research

Research is being carried out to ensure, and to demonstrate to the public, that geological storage will be safe. Public concerns interacting with politics and legislation have sometimes led to unrealistic requirements. Nevertheless, the large expenditures of public money have produced advances in rock science and technology. They have funded many full-scale experiments, leading to better constitutive laws, time-dependent and thermomechanical data, new testing and monitoring methods and instruments, and sophisticated numerical models for rock behavior and groundwater flow.

American repository research aims to demonstrate constructibility, retrievability, and containment. Numerical models and predictions are continually checked against the results of monitored in situ experiments (Heuze, 1981; Hustrulid, 1983; Munson and Fossum, 1986; Blejwas, 1989).

Test shafts and chambers explore the large-scale stress-strain-flow response of rock masses. Experimental facilities include the Underground Research Laboratory in Pinawa, Canada, the Asse salt mine in Germany, the Grimsel Rock Laboratory in Switzerland, and the Stripa Mine in Sweden (Côme et al., 1985; Simmons, 1985; Martin, 1989).

In underground experiments, *block tests* at a modest scale (the blocks measure typically 1 to 3 m on a side) investigate the behavior of elements of the rock mass containing joints (Barton and Lingle, 1982). The block boundaries are loaded by flat jacks installed in slots; in some experiments the blocks are also heated and injected with water. *Mine-by tests* such as conducted at the Nevada site make use of instrumented tunnels. Drifts are instrumented to either side of the alignment of a proposed tunnel heading, which is then excavated while taking readings.

Heater tests use electric heaters to simulate spent fuel canisters. In the spent fuel test at Climax, in the United States, instruments were installed in granite to monitor temperatures, stress changes, and rock displacements during 3 years of heating followed by 6 months of cooling. Finite element analyses modeled the behavior of the jointed rock mass (Stephannson et al., 1980; Butkovich and Patrick, 1986; Patrick, 1986).

6.6.7 The Swedish CLAB and SFR facilities

The *CLAB project* in Sweden is the world's first licensed underground facility for storage of nuclear spent fuel (Roshoff et al., 1983). It is a cavern 117 m long, 21 m wide, and 27 m high, smooth-blasted in granite 30 m below ground surface. Waste can be stored for 30 to 40 years before final disposal.

The granite was cement-grouted through a fan pattern of 6-m-long drillholes, then reinforced by pattern bolting, and lined with 100-mm-thick reinforced shotcrete in the roof and 50-mm shotcrete in the walls. The cost was about 50 percent more than for support of a conventional underground excavation in similar rock conditions. Free-standing inner walls of concrete and a roof of steel were erected within the cavern. The space between these and the rock is ventilated and drained to control humidity, temperature, and dust. Fuel in the cavern is stored in water-filled pools lined with concrete and stainless steel.

The *SFR Swedish Final Repository,* commissioned in 1988, is to receive all low and intermediate wastes from the 12 Swedish nuclear power plants up to the year 2010. The repository, 130 km north of Stockholm on the Baltic coast, and 65 m beneath the sea bed, is reached through parallel access tunnels extending 1 km from the harbor (Fig. 6.11). In creating the repository, a total of 430,000 m^3 of granite was excavated from 4.5 km of tunnels, caverns, and shafts. The radioactive waste will be stored in a concrete silo surrounded by bentonite clay in a rock chamber 31 m in diameter and 70 m high (Carlsson and Christiansson, 1988; Broch, 1989; IAEG, 1989).

6.7 Disposal of Chemical and Other Wastes

6.7.1 Toxic chemicals

The purpose of waste disposal is to protect human health and the environment. Wastes placed underground are soon forgotten, but over a long period, they can still be dangerous. Because it is never possible to ensure absolute safety, it is generally accepted that today's society should try to leave conditions for future generations no worse than it would accept for itself.

A recent study (IAEG, 1989) identified several main groups of waste including household, industrial, and mine waste; radioactive products; and other hazardous nondestructible wastes. Some products such as chemical warfare residues, neurotoxic pesticides, brominated hydro-

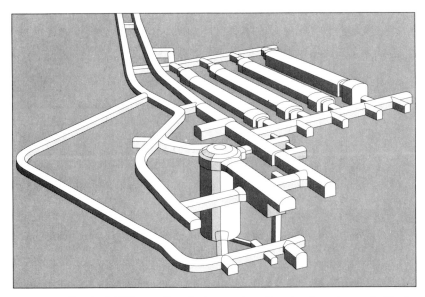

Figure 6.11 Sweden's SFR repository for low- and medium-level reactor waste (Carlsson & Christiansson, 1988; IAEG, 1989).

carbon liquids, or spent acids laced with heavy metals could be considered even more toxic than relatively insoluble radioactive wastes, yet environmental controls have for the most part been inadequate.

Waste products have been injected into porous rocks or placed in solution-mined or excavated caverns in various parts of the world. Currently, stable rock caverns are favored over other approaches, and *deep well disposal* is an alternative (Flak and Brown, 1988). Proponents of this method argue that wastes can always be placed so deep that the probability of them entering into a flow system that connects with shallow deposits becomes negligible, even for extremely long periods.

Chemical wastes have been stored in England since 1965 in an old mine at Walsall Wood colliery, at a depth of about 900 m. The mine is isolated environmentally by a geological graben with clay-filled faults to either side and shales above.

Chemical wastes are being stored in the 700-m-deep Herfa-Neurode mine in West Germany, where 2 million cubic meters of mined space is created annually (Maury, 1985). Two salt beds 3 m thick are being mined, separated by 60 m of sedimentary rock. The 15- by 20-m rooms have excellent stability. They are surrounded by an elaborate monitoring system to minimize the possibility of pollution.

In Ontario, Canada, chemical wastes were for many decades

pumped into solution caverns in salt at depths of 700 m. Concerns have put a stop to this method of disposal, at least temporarily. Questions include: Will the final pressure in the liquid waste be higher than in the surrounding groundwater, and might there be upward flow? Will cavern closure lead to hydraulic fracture by the wastes under high pressure? Will the cavities migrate upward through the salt?

In Norway, dumping of chemical process wastes in fjords has caused serious pollution problems, prompting new environmental restrictions. An alternative described by Aarvoll et al. (1987) uses purpose-mined chambers to store electrolytic wastes rich in heavy metals. The Norzink Company has been instructed by the environmental authorities to deposit its annual production of up to 60,000 m^3 of waste residues in rock caverns. The first was completed in 1985 (Broch, 1989).

6.7.2 Sewage and waste water

Waste water in urban environments is collected by underground pipes, but treatment is usually above ground, leading to problems of land zoning, odor, and aesthetics. Stockholm, however, has most treatment plants underground. The Water and Sewage Authority in Helsinki, Finland, is placing all current and future waste treatment plants in rock caverns; the Finns first used a rock chamber for this purpose in 1932. Häikiö and Kämppi (1987) point out some advantages, including the fact that digester gas is corrosive to concrete, but not to natural hard rocks.

The Tunnel and Reservoir Plan (TARP) in Chicago is the world's largest pollution and flood control project. A system of deep tunnels, reservoirs, and pumping stations collect, store, and treat a large portion of combined sewage produced in the Chicago area. TARP, excavated to depths of more than 100 m in dolomite rock, includes about 210 km of tunnel 2.7 to 11 m in diameter; 252 vertical shafts; several reservoirs of up to 102,000,000 m^3 at the surface and one underground mined storage reservoir of 2,500,000 m^3; and pump house caverns each 19 m wide by 29 m high by 84 m long (Cikanek and Goyal, 1986).

6.8 Cavern Design and Construction

6.8.1 Objectives

The basic goals of cavern design and construction are to minimize disturbance of the rock, to conserve or enhance natural stability of the cavern by installing a minimum of support, and to install permanent linings only sufficient to satisfy the function of the cavern. As in any complex project in rock, site data, test results, modeling information, construction control, and monitoring are essential elements for success. Useful reviews and case histories are provided by U.S.-NCRM (1978), Maury (1979), Saari (1986), Barla (1987), Dorso et al. (1987), and Liu et al. (1987).

6.8.2 Site investigation

Large-span caverns must be designed to remain stable for a long time. To achieve the necessary confidence in the design calls for an intensive program of investigation, testing, and numerical modeling of cavern behavior. Fortunately this is aided by the limited extent and depth of most caverns, such that a modest investment in drilling and testing goes a long way toward characterization of the ground (Franklin, 1989).

Detailed jointing statistics are needed, as are the locations and characteristics of individual lithologic contacts, shears, and faults (*Rock Engineering,* Chap. 3). Early information from rock outcrops is progressively updated by drill core data, by oriented core and television camera techniques, and by observations in exploratory shafts and in the chamber itself, during the early stages of construction. Direct site investigation by means of an exploratory tunnel in the crown of the cavern may be considered a necessity (Gysel, 1986). The same tunnel is often used for ground pretreatment (Sec. 6.8.4.1).

The localized nature of the cavern offers excellent opportunities for tomography between closely spaced exploratory drillholes, giving a map of rock quality variations (*Rock Engineering,* Sec. 6.4.2.3). Measurements are normally required of in situ stress, groundwater pressures, and rock mass permeabilities. Exploratory adits used initially for geological mapping later give access for plate load tests to determine rock mass deformability, and for direct shear tests on discontinuities (Cunha, 1987).

Field measurements of piezometric pressures and hydraulic conductivity of the rock mass serve as input for numerical models that simulate water inflows and the effects of joint water pressure on stability of the cavern wall and roof.

In addition to these conventional features of an investigation, special aspects have to be considered according to the application. Caverns for cryogenic and nuclear waste storage require particular attention to the thermal and hydrogeological characteristics of the site. Groundwater considerations take precedence in the design of caverns for hydrocarbon fuels. Investigations for storage in porous aquifers must determine relative mobility and saturation-desaturation curves of the rock; the reservoir must be free of channels, and a water base with good lateral seals is necessary. Investigations for industrial and commercial facilities focus on accurate costing of the work, and on methods for making the caverns safe, dry, and attractive for human occupancy.

Investigation for the Swedish undersea SFR radioactive waste storage facility (Sec. 6.6.7) started with seismic surveying, and included about 1300 m of core drilling from offshore platforms. The coreholes were geophysically logged and hydrogeologically tested. Much of the investigation, however, was conducted in a second phase, from underground pilot headings. The philosophy of carrying out complementary investigations below ground and having a flexible layout was found beneficial (Carlsson and Christiansson, 1988).

6.8.3 Design

6.8.3.1 Design philosophy. The designer must assess the project feasibility based on the site investigation data, then decide on the cavern location, orientation, construction procedures, and support and monitoring strategies.

The ideal site is one which minimizes contact with major zones that are weak, water-bearing, or highly stressed. The goal is never to build the largest possible cavern, but to exploit the particular rock mass properties in a fashion optimum for the project. One should be prepared, for example, to modify the traditional surface layout of a facility (e.g., nuclear power plant) to take into account the advantages and limitations of underground excavations. Much time and money can sometimes be saved by arranging the plant components in several smaller cavities rather than attempting to fit the conventional surface layout into a single large excavation.

6.8.3.2 Cavern shape. Although nature and numerical analyses indicate to us that the most stable caverns are egg-shaped, ease of excavation, end use, and habit cause most caverns to be designed and built with at least some portion of the walls vertical. The "key-hole" design so common in hydroelectric projects, the long low arch with small vertical walls, the circular cross section, and the standard vertical chamber are common geometries (Gysel, 1986).

Precise blasting, which depends greatly on precise drilling, can be achieved more easily when the geometry is simple, and vertical walls make sense for downward-drilled benches. In sedimentary rocks, the perimeter tends to break back to bedding planes, and there is little point in resisting this tendency, as was demonstrated, for example, in the TARP project (Sec. 6.8.4.2). On the other hand, an elliptical or egg-shaped cross section may prove beneficial to avoid bursting or squeezing in highly stressed or very broken or weak ground.

6.8.3.3 Numerical and physical models

Numerical modeling methods are most useful when applied to the design of major permanent underground openings such as hydropower and storage caverns. Examples abound in the published literature. Sugawara et al. (1988) combined boundary element and characteristics methods in an analysis of the yielding of rock around large caverns. Sousa (1983) describes combinations of the finite element and boundary element methods for three-dimensional analysis. The approach was applied to a power station in Mozambique. Deformability and shear strength of joints in the granitic gneiss were measured by in situ testing, and ground stresses were determined using small flat jack tests and the LNEC overcoring procedure. A physical model of the same powerhouse configuration was constructed and tested using blocks of gypsum plaster and diatomite.

In any cavern, the stress state varies from biaxial loading in the extensive rock walls to complex three-dimensional conditions in corners. During staged excavations, some parts are first highly stressed, then unloaded, and loaded again. Cyclic loading also occurs during the filling and emptying and heating and cooling of storage caverns. We do not yet have a generalized stress path constitutive law for rock; therefore even when the numerical model is the best and most sophisticated available, empirical design and field verification remain essential (Amadei et al., 1987; Starfield and Cundall, 1988).

6.8.3.4 Empirical methods.

Empirical methods have an important place alongside the numerical models, and much is to be gained by review of the behavior of existing facilities (Maury, 1979; 1987). An early review by Cording et al. (1971) compares support measures and monitored displacements in a number of large caverns, and gives typical crown and sidewall support pressures (applied by tensioned anchors and bolts) as functions of the width and height of opening (Fig. 6.12). As the size of the opening increases, the support pressures required to maintain stability also increase, and longer bolts and tensioned anchors are needed.

Bolt lengths in arched crowns typically range from 0.2 to 0.4 times the cavern width B. In planar walls, bolt lengths range from 0.1 to 0.5 times the cavern height H. The following empirical relationships apply for support pressures in the crown and sidewalls, expressed as a function of rock unit weight and cavern width and height:

$$P_v = nB\gamma$$

and

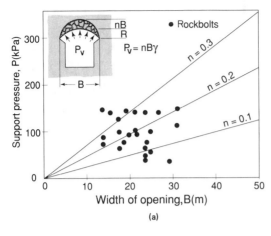

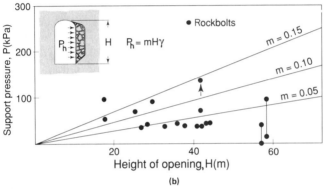

Figure 6.12 Support pressures applied by rockbolts and anchors in caverns (from Cording et al., 1971). (a) Crown of cavern; (b) cavern walls.

$$P_h = mH\gamma$$

P_v and P_h are the crown and sidewall pressures to be provided by rockbolting, γ is the unit weight of rock, and n and m are empirically determined coefficients. They vary according to rock quality, n from 0.1 to 0.3, and m from 0.05 to 0.15. The empirical method is based on observations in caverns in moderate to excellent rock, and at moderate levels of ground stress; results should not be applied to conditions of poor rock or high ground stress.

Kaiser (1986) discusses a more recent approach to empirical design of caverns based on the Q system of rock mass quality classification. This is an extension of the tunnel design method reviewed in Sec.

5.5.3.6. Because there is less experience with caverns than with tunnels, approaches such as this should be used with caution.

6.8.3.5 Conserving the groundwater barrier. In unlined caverns for liquid and gas storage, a gradient in the regional groundwater table toward the cavern is always necessary. Bérest et al. (1982) have proposed the following equation to determine the maximum permissible gas or vapor pressure P:

$$P < P_h - P_{dz} - P_{sf} - F$$

where P_h = the head of groundwater above the roof
P_{dz} = the thickness of the blast-damaged or bolted zone, whichever is greater
P_{sf} = a head reduction necessary to account for dynamic gradient effects
F = a security factor (head of water) empirically chosen by considering the nature of the rock mass, the type of fluid, and case histories

Maintaining hydrogeological security also means sealing the shaft leading to the storage cavern. In the United States, shafts are usually fully cemented, except for the service pipes. In Europe, more often the shaft is plugged only above the cavern, and lined and water-filled above that. Although more expensive, the European method gives better control of the water barrier. Bérest (1989) reports a case where a salt solution cavern for gas (70 percent ethane, 30 percent butane) lost pressure integrity because of a leak along the access casing, which was cemented in place.

Water tables close to the surface present no problem; however, in dry climates or in rocks with low recharge capacity, the hydraulic barrier must be maintained by active injection, using *recharge wells*. Well fields must be designed to guarantee that flow is always toward the cavern. If the rock mass is moderately to highly permeable, the chamber must be partially isolated by grouting, and the injected groundwater curtain is maintained inside the grouted zone.

6.8.3.6 Solution cavern design. The design of leached cavities is based on models of the solution process and closure rates. For storage, the cavern must be structurally sound, preferably cylindrical with a domed or conical roof. A spherical geometry is more resistant to closure in the isotropic stress fields typical of salt deposits. The cover of salt between the crown of the cavern and the overlying nonsalt beds should be at least 30 percent of the cavity diameter to ensure integrity of long-term seal in the vertical direction.

The chemical plants northeast of Edmonton, Alberta, make use of solution-mined cavities in salt beds at a depth of 1800 m (the Lotsberg Formation of the Prairie Evaporites). Their spacing and size have been selected according to the estimated closure rates. Solution caverns are maintained full of liquid at a pressure equal to the hydrostatic head of a saturated brine column, which forms the immiscible liquid beneath the product, and is recharged from a surface brine pond. The brine provides stability, and also a positive head for expulsion of the lighter, immiscible petrochemicals. Because the brine in the surface pond becomes diluted by rainfall, the cavern is being dissolved slowly, counteracting closure rates.

6.8.4 Construction

6.8.4.1 Ground pretreatment.
Grouting is used to reduce groundwater pressure on a cavern lining, or to reduce infiltration or exfiltration (Fig. 6.13). Unusual end-use requirements or poor ground conditions will lead to full *pregrouting* from pilot tunnels, or even *freezing* of the ground before construction begins.

The Mingtan pumped storage cavern, one of the largest in Taiwan, is 23 m wide, 47 m high, and 159 m long. It contains six reversible pump-turbines supplying 1600 MW of peak power. The sandstone and siltstone rocks are faulted. Clay seams in the cavern roof were treated from small adits in advance of general excavation, by *water jetting* at pressures from 20 MPa to as high as 240 MPa, then backfilling with nonshrink mortar. Thick seams were removed by mechanical excavation and then jetted clean. The cavern roof was pre-reinforced by high-strength tendons protected by corrugated sheathing against corrosion. During excavation, a flexible lining of rock bolts and steel fiber reinforced shotcrete was applied.

6.8.4.2 Blasting and support sequence.
Large vertical caverns and silos may be developed by *vertical crater retreat* (VCR) and other methods of open stope mining (Chap. 7). Upper tunnels give access for drilling and loading of blastholes, and tunnels at the base provide drawpoints for removal of broken rock.

More often, the cavern is excavated from the top down using a *benching method*. A typical sequence begins with excavation of an access tunnel by conventional methods, where blastholes are drilled parallel to the line of advance. This is expanded laterally to form the cavern roof, which is reinforced while it is easily accessible. The heading is then deepened, again using horizontal drilling (*breasting*) but with

Figure 6.13 200-L/s groundwater inflow encountered during the construction of a Norwegian hydropower machine hall. (*Photo, courtesy E. Broch*)

more widely spaced holes (Fig. 6.14). With the increased headroom, vertical blastholes are now more convenient for removing the remaining deeper benches, using a procedure similar to bench blasting in quarries.

Because cavern sidewalls are often vertical and high, care must be exercised in assessing which blocks may slide. The necessary supports, usually high-capacity tensioned anchors, can be installed at the appropriate level in the staged excavation to prevent loosening.

Pumphouse caverns for the Chicago TARP wastewater project (Sec. 6.7.2) were explored with an initial 3.7- by 6.1-m adit along the crown. Polyester-resin-grouted rockbolts 9 m and 6 m long were installed vertically to pin together the crown beds. The sides were slashed to

Figure 6.14 Drilling blastholes in the bench of a rock cavern with a hydraulic jumbo (Tamrock Paramatic).

full cavern width. The crown was repaired by dental treatment and grouting. A cast-in-place concrete arch was installed, nominally 0.2 m thick but actually 1.2 to 2.0 m thick in many places because of overbreak (the arched roof design proved itself impractical in horizontally bedded rock). The remainder of the pumphouse was then excavated in five bench lifts; blastholes were downdrilled using pneumatic air tracks. Sidewall shotcrete was placed to a thickness of 50 mm, and the walls were further reinforced with 3-m-long 35-mm-diameter rockbolts at 1.5-m centers (Cikanek and Goyal, 1986).

Johansson and Lahtinen (1976) describe construction of two parallel 16- by 16.5- by 366-m caverns with total capacity of 300,000 m^3, for storing light oil 45 m below the surface in granodiorite rock at Nokia in Finland. The caverns were separated by a 35-m-wide pillar. Excavation of the 7-m-high top heading was followed by horizontal benching, also to a height of about 7 m and, finally, by excavation of a lower bench 12.5 m high using vertical blastholes. Extensive scaling was needed, and the roof was bolted, meshed, and lined with shotcrete.

Gagne (1972) gives details of drilling, delay patterns, and other aspects of blast design for the Churchill Falls underground power complex in Labrador, Canada. More than 1.76 million cubic meters of rock were excavated at rates of 7600 to 29,000 m^3 per week during a 34-month period.

Figure 6.15 Controlled blasting at Atomic Energy of Canada Limited's Underground Research Laboratory. Careful blasting in the shaft brow at a depth of 300 m reduced damage to the walls of the excavation, and the requirement for rock reinforcement and support (Kuzyk et al. 1987) (*Photo courtesy G. Kuzyk and AECL*).

6.8.4.3 Controlled perimeter blasting. Caverns must be carefully excavated to avoid rock damage, and to minimize the support needed. Controlled perimeter blasting methods are employed, with accurate drilling of closely spaced perimeter holes, a minimum of burden, and light charges (Fig. 6.15). Often the perimeter is presplit, or even line-drilled and then slashed in thin layers (*Rock Engineering,* Secs. 13.4 and 13.5). The top heading, even in hard rock, may sometimes be excavated mechanically by a mini-tunneling machine.

Overblasting is particularly undesirable when caverns are required to store fluids. Open fissures promote desaturation of the ground, which often is difficult to resaturate. Exfiltration and infiltration are greatly aggravated if a direct connection is established with nearby aquifers.

6.8.4.4 Mechanical excavation. Whereas blasting is the most common excavating method, particularly for large caverns in hard rock, mechanical methods can be attractive in softer rocks, and full-face boring machines can be used in both hard and soft conditions if a network of tunnels is an acceptable alternative. Some types of caverns, such as those for underground hydropower stations, have a geometry closely defined by the machinery and other facilities to be accommodated, whereas others, such as those for fluid storage, permit more flexibility of shapes and sizes.

Hodgson (1986) describes the use of mechanical excavators to construct an underground pumped storage facility in shale and chalk. To minimize disturbance to the wall rock, these soft formations were mined by a combination of TBM, roadheader, and mechanical ripping. A full-face TBM was selected for excavating the 10.2-m-diameter circular tunnels. Roadheaders were employed for the 5.5-m-diameter upstream and downstream tunnels, the crown drifts of the powerhouse and service vault chambers, and the arches of the smaller chambers in the system. Rippers were chosen for excavating all shafts and the core of the powerhouse, the service vault, valve chamber, access chamber, and connecting chamber.

6.8.4.5 Control of difficult ground. Long tensioned anchors are installed to guard against wedge failures in the wide-span roof and high walls of rock caverns. Shotcreting is employed in most caverns in moderate to poor rock as a precaution against raveling of the crown, which becomes inaccessible after benching (Fig. 6.16). In particularly bad ground, a *core support method* is used in which small sections of the cavern are excavated one by one and reinforced and lined before the central core of rock is removed. Gysel (1986) reports one case of a Pol-

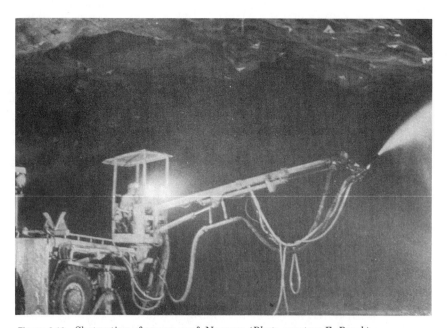

Figure 6.16 Shotcreting of cavern roof, Norway. (*Photo courtesy E. Broch*)

ish cavern, where the benching method could not be used in very friable rock. Leaving a central core has the advantage of allowing temporary bracing between the core and the cavern walls.

The Fortuna hydroelectric project in Panama encountered particularly poor conditions in faulted volcanic and pyroclastic rocks. These conditions led to the adoption of a wall support system consisting of soldier piles, horizontal walers, and prestressed anchors, similar to the common method for supporting excavations in near-surface soils (Deere at al., 1986). The powerhouse was completed without incident, with additional support placed at different stages in response to an intensive monitoring program as outlined in Sec. 6.8.5.1.

Lindblom (1986) recalls two cases where unexpectedly *high horizontal stresses* caused substantial problems in cavern construction. In Finland, three parallel caverns were excavated at a depth of 50 m in granite. Horizontal rock stresses of 15 MPa acting perpendicular to the cavern axis generated tension in the walls and heavy compression in the roof sections. This gave rise to rockbursts in the compressed roof, and horizontal fracturing of pillars. Heavy inflows of water required grouting. The second case, in Sweden, related to six caverns for crude oil storage. No support had been planned except spot bolting and thin shotcreting in the roof. However, serious rockbursts occurred in the roof. The ejection of disks of rock posed a serious spark hazard in a crude oil storage facility. Rock stresses were measured after the incident, and amounted to 15 MPa at a depth of 60 m. The roof support was changed to systematic 2-m-long bolting and wire mesh.

6.8.4.6 Solution mining.
Cavern volumes of up to 1 million cubic meters can be mined in salt formations by leaching. Wells are drilled, and the desired cavern configuration is obtained by adjusting the locations of leaching. The tubing can later be employed for the pumping of stored products, either by water or brine displacement or by submersible pump. Brine displacement is usually preferred, because the cavern's internal pressures can be maintained high enough to slow down closure rates.

About 7 m^3 of water are needed for every cubic meter of salt removed. Cavern size can be doubled in four to five cycles of fresh water displacement.

6.8.5 Monitoring

6.8.5.1 Monitoring during construction.
Design assumptions for mined caverns and converted mines should be checked by close monitoring of rock and support conditions during construction, and by a comprehensive instrumentation program (Sakurai, 1982; Kaiser, 1987).

Typically, rock movements are small and must be measured by sensitive multiple extensometers and inclinometers. Pore water pressures and tensions in anchors and rockbolts are monitored as a further check. Access is initially from surface, then from galleries or shafts, later from within the cavern. Remote readings are necessary in view of the size and geometry of the cavern. Instruments must continue to function throughout the various stages of cavern excavation, and may even be designed to play a useful role during the operating life of the facility.

For a structure as important as a cavern, the early phases of construction should proceed slowly under close surveillance, and may be planned as a large-scale·test of the design. Monitored trials give a check on the assumptions and predictions, and allow adjustments to methods of blasting and rock stabilization.

The Drakensberg pumped storage scheme in South African sedimentary rocks (Sharp et al., 1979) provides an outstanding example of this approach. The crown and walls of the machine hall were excavated in four closely monitored stages, using sidewall "slots" to a depth equal to about half the final sidewall height, and a length equal to about 1.5 times the span (Fig. 6.17). The results led to optimization of smooth-wall blasting methods and support patterns not only for the machine hall, but also for the transformer and valve halls and the control block. The trials led to significant reductions in support requirements that more than offset the cost of the experimentation.

Subway stations tend to be built close to surface so as to minimize lengths of escalator and shaft. They are often beneath cities, and in

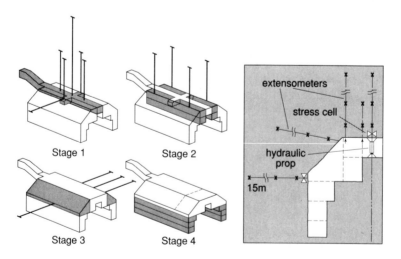

Figure 6.17 Trial enlargement of the machine hall for the Drakensberg pumped storage scheme (Sharp et al., 1979).

less than ideal ground. The combination of poor ground, shallow depth, and sensitive surface structures calls for careful monitoring. An example is the Taicoo cavern, a major station of the Hong Kong mass transit railway. It was excavated in weathered rock with extensive monitoring to ensure stability of the cavern and to safeguard nearby buildings (Sharp et al., 1986).

Georgia Power's 800-MW Rocky Mountain hydropower project provides a further example of the role of instrumentation during the construction phase. Movements measured seldom exceeded 5 mm, and they stabilized without additional ground reinforcement. At the few locations where movements exceeded 25 mm, additional 10-m-long tensioned rock anchors were installed (Prager et al., 1986).

The cavern of the Fortuna hydropower project in Panama (Sec. 6.8.4.2) suffered a major change of layout because of unexpectedly chaotic geology. Deere et al. (1986) describe the elaborate monitoring system used, which included a network of 180 extensometers, load cells, and convergence points. The heterogeneity of the rock mass made numerical modeling impractical. The support system adopted was based on an observational method of design, so that instrumentation assumed particular importance.

The *NATM observational method,* so effective in tunnels (Sec. 5.5.1.3), can be applied also to caverns. The Cirata hydroelectric power cavern in Japan is one of the largest in the world: 253 m long, 35 m wide, and 50 m high, with an excavated volume of 320,000 m^3 (Kamamura et al., 1986). Walls of the egg-shaped chamber, excavated in shale and pyroclastic rock containing clay seams, were reinforced by anchors, rockbolts, and shotcrete. Twenty-two cross sections as well as end walls were instrumented, and the system was scanned by microcomputer. Additional anchors and bolts were installed at locations of excessive ground movements, and the monitoring system confirmed the effectiveness of these measures.

Volumetric monitoring of solution-mined *salt caverns* is necessary to control solution, and to obtain a cavern of the required geometry. The shape of the cavity can be determined by a rotating sonar pulse, or from the period of oscillation of an induced pressure pulse (Bérest et al., 1983; Bérest, 1985).

6.8.5.2 Monitoring caverns in service. Caverns for such purposes as hydropower facilities are usually thoroughly stabilized by the time they are commissioned, so ongoing measurements, for example of rock displacements, are needed only under special circumstances. Long-term monitoring is much more a requirement when caverns are used to store oils, gases, and toxic or radioactive wastes. Here groundwater is the medium of greatest concern, either in its role as a barrier to escape of fluids, or as a potential medium of contaminant migration. A system of piezometers and wells is installed early and routine measurements are made of both water quality and pressure fluctuations.

Further instrumentation is needed in and around fluid storage chambers to measure pressure fluctuations in the stored liquids, stresses and displacements in the rock roof and walls, and evolution of the thermal and pore pressure regimes. For example, the May sur Orne mine, converted to a hydrocarbon storage facility, was instrumented with 60 vibrating wire stress-change measuring devices at depths of 3, 6, and 9 m from the free surface of the cavern (Maury, 1977a). Acoustic sensors were employed to detect fracturing, and many pressure transducers and thermocouples were installed. More than 10 km of wiring was needed to connect the instruments. The monitoring system has given many years of trouble-free service.

Caverns in salt are monitored periodically to assess loss or gain of storage volume. A liquid-filled cavern can be measured by closing the exit ports, and calculating the rate of pressure build-up knowing the volumetric compressibilities of the liquid and the surrounding rock.

References

Aarvoll, M., Barbo, T. F., Hansen, R., and Lövholt, J., 1987. "Storage of Industrial Waste in Large Rock Caverns." In: *Large Rock Caverns. Proc. Int. Symp.*, Helsinki, Finland, Pergamon Press, Oxford, vol. 1, pp. 759–770.

Amadei, B., Robison, M. J., and Yassin, Y. Y., 1987. "Rock Strength and the Design of Underground Excavations." In: *Large Rock Caverns. Proc. Int. Symp.*, Helsinki, Finland, Pergamon Press, Oxford, vol. 2, pp. 1135–1146.

ASCE, 1971. *Proc., American Society of Civil Engineers Symp. Underground Rock Chambers*, Phoenix, Arizona.

Barla, G., 1987. "General Report on Case Histories, Mines and Large Tunnels." In: *Large Rock Caverns. Proc. Int. Symp.*, Helsinki, Finland, Pergamon Press, Oxford, vol. 3, pp. 1715–1764.

Barton, N., and Lingle, R., 1982. "Rock Mass Characterization Methods for Nuclear Waste Repositories in Jointed Rock." *Proc. Int. Symp. Rock Mech.: Caverns and Pressure Shafts*, Balkema, Rotterdam, vol. 1, pp. 3–18.

Bawden, W. F., and Roegiers, J. C., 1985. "Gas Escape from Underground Mined Storage Facilities, a Multiphase Flow Phenomenon." *Proc. Int. Symp. Fundamentals of Rock Joints*, Bjorkliden, Sweden, pp. 503–514.

Bennett-Smith, L., 1977. "Examples of Storage in Mined Space in the U.S. and Canada." *Proc. 1st Int. Symp. Storage in Excavated Rock Caverns*, Stockholm, Sweden, pp. 641–643.

Bérest, P., 1981. "Stabilité des Cavités de Stockage d'Hydrocarbure dans le Sel." *Revue Française de Géotechnique*, no. 16, pp. 5–10.

———, 1985. "Phénomènes Vibratoires dans les Colonnes Pétrolières—Application au Calcul du Volume des Cavités Souterraines." *Revue Française de Géotechnique*, no. 32, pp. 5–17.

———, 1989. "Accidents of Underground Oil and Gas Storage—Case Histories and Prevention." In: *Storage of Gases in Rock Caverns*, Balkema, Rotterdam, pp. 289–301.

———, Habib, P., Boucher, M., and Pernette, M., 1983. "Détermination du Volume d'une Cavité Souterraine Remplie de Liquids par Mésure d'une Periode d'Oscillation."

Proc. Int. Symp. Soil and Rock Investigations by In Situ Testing, Paris, vol. 2, pp. 443–447.

———, Ghoreychi, M., Fauveau, M., and Lebitoux, P., 1986. "Mechanisms of Creep in Gas Storage Caverns. Effect of Gravity Forces." *Proc. 27th U.S. Symp. Rock Mech.,* Tuscaloosa, Ala., pp. 789–794.

———, ———, Roman, J., and Bazargan-Sabet, B., 1989. "Comportement Mécanique du Sel Broyé Utilisé comme Remblai dans un Massif Salifère." *Proc. Int. Conf. Rock at Great Depth,* V. Maury and D. Fourmaintreaux (eds.), Balkema, Rotterdam.

———, Ledoux, E., and Tillie, B., 1982. "Etanchéité des Stockages d'Hydrocarbures Liquéfiés en Galeries non Revêtues dans un Milieu Aquifère." *Revue de l'Institut Français du Pétrole,* vol. 37, no 3.

Bergman, M. (ed.), 1977. *Storage in Excavated Rock Caverns, Proc. Rockstore 77, Int. Symp.,* Pergamon Press, Oxford, 3 volumes.

———, (ed.), 1980. *Subsurface Space, Rockstore 80, Proc. Int. Symp.,* Pergamon Press, Oxford, 3 volumes.

Blejwas, T. E., 1989. "Experiments in Rock Mechanics for the Site Characterization of Yucca Mountain." *Proc. 30th U.S. Symp. Rock Mech.,* Morgantown, W. Va., pp. 39–46.

Bogdanoff, I., 1986. "Block Filled Rock Caverns—a Concept for Large Thermal Oil or Gas Storage under Bad Geological Conditions." *Proc. Int. Symp. Large Rock Caverns,* Helsinki, Finland, vol. 1, pp. 467–478.

Boivin, D. J., 1985. "Les Aménagements Souterrains des Grandes Villes du Canada." *Annales de l'Institut du Bâtiment et des Travaux Publics,* no. 433, pp. 83–96.

Brekke, T. L., and Jørstad, F. A. (eds.), 1970. *Large Permanent Underground Openings. Proc. Int. Symp.,* Oslo, 1969, Universitetsforlaget, Oslo.

Broch, E., 1989. "The Technical, Economical, Environmental Disclosures of the Underground Employment in the Future." *General Rept., Suolosottosuolo; Int. Cong. Geoengineering,* Turin, Italy, 12 pp.

———, and Odegaard, L., 1983. "Storing Water in Rock Caverns." *Underground Space,* vol. 7, pp. 269–272.

Brown, E. T., 1987. "Research and Development for Design and Construction of Large Rock Caverns. In: *Large Rock Caverns. Proc. Int. Symp.,* Helsinki, Finland, Pergamon Press, Oxford, vol. 3, pp. 1937–1948.

———, and Hudson, J. A. (eds.), 1984. *Design and Performance of Underground Excavations.* Int. Symp., British Geotechnical Society, London, 518 pp.

Butkovich, T. R., and Patrick, W. C., 1986. "Thermo-Mechanical Modelling of the Spent Fuel Test—Climax." *Proc. 27th U.S. Symp. Rock Mech.,* Tuscaloosa, Ala., pp. 898–905.

Carlsson, A., and Christiansson, R., 1988. "Site Investigations for the Swedish Undersea Repository for Reactor Waste." *Proc. Int. Symp. Rock Mech. and Power Plants,* Madrid, pp. 561–569.

Cikanek, E. M., and Goyal, B. B., 1986. "Experiences from Large Cavern Excavation for TARP." *Proc. Int. Symp. Large Rock Caverns,* Helsinki, Finland, pp. 35–46.

Côme, B., 1984. "Conception de Dépôts Définitifs Profonds pour Déchets Radioactifs: Exigences, Exemples Actuels et Idées Nouvelles." *Annales de l'Institut du Batiment et des Travaux Publics,* no. 430, pp. 90–115.

———, Johnston, P., and Müller, A. (eds.), 1985. *Design and Instrumentation of In Situ Experiments in Underground Laboratories for Radioactive Waste Disposal, Proc. of a CEC-NEA Workshop,* Balkema, Rotterdam, 474 pp.

Cording, E. J., Hendron, A. J., and Deere, D. U., 1971. "Rock Engineering for Underground Caverns." *Proc. Symp. Underground Rock Chambers,* Phoenix, Ariz., American Society of Civil Engineers, New York, pp. 567-600.

Cunha, A. P., 1987. "A Geotechnical Investigation Methodology for Caverns in Rock." *Proc. Int. Symp. Rock Mech.: Caverns and Pressure Shafts,* Balkema, Rotterdam, vol. 1, pp. 19–25.

Curtis, G., and Rutherford, J., 1981. "Expansive Shale Damage, Theban Royal Tombs." *Proc. 10th Int. Conf. Soil Mech. and Found. Eng.,* Stockholm, Balkema, Rotterdam, vol.3, pp. 71–74.

Daerga, P. A., Stephansson, O., and Sagefors, I., 1986. "Funnel Store—New Concept for Large Rock Caverns." *Proc. Int. Symp. Large Rock Caverns,* Helsinki, Finland, vol. 1, pp. 479–488.

Deere, D., Isaza, E., Fellor, E., and Giussani, L., 1986. "Monitoring of Power House Cavern for Fortuna Hydroproject." *Proc. Int. Symp. Large Rock Caverns,* Helsinki, Finland, vol 2, pp. 907–920.

de Laguérie, P. V., 1985. "Deux Expériences de Laboratoires Souterrains en Conditions Difficiles." In: *Design and Instrumentation of in Situ Experiments in Underground Laboratories for Radioactive Waste Disposal,* eds., B. Côme, P. Johnston, and A. Müller (eds.), Balkema, Rotterdam, pp. 21–32.

Dolezalová, M., Hejda, R., and Leitner, F., 1982. "FEM-Analysis of an Underground Cavern with Large Span." *Proc. Int. Symp. Rock Mech.: Caverns and Pressure Shafts,* Balkema, Rotterdam, vol. 1, pp. 207–219.

Dorso, R., del Rio, J. C., de la Torre, D., Sarra Pistone, R., 1987. "Powerhouse Caverns of the Hydroelectric Complex "Rio Grande No. 1." *Proc. Int. Symp. Rock Mech.: Caverns and Pressure Shafts,* Balkema, Rotterdam, vol. 1, pp. 221–238.

Dreyer, W., 1982. *Underground Storage of Oil and Gas in Salt Deposits and Other Non-Hard Rocks.* Halsted Press, New York, 207 pp.

Duffaut, P., 1980. "Past and Future of the Use of Underground Space in France and Europe." *Underground Space,* vol. 5, pp. 86–91.

———, 1982. "Les Facteurs qui Limitent la Taille des Cavités." *Proc. Int. Symp. Rock Mech.: Caverns and Pressure Shafts,* Balkema, Rotterdam, vol. 1, pp. 245–253.

———, 1985. "Les Laboratoires Souterrains de Mécanique des Roches avant les Déchets Nucléaires." In: *Design and Instrumentation of In Situ Experiments in Underground Laboratories for Radioactive Waste Disposal,* B. Côme, P. Johnston, and A. Müller, (eds.), Balkema, Rotterdam, pp. 8–20.

Dusseault, M. B., 1989. "Saltrock Behavior as an Analogue to the Behavior of Rocks at Great Depth." *Proc. Int. Conf. on Rock at Great Depth,* V. Maury and D. Fourmaintreaux (eds.), Balkema, Rotterdam, pp. 10–17.

Eriksson, L. G., 1989. "Underground Disposal of High Level Radioactive Waste in the United States of America—Program Overview." *Bul., Int. Assoc. of Eng. Geol.,* no. 39, pp. 35–51.

Finnish Tunnelling Assoc., 1986. *Rock Engineering in Finland.* Rakentajain Kustannus Oy, Helsinki, 188 pp.

Flak, L. H., and Brown, J., 1988. "Case History of Ultra Deep Disposal Well in Western Colorado." *Proc. Int. Ass. of Drlg. Cont. and Soc. Pet. Eng. Drilling Conf.,* Dallas, Texas, pp. 381–394.

Fordham, C. J., 1989. "Use of Halite Backfill in Potash Mines." Ph.D. Thesis, University of Waterloo, Waterloo, Ontario, 150 pp.

———, Dusseault, M. B., Mraz, D., and Rothenburg, L., 1989. "The Use of Relaxation Tests to Predict the Compaction Behavior of Halite Backfill." *Proc. 4th Int. Symp. on Mining with Backfill,* Montreal, Balkema, Rotterdam, pp. 327–334.

Franklin, J. A., 1989. "Stability of Shallow Caverns." In *Rock Caverns, Hong Kong,* A. W. Malone and P. G. D. Whiteside (eds.), Inst. Min. Met. (U.K.), pp. 203–212.

Gagne, L. L., 1972. "Controlled Blasting Techniques for the Churchill Falls Underground Complex." *Proc. North American Rapid Excavation and Tunneling Conf.,* Chicago, Ill., vol. 1, pp. 739–764.

Goodall, D. C., Åberg, B., and Brekke, T. L., 1988. "Fundamentals of Gas Containment in Unlined Rock Caverns." *Rock Mech. and Rock Eng.,* vol. 21, pp. 235–258.

Gyenge, M., 1980. "Nuclear Waste Vault Sealing." *Proc. 13th Cdn. Rock Mech. Symp.,* Toronto, Ontario, pp. 181–192.

Gysel, M., 1986. "Design and Construction of Large Caverns: Developments and Trends over the Past 25 Years—a Swiss Experience." *Proc. Int. Symp. Large Rock Caverns,* Helsinki, Finland, vol 1, pp. 81–95.

Häikiö, E., and Kämppi, A., 1987. "Use of Rock Caverns in the Helsinki Water Supply and Wastewater Disposal System." In: *Large Rock Caverns. Proc. Int. Symp., Helsinki, Finland,* Pergamon Press, Oxford, pp. 489–495.

Haimson, B. C., Doe, T. W., and Fu, G. F., 1978. "Geotechnical Investigation and Design of Annular Tunnels for Energy Storage." In: *Storage in Excavated Rock Caverns—Rockstore 77*, Pergamon Press, Oxford, pp. 275–282.

———, Ma, M. S., O'Donnell, M., and Ren, N. K., 1978. "Site Investigation for Energy Storage Caverns at Waterloo, Wisconsin." *Proc. 3d Int. Cong., Int. Assoc. Eng. Geol.*, sec. 3, vol. 2, pp. 76–86.

Hardy, H. R., and Langer, M. (eds.), 1982, 1988. *Proceedings, 1st (1981) and 2d (1984) Conf. Mechanical Behavior of Salt.* Transtech Publications, Clausthal, Germany.

Heuze, F. E., 1981. "Geomechanics of the Climax Mine-By, Nevada Test Site." *Proc. 22d U.S. Rock Mech. Symp.*, Cambridge, Mass., pp. 458–464.

Hodgson, J. L., 1986. "A Construction Planning Study for an Underground Hydroelectric Pumped Storage Facility in Weak Rock." *Proc. 27th U.S. Symp. Rock Mech.*, Tuscaloosa, Ala., pp. 795–799.

Hudson, J. A., 1983. "U.K. Rock Mechanics Research for Radioactive Waste Disposal." *Proc. 5th Int. Cong. Rock Mech.*, Melbourne, Australia, vol. E, pp. 161–165.

Hustrulid, W., 1983. "Design of Geomechanical Experiments for Radioactive Waste Disposal—A Rethink." *Proc. Int. Symp. Field Measurements in Geomechanics*, Zurich, Switzerland, pp. 1381–1408.

IAEA (Int. Atomic Energy Agency, Vienna, Austria), 1980. *Proc. Int. Symp. Underground Disposal of Radioactive Wastes*, Otaniemi, Finland.

———, 1983. *Proc. Int. Conf. Radioactive Waste Management.* Vienna.

IAEG, 1989. "Problems of Underground Disposal of Waste: Report of IAEG Commission 14." *Bul. Int. Assoc. Eng. Geol.*, no. 39, pp. 3–58.

Inada, Y., and Taniguchi, K., 1986. "A Theoretical Analysis of the Range of the Plastic Zone Around Openings due to Storage of L.N.G." *Proc. 27th U.S. Symp. Rock Mech.*, Tuscaloosa, Ala., pp. 782–788.

Jansson, G., 1974. "Rock Cave Storage Can Be Cheaper Option." *Oil and Gas Journal*, October 28, pp. 74–82.

Johansson, S., and Lahtinen, R., 1976. "Oil Storage in Rock Caverns in Finland." *Tunneling 76*, Publ. Institution of Mining and Metallurgy, Great Britain, M. J. Jones (ed.), pp. 41–82.

Kaiser, P. K., 1986. "Excavation and Mining Methods: General Report," In: *Proc. Int. Symp. Large Rock Caverns*, Helsinki, Finland, vol 3, pp. 1877–1907.

———, 1987. "General Report on Excavation and Mining Methods." In: *Large Rock Caverns. Proc. Int. Symp.*, Helsinki, Finland, Pergamon Press, Oxford, vol. 3, pp. 1877–1913.

Kerisel, J., 1987. *Down to Earth*. Balkema, Rotterdam, 149 pp.

Kamemura, K., Homma, N., Shibata, K., Harada, T., and Soetomo, S. W., 1986. "Observational Method on Large Rock Cavern Excavation." *Proc. Int. Symp. Large Rock Caverns*, Helsinki, Finland, vol 2, pp. 1503–1512.

Kuzyk, G. W., Babulic, P. J., Lang, P. A., and Morin, R. A., 1987. "Blast Design and Quality Control Procedures at Atomic Energy of Canada Limited's Underground Research Laboratory." *Proc. 13th Ann. Conf. Soc. Explosives Engrs.*, Miami, Fla., 13 pp.

Langer, M., 1989. "Waste Disposal in the Federal Republic of Germany: Concepts, Criteria, Scientific Investigations." *Bul. Int. Assoc. Eng. Geol.*, no. 39, pp. 53–58.

Liljestrand, B., 1986. "A Technical Economical Comparison of Sports and Leisure Facilities Built in Rock Caverns and Equivalent Facilities Above Ground." *Proc. Int. Symp. Large Rock Caverns*, Helsinki, Finland, vol. 1, pp. 523–534.

Lindblom, U. E., 1986. "Developments in Design Methods for Large Rock Caverns. General Report." In: *Proc. Int. Symp. Large Rock Caverns*, Helsinki, Finland, vol. 3, pp. 1835–1867.

Liu, S. C., Cheng, R. Y., and Hsieh, C. S., 1987. "Design and Construction of Underground Power Caverns in Taiwan." *Proc. Int. Symp. Rock Mech. Caverns and Pressure Shafts*, Balkema, Rotterdam, vol. 1., pp. 329–346.

———, Cheng, Y., and Chang, C. T., 1988. "Design of Mingtan Power Cavern." *Proc. Int. Symp. Rock Mech. and Power Plants*, Madrid, Spain, pp. 199–208.

Martin, D. C., 1989. "Failure Observations and in Situ Stress Domains at the Underground Research Laboratory." *Proc. Int. Conf. Rock at Great Depth,* V. Maury and D. Fourmaintreaux (eds.), Balkema, Rotterdam, pp. 719–726.

Maury, V., 1977a. "Environmental Protection, Monitoring and Operation at the May sur Orne Underground Oil Storage Facility." *Proc. Rockstore 77, Int. Symp.,* Pergamon Press, Oxford, vol. 1.

———, 1977b. "An Example of Underground Storage in Soft Rock (Chalk)." In: *Storage in Excavated Rock Caverns, Proc. Rockstore 77,* Int. Symp., Pergamon Press, Oxford, vol 2, pp. 681–689.

———, 1979. "Utilisation des Essais et Mésures en Laboratoires et In Situ dans Cinq Projets de Stockage Souterrains." *Proc. 4th Int. Cong. Rock Mech.,* Montreux, Balkema, Rotterdam, vol. 2, pp. 417–428.

———, 1985. "Rock Mechanics: A Basic Discipline for Underground Storage. Gen. Rept.," *Proc. Int. Symp. Role of Rock Mech. in Excavations for Mining & Civil Works,* Zacatecas, Mexico, 48 pp. (in French, title: "La Mécanique des Roches: Une Discipline de Base pour le Stockage Souterrain").

———, 1987. "Observations, Research, and Recent Results about Failure Mechanisms around Single Galleries." *Proc. 6th Int. Cong. Rock Mech.,* Montreal, (in French), Balkema, Rotterdam, vol. 2., pp. 1119–1128.

———, and Fourmaintreaux, D. (eds.), 1989. *Rock at Great Depth. Proc. Int. Symp.,* Balkema, Rotterdam, 3 volumes.

McCreath, D. R., and Mitchell, D. E., 1978. "Build Underground and Conserve Energy: Fact or Fiction?" *Eng. Journal,* April, pp. 14–16.

Morfeldt, C. O., 1972. "Storage of Oil in Unlined Caverns in Different Types of Rock." *Rock Mech.,* pp. 409–421.

Munson, D. E., and Fossum, A. F., 1986. "Comparison between Predicted and Measured South Draft Closures at the WIPP Using a Transient Creep Model for Salt." *Proc. 27th U.S. Symp. Rock Mech.,* Tuscaloosa, Ala., pp. 931–939.

Nilsen, B., and Olsen, J. (eds.), 1989. *Storage of Gases in Rock Caverns.* Conf Proceedings, Balkema, Rotterdam.

Nordell, B., 1988. "A Large Scale Borehole Heat Store during Five Years of Operation." *Proc. Int. Symp. Rock Mech. and Power Plants,* Madrid, Spain, pp. 583–587.

OECD-NEA (Organisation for Economic Co-operation and Development, Nuclear Energy Agency), 1978. *Proc. NEA Workshop, in Situ Heating Experiments in Geological Formations.* Ludvika, Sweden.

———, 1979. *Proc. NEA Workshop, The Use of Argillaceous Materials for the Isolation of Radioactive Waste.* Paris, France.

———, 1982. *Geological Disposal of Radioactive Waste: Research in the OECD Area.* OECD, Paris, 54 pp.

———, 1983. *Proc. Conf. Geological Disposal of Radioactive Waste—In Situ Experiments in Granite.* OECD, Paris.

Patrick, W. C., 1986. "Spent Fuel Test—Climax: An Evaluation of the Technical Feasibility of Geologic Storage of Spent Nuclear Fuel in Granite." *Final Report UCRL 53702,* Lawrence Livermore National Laboratory, 299 pp.

Pilebro, H., Brunström, C., and Larsson, M., 1987. "The Lyckebo Project—Thermal Energy Storage in a Rock Cavern." In: *Proc. Int. Symp. Large Rock Caverns,* Helsinki, Finland, Pergamon Press, Oxford, vol. 1, pp. 549–556.

Pincus, H. J., 1978. "Geological Aspects of Storage of Compressed Air in Porous, Permeable Rocks." *Proc. 3d Int. Cong., Int. Assoc. Eng. Geol.,* Madrid, Spain, pp. 107–116.

Prager, R. D., Kendorski, F. S., and Lundell, C. M., 1986. "The Design and Construction of the Excavation for the Underground Works at the Rocky Mountain Pumped Storage Project." *Proc. 27th U.S. Symp. Rock Mech.,* Tuscaloosa, Ala., pp. 968–974.

Pusch, R., 1985. "The Stripa Buffer Mass Test Instrumentation for Temperature, Moisture, and Pressure Measurements." In: *Design and Instrumentation of In Situ Experiments in Underground Laboratories for Radioactive Waste Disposal,* B. Côme, P. Johnston, and A. Müller, (eds.), Balkema, Rotterdam, pp. 303–314.

Rometsch, R., 1985. "An Overview of Strategies for In-Situ Investigations for Radioactive Waste Disposal." In: *Design and Instrumentation of In Situ Experiments in Un-*

derground Laboratories for Radioactive Waste Disposal, B. Côme, P. Johnston, and A. Müller, (eds.), Balkema, Rotterdam, pp. 33–39.

Roshoff, K., Stephansson, O., Larsson, H., Stanfors, R., and Eriksson, K., 1983. "KLAB an Intermediate Storage for Spent Nuclear Fuel in Sweden." *Proc. 5th Int. Cong. Rock Mech.,* Melbourne, Australia, vol. E., pp. 151–159.

Saari, K. H. O. (ed.), 1987. *Large Rock Caverns. Proc. Int. Symp.,* Helsinki, Finland, Pergamon Press, Oxford, 3 volumes, 2084 pp.

St. John, C. M., 1982. "Repository Design." *Underground Space,* vol. 6, pp. 247–258.

Sakurai, S., 1982. "Monitoring of Caverns during Construction Period." *Proc. Int. Symp. Rock Mech.: Caverns and Pressure Shafts,* Balkema, Rotterdam, vol. 1, pp. 433–441.

Schneider, H. J., 1988. "Geotechnical Requirements of Subsurface Repositories in Mines, Rock and Salt Caverns for Interim or Final Disposal of Hazardous Wastes." *Bul. Int. Assoc. Eng. Geol.,* no. 37, pp. 71–76.

Sharp, J. C., Pine, R. J., Moy, D., and Byrne, R. J., 1979. "The Use of a Trial Enlargement for the Underground Cavern Design of the Drakensberg Pumped Storage Scheme." *Proc. 4th Int. Cong. Rock Mech.,* Montreux, Switzerland, vol. 2, pp. 617–626.

———, Smith, M. C. F., Toms, I. M., and Turner, V. D., 1986. "Taikoo Cavern, Hong Kong—Performance of a Large Metro Excavation in a Partially Weathered Rock Mass." *Proc. Int. Symp. Large Rock Caverns,* Helsinki, Finland, vol. 1, pp. 403–423.

Simmons, G. R., 1985. "In Situ Experiments in Granite in Underground Laboratories: A Review." In: *Design and Instrumentation of in Situ Experiments in Underground Laboratories for Radioactive Waste Disposal,* B. Côme, P. Johnston, and A. Müller, (eds.), Balkema, Rotterdam, pp. 56–81.

Sousa, L. R., 1983. "Three-Dimensional Analysis of Large Underground Power Stations." *Proc. 5th Int. Cong. Rock Mech.,* Melbourne, Australia, vol. F., pp. 169–174.

Spiers, C. J., (and four coauthors), 1986. "The Influence of Fluid-Rock Interaction on the Rheology of Salt Rock." *Final Report, Nuclear Science and Technology,* vol. EUR 10399 EN, 131 pp.

———, (and five coauthors), 1989. "Long-Term Rheological and Transport Properties of Dry and Wet Salt Rocks." *Final Report, Contract Fl1W-0051-NL.* Office for Official Publications of the European Communities, Luxembourg, 161 pp.

Starfield, A. M., and Cundall, P. A., 1988. "Towards a Methodology for Rock Mechanics Modeling." *Int. Jour. of Rock Mech., Min. Sci., and Geomech. Abstr.,* vol. 25, no. 3, pp. 99–106.

Stephansson, O., Blomquist, R., Groth, T., Jonasson, P., and Terandi, T., 1980. "Modelling of Temperature Fields and Deformations for Radioactive Waste Repositories in Hard Rock." *Underground Disposal of Radioactive Wastes,* Int. Atomic Energy Agency, Vienna, IAEA-SM-244/164, vol. 2, pp. 121–135.

Sugawara, K., Aoki, T., and Suzuki, Y., 1988. "A Coupled Boundary Element-Characteristics Method for Elasto-Plastic Analysis of Rock Caveran." *Proc. Int. Symp. Rock Mech. and Power Plants,* Madrid, Spain, pp. 249–258.

Svemar, C., and Sagefors, I., 1987. "Some Aspects on Forming a High-Level Nuclear Waste Repository which Appeals to Public Opinion." In: *Large Rock Caverns. Proc. Int. Symp.,* Helsinki, Finland, Pergamon Press, Oxford, vol. 1, pp. 847–854.

U.S.-NCRM (U.S. National Committee for Rock Mech.), 1978. *Rock Mechanics Problems Related to Underground Construction.* 1977 Report US-NCRM. National Academy of Sciences, Washington, D.C., pp. 7–14.

Vasilescu, M. S., Benziger, C. P., and Kwiatkowski, R. W., 1971. "Design of Rock Caverns for Hydraulic Projects." *Proc. Symp. Underground Rock Chambers,* Phoenix, Ariz., American Society of Civil Engineers, New York, pp. 21–50.

Wittke, W., (ed.), 1982. "Caverns and Pressure Shafts." *Proc. Int. Symp. Rock Mech., Aachen.* Balkema, Rotterdam, 3 volumes, 1356 pp.

Woodard, D. R., 1980. "The Kansas City Underground Experience." In: *Rockstore '80: Subsurface Space,* Pergamon Press, Oxford, vol. 1, pp. 63–75.

Chapter 7

Underground Mining

7.1 Introduction to Mining

7.1.1 Types of orebody

Understanding mining methods and sequences requires an appreciation of the nature and geometry of orebodies, and this in turn requires some knowledge of the geological origins of the ore. Different methods of extraction are selected depending on whether the orebody is tabular or massive, thin or thick, horizontal or steeply inclined. This together with the strengths and jointing characteristics of the ore and "country" rock will help dictate the optimum mining approach.

Ores are found in all three major rock types: sedimentary, igneous, and metamorphic. Mineral deposits in *sedimentary rocks* are usually *stratiform* (in the form of beds or seams), often horizontal and interbedded with barren horizons of shale or sandstone. Stratiform orebodies include evaporites such as potash, bedded salt, and gypsum, and organic deposits such as coal, oil shale, and lignite. Other stratiform orebodies include replacement lead-zinc deposits of the Mississippi Valley type and sideritic iron ores. Marine precipitation forms accumulations of iron and manganese in stratiform bodies and nodules on the sea bed.

Denser minerals eroding from igneous or metamorphic parent rocks often become concentrated in sedimentary strata by water flowing in rivers and streams. Examples are the well-known *placer deposits* of gold in California, the Yukon, and Alaska, diamonds in Namibia, and tin in Malaysia.

Groundwaters percolating through porous rocks may dissolve and concentrate valuable minerals. The Pine Point lead-zinc orebody in the Northwest Territories, Canada, was created by long-term basin-scale processes of solution, flow, and precipitation when the rocks were at greater depths than today. The uranium ores of Saskatchewan, found at the contact between ancient sediments and the underlying metamorphic and igneous rocks, are precipitates which developed when moving groundwater came into contact with rocks of different chemistry.

Weathering and leaching of igneous and metamorphic rocks in tropical climates by groundwater produce bauxite, a rock rich in alumina, and iron oxide ores from rocks rich in iron silicate and carbonate. *Biological decomposition* generates deposits of coal, lignite, and the hydrocarbons, and temperatures and pressures may metamorphose these deposits. High-grade *metamorphism* leading to recrystallization of carbon has formed the graphite deposits of Texas and Mexico. *Diapiric processes* create *salt domes* capped with sulphur and anhydrite.

Hydrothermal activity, the circulation of hot aqueous solutions charged with mineral matter, creates orebodies in sedimentary, metamorphic, or igneous rocks. Hydrothermal solutions are generated deep in the crust, and slowly work their way upward to regimes of lower temperatures and pressures, where the valuable minerals are precipitated. Usually, the deposits are in the form of veins or lenticular bodies. Hydrothermal activity was responsible for low-grade disseminated deposits of crystalline copper-molybdenum minerals at Bingham, Butte, and Climax, in the United States, and for copper and gold at Noranda, Canada.

Orebodies in igneous and metamorphic formations can be *finely disseminated* throughout a large volume of the rock mass or, more often, are concentrated as *vein deposits* or as alteration products along major joints and faults. *Magmatic differentiation* of molten rock into fractions containing different concentrations of mineral is responsible for layered deposits of chromite in Zimbabwe and South Africa. Intrusions of magma from the earth's mantle followed by magmatic differentiation formed the nickel ores of Sudbury, Canada.

7.1.2 Exploration, development, and production

The earliest phase of mining is *exploration,* followed by *development* (mining of underground access ways and drawpoints) and *production* (extraction of ore from the stopes). These three activities occur at the same time at different levels and locations in the mine (Fig. 7.1). Typically, the deeper levels are being explored while intermediate levels are being developed, and upper ones mined.

The continuation of profitable mining relies on exploratory drilling to map mineral grades and delineate the geometries of future stopes. The aim is to maintain or increase proven reserves and supplies of ore

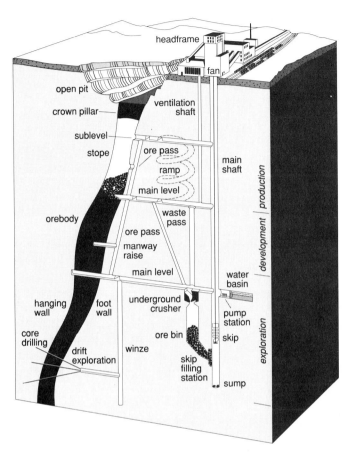

Figure 7.1 Schematic layout of an underground mine (after Hamrin, 1982).

to the mill, and to complete planning and design of new stopes well in advance of the currently active ones.

The search for new mineral deposits makes extensive use of remote sensing and airborne geophysical methods for investigating large tracts of land at minimal cost. Potential metallic orebodies can be identified in surrounding barren country rock by differences in their density, conductivity, and magnetism. Anomalies are investigated from assays on streambed sediments and exploratory cores obtained by drilling. Exploratory drilling from the surface is followed by the excavation of shafts and adits, giving access for drilling from underground.

The most common pitfalls in evaluation of a mining property are an orebody insufficiently defined to obtain reliable estimates of reserves;

incorrect estimates of grade, dilution, and mining recovery; optimistic assessments of ground quality; and unanticipated problems of ground instability or groundwater inflow. Too often exploratory drilling is dedicated to defining the orebody grade without considering the mechanical competence of the rock that will form the walls of a future stope. Growing awareness of geomechanics, however, is leading to a better integration of mineral exploration with rock quality assessment.

7.1.3 Mine openings—terminology and classification

A layperson might be excused for thinking that miners have invented a secret language that bears only a passing resemblance to English. An underground passage for access of personnel and supplies, for transportation or for ventilation, is called a *drift, drivage, level,* or *roadway,* not a tunnel. The rock overhead is the *roof* or the *back,* depending on whether you are in a coal or a hard rock mine, less often the *crown* as in a tunnel. Many more colloquial and local terms (goaf, gob, inby, outby, etc.) reflect the fact that mining is one of the oldest industrial activities, with vestiges of dialects from far away and long ago.

Of greater practical importance is a classification of mine openings into three types according to function and degree of permanence: Service openings give long-term access throughout the mine, development openings give immediate access to a stope, and stopes are chambers where ore is extracted from the orebody.

Surface facilities include the headframe to raise and lower personnel, ore, and mining equipment and materials (Fig. 7.2). The network of *service openings* includes near-horizontal drifts excavated in a series of *levels* at regular depth intervals, and connected by *shafts, raises, inclines,* and *ramps.* A shaft is excavated steeply or vertically downward, whereas a raise is excavated upward. Inclines and ramps are at low angles, usually less than 15°; an incline is usually straight, whereas a ramp usually follows a spiral. Near-horizontal adits provide access to mines in hilly or mountainous areas. Shafts give access to deep levels of the mine, whereas ramps are limited to shallow depth or connect locally between levels.

Service openings are designed to remain open and secure for the life of the mine. Shafts and haulage drives must continue to allow high-speed operation of cages and skips, ore vehicles, conveyors, and trucks for transporting personnel. They are located in the most stable rock available, often in the footwall where the ground is least likely to be affected by extraction, and if possible, where minerals are scarce or

Figure 7.2 Surface facilities for a modern underground salt mine in Louisiana (depth 400 m). The hoist frame on the left is for ore, and the one on the right is for personnel.

absent. Openings have to provide adequate ventilation to the mine, and special ventilation shafts and raises are common.

Service openings are increasingly excavated by boring machine (Handewith, 1980; Yuen et al., 1987); 4 to 6 km of tunnels, development headings, and haulage ways can be enough to justify use of a full-face TBM (*Rock Engineering,* Sec. 15.2). Support methods are similar to the primary support used in tunnels, that is, rockbolts, mesh, and shotcrete (Chap. 5).

Development openings such as *sublevels* and *drawpoints* are used in mining for a period only a little longer than the life of the stope that they serve. They are designed as part of the overall stoping layout, so that they will continue to function as long as adjacent mining is in progress. They are excavated by blasting or by roadheader (Fig. 7.3), and often are stabilized by rockbolts and wire mesh.

Stopes are the largest of the three classes of opening and have the greatest radius of influence and require the most attention in design; their performance may affect the entire mine. They are the least permanent, require only short-term stability, are the least accessible, and often are difficult to monitor and impossible to fully stabilize.

7.2 Mining Methods

7.2.1 Stable and caving alternatives

To suit the various shapes and qualities of orebody there are two distinct categories of mining method: *stable stope* and *caving,* with a com-

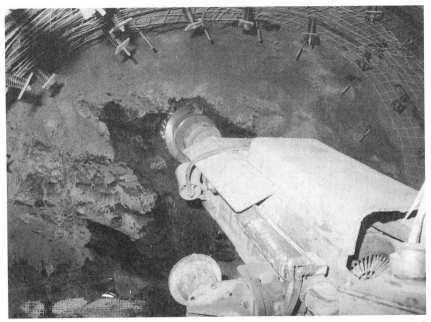

Figure 7.3 Excavating a ventilation tunnel through soft sandstones, Ping Zhung coal mine, Inner Mongolia, China. (*Photo courtesy Lu Haichang*)

plete spectrum of intermediate methods between these two extremes (Table 7.1 and Fig. 7.4). Comprehensive descriptions of various methods are given by Hamrin (1982), Budavari (1983), Brady and Brown (1985), and Jeremic (1985).

In *fully stable* methods, mine openings are designed to remain stable with minimal displacements, often in the elastic stress range, at

TABLE 7.1 Spectrum of Mining Methods from Fully Stable (top) to Fully Caving (bottom)

Method	Support medium	Span	Access
Fully Stable			
Open stoping	None or pillars	Large	None
Room and pillar	Pillars	Small	Good
Cut and fill	Fill	Small	Good
Shrinkage	Broken ore	Small	Good
Vertical crater retreat	Broken ore	Large	None
Long wall	None (caving)	Large	None
Sublevel caving	None (caving)	Large	None
Block caving	None (caving)	Large	None
Fully Caving			

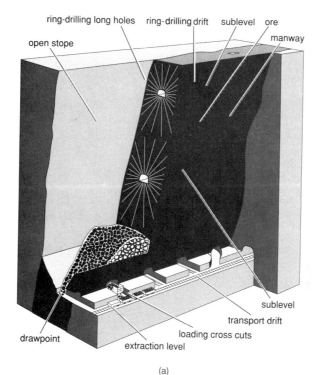

Figure 7.4 Mining methods. (a) Sublevel open stoping with ring-drilled blastholes; (b) room and pillar (After Hamrin, 1982 in: *Underground Mining Methods Handbook*, AIME).

320 Rock Engineering Underground

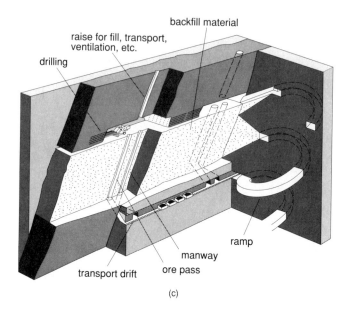

(c)

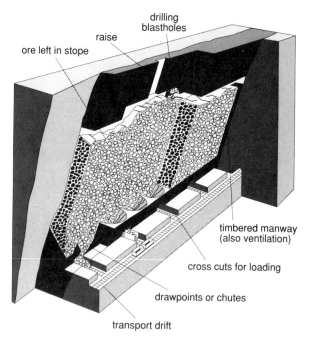

(d)

Figure 7.4 (*Continued*) Mining methods. (c) Overhand cut and fill; (d) shrinkage stoping (After Hamrin, 1982 in: *Underground Mining Methods Handbook*, AIME).

Underground Mining 321

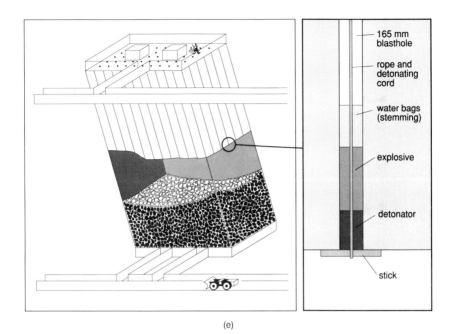

(e)

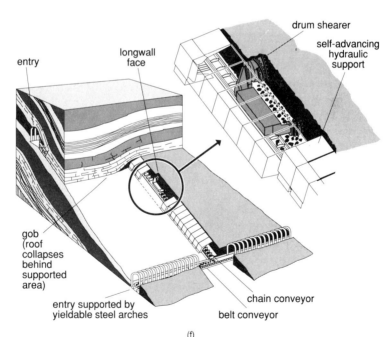

(f)

Figure 7.4 (*Continued*) Mining methods. (*e*) Vertical crater retreat; (*f*) longwall mining in coal and other soft rocks (After Hamrin, 1982 in: *Underground Mining Methods Handbook*, AIME).

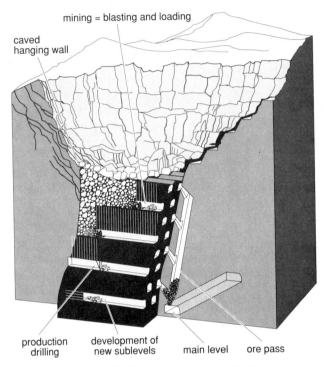

Figure 7.4 (*Continued*) Mining methods. (*g*) Sublevel caving (After Hamrin, 1982 in: *Underground Mining Methods Handbook*, AIME).

least for the life of the stope. They may be naturally stable without support (as in open stoping), or with pillar support (as in room-and-pillar mining). Alternatively they may be given artificial support either by backfill (as in cut-and-fill mining) or by broken ore (as in shrinkage stoping and the vertical crater retreat, or VCR, method).

In *caving* methods, the ore block is undercut to encourage collapse. Only temporary local support is provided to protect personnel and equipment, and the mined space fills spontaneously with fragmented ore and waste rock that is induced to rupture (cave) and flow. As extraction proceeds, the caving front migrates through the orebody in a process that may self-stabilize by bulking, or, if the workings are shallow, will eventually lead to surface subsidence and sinkholes (Fig. 7.5). The objective is to maintain a steady flow of ore that keeps pace with extraction, and to avoid blockages, bridges, and unstable voids that delay production and are a safety hazard.

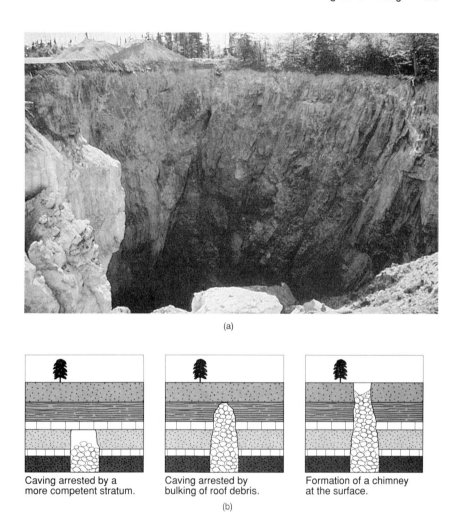

(a)

Caving arrested by a more competent stratum.   Caving arrested by bulking of roof debris.   Formation of a chimney at the surface.

(b)

Figure 7.5 Chimney subsidence. (*a*) Cave to surface at Heath Steele Mines Ltd., New Brunswick, Canada (*Photo courtesy Noranda Minerals Inc.*); (*b*) mechanism of progressive caving (Karfakis, 1986).

7.2.2 Open stoping

In open stoping (also known as *sublevel stoping* or *blasthole stoping*), the ore is broken and removed from drawpoints, leaving an open stope, often with a substantial unsupported span (Figs. 7.4*a* and 7.6). The method can be used in a massive orebody or in a tabular one having a dip steeper than the angle of repose of the broken ore, which must flow by gravity.

The dimensions of stopes and pillars are designed to match the ge-

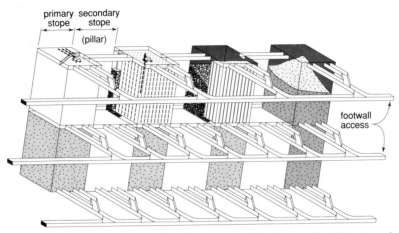

Figure 7.6 Open stope mining using the transverse blasthole method (Potvin and Hudyma, 1989).

ometry of the orebody, and many variations are possible as discussed by Potvin and Hudyma (1989). An important condition for economic open stoping is a stable hangingwall for the required spans and period of mining. Hangingwall quality governs the maximum length and height that can be permitted in any given width of open stope. In poor quality rock, open stoping is not possible.

The stope block is first developed by mining an extraction level, with drawpoints placed to ensure maximum recovery of broken ore. One or more *slot raises* are mined alongside the ends or up the middle of the ore panel, and these are developed into slots to make room for the *bulking* of ore when it is blasted. Fragmented ore occupies 20 to 50 percent more volume than the same ore in place (Sec. 7.6.2.1). Sublevels are developed to give access for blasthole drilling. The ore is then blasted progressively into the stope, and flows downward to the drawpoints for extraction. Continuous draw normally helps to minimize the risk of blockages. The ore can be drawn until the stope is empty, provided that the hangingwall is sufficiently competent to stand unsupported. When the hangingwall is less stable, the stope is left at least partially full to prevent hangingwall collapse and avoid problems of dilution and blockages at the drawpoints caused by slabs of hangingwall rock.

The method of *longhole fan drilling* (Fig. 7.4a) uses conventional small diameter holes with high explosive. Current mining technology favors the alternative of *large-diameter blastholes* drilled by in-the-hole or rotary drills (Fig. 7.6). Long, straight holes reduce stope development costs because fewer sublevels are required. Blastholes are most commonly drilled in vertical rings or fans. Less frequently, they are drilled in a downward parallel array or in a horizontal fan pattern from corner raises. The explosive consists of pneumatically loaded AN/FO or a slurry. Holes for the production blast are detonated one or two rows at a time using one of several techniques: full column loading; decking, where individual charges are separated by about 1.5 m of sand and individually primed to ensure complete detonation; delay decking using detonating cords to initiate the charge sequentially upward; or bench slicing, in which only the bottom of the slice is loaded.

Backfilling, if used at all, can be delayed until extraction in a given stope is complete. It may be needed for long-term stabilization of walls and pillars, but in any case, the open stopes provide a convenient place to dispose of waste rock and tailings. Country rock is at first supported by ore pillars left between the stopes. These are often recovered later, for example, by backfilling the stopes with cemented tailings, and then by mining the pillars; the ore becomes the stope, the fill now acts as a pillar. Alternatively, recovery may proceed without backfilling; the ore remnants are blasted into the open stopes, relying on delayed collapse of the overextended stopes and continued ability to draw fragmented ore with low dilution from beneath the collapsed, larger-sized, and less mobile waste rock.

Ground control systems for open stoping require careful design (Stillborg, 1984). The drilling drift at the upper level usually becomes the crown of the stope after blasting and may require reinforcement, not only to stabilize the drift itself (support required for this is minimal), but also to stabilize the hangingwall and prevent chimneying in the stope. Stope dimensions can sometimes be increased by installing a system of rockbolts and reinforcing cables from the crown drift and sublevels. Unfortunately, sublevel drifts provide minimal access for installation of stabilizing systems. If chimneying occurs in spite of these precautions, it can be arrested by allowing the stope to fill with broken ore and bulked waste rock or if necessary by introducing backfill. Delayed backfilling in this manner is costly because any ore subsequently mined is diluted. Upward migration of caving ground can only be tolerated if the stope is deep so that caving will not become hazardous (Sec. 7.5.2). If uncontrolled caving continues, open stoping has to be abandoned for a method that places less reliance on large-scale ground stability.

7.2.3 Room-and-pillar mining

Ore is mined from *rooms* or *entries*. Remnants of ore are left behind in a regular pattern of "pillars" to maintain stability of the immediate roof and of the workings as a whole. If the thickness of orebody is greater than about 6 m, it is worked in multiple passes, usually by mining a top slice and then a bench (Fig. 7.4*b*).

Room-and-pillar mining is the most common and least expensive method when the orebody is stratiform or lenticular and dips at no more than 30°. High-grade pockets of ore can be mined selectively, leaving behind low-grade ore as irregularly distributed pillars. However, to leave pillars in an orebody means permanent sterilization of a fully-proven and developed mining reserve.

In a room-and-pillar mine, room width is limited only by the maximum span that can stand unsupported for as long as needed to extract the ore. The room-and-pillar method works best when the ore is near surface; deeper ore means higher vertical stresses, requiring that more ore be left behind in pillars, in other words, a lower *extraction ratio* (defined as ore removed divided by original ore in place). The roof rock must be strong, and the vertical jointing widely spaced, because the major hazard is premature roof falls (Heuze and Goodman, 1971). The mined-out rooms provide a convenient disposal area for waste rock.

Room-and-pillar mining is used in nine out of ten coal mines in the United States (Bieniawski, 1983), and in almost all potash, salt, and gypsum mines. In coal mines, generally about half the coal is left behind as pillars, although an extraction ratio of 70 percent or more is achieved in about one-quarter of the mines practicing the *retreat mining* method, in which rooms and pillars are developed as mining advances but subsequently the pillars are recovered or "robbed" on retreat. Bieniawski gives typical dimensions for room-and-pillar coal mines in the United States as follows:

Depth below surface: 150 (25 to 480) m

Pillar height (seam thickness): 2 (1 to 4.5) m

Entry width (roof span): 4.8 to 6.1 (3.7 to 8.2) m

Pillar width: 15 (5 to 23) m

Pillar length: 20 (6.1 to 27) m

Width/height ratio: 8 (2 to 16)

Length/width ratio: 1.25 (1 to 3)

Extraction: 50 percent (25 to 85 percent)

In room-and-pillar potash mining in Saskatchewan, roof collapse cannot be allowed to occur because of water inflow, and a mine is often developed in panels, typically 1.6 by 1.6 km in size. Barrier pillars separate the panels, and provide protected conditions for permanent travelways. For panel development, rooms are excavated in several passes of a continuous mining machine (Fig. 7.7), leaving long continuous wide pillars to give better confining stress development within the pillars. Rooms may be developed to their full width using multiple parallel passes of a mining machine, or lateral chevron cuts from a central drift. Developed dimensions are typically as follows:

Depth below surface: 1000 m

Pillar height (seam thickness): 3.5 to 4.2 m

Pillar width: 26 to 35 m

Pillar length: >200 m, generally full panel width (1.5 km)

Width/height ratio: 8 to 10

Room width: 18 to 26 m

Pillar to room width: 1.1 to 1.6

Extraction: 28 to 48 percent

In shallow salt mining, square or rectangular pillars are commonly left behind. The Goderich mine in Ontario in a bedded salt uses drill and blast methods, pulling the entire face above an undercut slot, which ensures a flat roadway. Domal salt mines often use an initial entry development phase for a section, then continue development by one or more blasted benches. In these mines, typical development data could be:

Depth below surface: 200 to 500 m

Pillar height: 15 to 26 m

Pillar width: 20 to 35 m, often square

Width/height ratio: 0.85 to 1.5

Room width: 10 to 20 m

Pillar to room width: 1.2 to 2.0

Extraction: 30 to 50 percent (not an issue for salt domes)

To recover coal left in the wall of open-pit mines in hilly areas, parallel entries are driven into the hillside normal to the highwall (often called *contour mines*) to extract as much coal as possible, often using auger techniques. Fully automated remote controlled systems are being designed to extract more coal by deeper drives. In these cases, slope stability considerations are combined with stability considerations for the pillars, a complex interaction problem.

(a)

(b)

Figure 7.7 Potash mining in New Brunswick, Canada. (a) Marietta continuous mining machine; (b) checking roof conditions. (*Photos courtesy DPPC Mine, Sussex*)

7.2.4 Cut-and-fill stoping

Wyllie (1969) reviews mechanized cut-and-fill stoping. The method was first used in coal mining, and in 1959 was adapted to hardrock mining in Canada. In the more common *overhand cut-and-fill* method, mining advances up-dip or vertically (Fig. 7.4c). The method is suitable for dips in the range 35° to 90° and spans of 1.5 to perhaps 15 m, and can be employed in both shallow and deep orebodies. The stope crown is in ore, and the floor is on backfill, usually cemented tailings, often hydraulically placed. Miners work continuously in the stope, drilling and blasting a slice of rock about 3 m thick from the advancing crown, and replacing it with backfill. Production drilling makes use of either horizontal holes drilled into a *breast,* or *uppers* inclined at 50° to 65° from the horizontal and spaced at 1.25- to 1.50-m centers. The drills are mounted on jumbos (Dickhout, 1973). Load-haul-dump (LHD) vehicles dig the broken ore and transport it to the ore passes, from where it flows by gravity to a lower transport drift.

In softer ores, such as potash in the Potash Corporation of America mine in New Brunswick, Canada, the ore is mined by a large roadheader working on a platform of granular halite extraction wastes returned to the mine face by conveyors.

Cut-and-fill stoping is ideal for stratiform orebodies with seams of uniform thickness at high angles. This method is used at the Mount Isa Racecourse orebody, Australia, where rich lead-zinc-silver-sulphide seams are several meters thick, and dip at about 65° to 70° (Lee and Bridges, 1980). Long grouted cable bolts are used to improve roof conditions; after each blast the protruding portion is cut off.

The *post pillar cut-and-fill* method is a modification where pillars between adjacent stopes in an orebody are reduced to posts by extracting rooms in them. The posts are not recovered; with a low safety factor, they rely on the constraint of the surrounding fill for stability.

In contrast, the *undercut-and-fill* method maintains a fill crown and mining proceeds downward through the orebody. Cemented fill is supported by a mat suspended by ties from the crown pillar. Once the fill has drained, mining of the underlying slice proceeds and progresses downward in successive slices.

Cut-and-fill stoping is labor intensive and relatively slow, so for the method to be economic, the value and grade of the orebody must be high. Personnel and equipment are required for processing and handling the backfill materials. However, the method can be highly selective and it permits close control of grades because barren lenses can be left unmined.

Cut and fill is the most common method for mining narrow

orebodies, orebodies, and in weak, broken, and comparatively unstable ground. Openings of minimum span can be created and advanced while immediately filling the mined-out areas with waste rock or tailings backfill. Improved stability leads to a general improvement in ventilation and drainage, and to a reduced risk of caving to surface.

7.2.5 Shrinkage stoping

The method is similar to overhand cut and fill, except that the stope walls are supported by broken ore, not backfill. About 70 percent of the ore (bulked to a porosity of about 30 percent by blasting) is left in the stope and the remaining 30 percent is drawn from the base of the stope after each blast. The miners work directly beneath the stope crown, on top of the broken ore. Mining proceeds continuously along the length of the stope and upward, removing first one lift and then another (Fig. 7.4d). Once the stope has been mined to its full design height, the ore is drawn from below until the stope is empty or until dilution caused by stope wall collapse becomes excessive.

For shrinkage stoping, the ore must resist crushing and compaction so that it continues to flow during its prolonged residence time in the stope. The hangingwall needs to be sufficiently competent to stand with the support from the broken ore, although it can be reinforced by bolting or cabling when needed.

As in the cut-and-fill method, blast vibrations and damage to the hangingwall are minimized because small volumes are blasted at any one time. Shrinkage stoping, like cut-and-fill mining, lends itself to improved stability in weak and broken ground, not only because opening spans are kept small, but also because there is always ready access for inspection and monitoring, and for installation of bolts and other support systems when needed.

7.2.6 Vertical crater retreat (VCR) method

VCR mining also maintains the stope nearly full of broken ore but, unlike shrinkage stoping, miners do not work in the stope. Long, large-diameter, parallel blastholes are drilled downward from a sublevel. The stope is developed by taking horizontal slices from the bottom up, by a series of *cratering blasts* using short "spherical" charges concentrated in the crown region (Fig. 7.4e). After each episode of blasting, sufficient ore is drawn to provide an expansion void for the next blast, leaving the stope almost full of broken ore.

The typical procedure in VCR mining is to attach a wooden stick to the end of a rope, lower it down the hole, and pull it back against the bottom opening. A plug higher up the rope supports the bags of slurry explosive and stemming sand or water bags. A 1-m column of explosive (17 kg of AN/FO) in a 150- to 180-mm-diameter hole is typical. Initiation is by detonating cord.

The main advantage of VCR is that it requires much less stope development. It is workable in many orebodies where shrinkage stoping is not, although because of drilling inaccuracies, seam widths narrower than about 3 m may present a problem. The method is particularly suitable for pillar recovery operations in massive orebodies. Although there is more work in loading, handling, and firing of explosives, the small spherical charges tend to give a fragmentation at least as good as with small-diameter blastholes. They result in a much reduced powder factor, and keep vibrations and the need for secondary breakage to a minimum. Missed holes can be reblasted safely. In spite of these advantages, VCR blasting can, under adverse circumstances, result in damage to the walls of stopes and in dilution of the ore by waste rock.

7.2.7 Longwall mining

Longwall mining is the most popular of the caving methods, that is, those that encourage the collapse of waste rock soon after mining. Longwall mining is the preferred method in a flat-lying (less than 20° dip), uniform, and continuous stratiform orebody when a high extraction ratio is required. It accounts for almost the entire European output of deep-mined coal and is being applied increasingly in the United States, Canada, Australia, and China (Smart and Redfern, 1986). The Alsatian potash deposits in France are also mined largely by longwalling, as are the hardrock gold reefs of South Africa.

In *soft rock longwalling,* twin parallel *entries* defining a panel to be mined are driven, usually with a roadheader boring machine. The entries accommodate conveying, service, personnel, and ventilation requirements.

The longwall face is advanced between these entries by repeated passes of a mining machine. A *plough* or a *rotating drum shearer* is moved back and forth between the entries. A plough can remove only a few centimeters, whereas drum shearers can remove from 25 to 100 cm with each pass (Fig. 7.8).

An armored *face conveyor* beneath the mining machine catches and transports the broken ore to a conveyor system in one of the entries. All modern longwall faces are equipped with powered, self-advancing *hydraulic roof supports* (*chocks* or *shields*) which protect the working area, and, together with the conveyer system, are advanced to keep pace with progress of the shearer. Roof support is provided by hydraulic legs acting between a base and a roof canopy or rear shield, and

(a)

(b)

Figure 7.8 Longwall coal mining. (*a*) Typical longwall face, height 2 m; (*b*) shearer and operator. (*Photos courtesy Kelly Bentham/Bill Gallant, CANMET Cape Breton Coal Research Laboratory*)

thrust for advancing the supporting system is provided by a horizontal ram. The powered support is only active when its position is moved after a slice of coal has been taken off the face. Thereafter, it becomes passive, developing an increased resistance to convergence as the hydraulic legs absorb closure. Overload protection is provided by a pressure release system (Smart and Redfern, 1986).

The hangingwall must deflect and cave as mining advances. Otherwise, excessive pressure may be carried by the hydraulic supports or by the face, which can lead to face bursts and poor control of mining. Openings are maintained only in the active working area and for the entries (Fig. 7.9b). The ideal ground characteristics for this method are weak and readily caving rock in the immediate roof, with stronger rock above this to bridge between the mined face and the consolidating bed of caved roof rock (the *gob*). There should be no aquifer beds in the roof close enough to be breached by the caving. The seam floor must have sufficient bearing capacity to carry the hydraulic roof support system.

A characteristic of longwall mining is large-scale displacement of country rock. Access to the longwall face is maintained through entries supported by appropriately dimensioned pillars (Mark, 1990). As the gob consolidates, displacements at the level of the seam migrate upward, and consequent regional subsidence can preclude or place restrictions on use of this method in urban areas (Sec. 7.6).

Longwall mining is practiced in an *advance* or *retreat* mode. In the retreat longwall mining method, lateral entries are initially developed to the full length of the panel, and the face then retreats back to the main entries. This requires considerable development before panel mining, but the advantage is that the longwall face is never separated from the main travelways by a large collapsed area where ground control problems may develop. Ventilation is also easier. In advance longwalling, the face moves away from the main entry during mining, and entry development keeps pace. Thick safety pillars may be left between panels to guarantee good access conditions for the panel development period.

In *hard rock longwall mining,* such as used in the South African gold mines, the face is advanced by blasting. Ore is drawn by a scraper down dip into a transport gulley and from there into an ore pass. Yielding props are used near the face, and resilient timber or concrete cribs (or *sandwich packs,* Fig. 7.10 and *Rock Engineering,* Sec. 17.2.1) are constructed in the void behind the face. The resilient support is designed to permit closure under conditions of controlled energy release, reducing the risk of rockbursting.

Figure 7.9 Lingan Colliery, Nova Scotia, Canada. (a) Room and pillar entry; (b) severely deformed gateroad, supported by steel arches and timber lagging. (*Photos courtesy Bill Gallant, CANMET Cape Breton Coal Research Laboratory*)

Negligible convergence

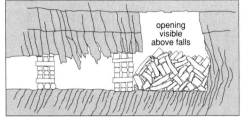

Appreciable convergence

Severe convergence

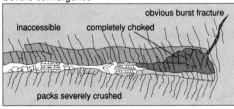

Figure 7.10 Rockbursts accompanied by varying amounts of roof-to-floor convergence in a South African hard rock longwall mining setting (Ortlepp and Steele, 1972). Note the combination of timber or sandwich pack and pipe stick support.

7.2.8 Sublevel caving

Sublevel caving was developed in Sweden at the Kiruna iron ore mine during the mid-1950s, and in Australia, Zambia, the Congo, and Canada in the 1960s. Ore is fragmented using blastholes drilled in fans from closely spaced sublevels, with the object of inducing downward flow of both the ore and the enclosing waste rock. Ore is extracted with an LHD vehicle operating at a number of drawpoints, and is conveyed to an ore pass outside the boundary of the orebody. Mining progresses downward as each sublevel is eliminated (Fig. 7.4g).

Sublevel caving is feasible only in orebodies dipping more steeply than 70°. Success depends on the fragmented ore flowing more easily (being more mobile) than the waste. The grade of ore must be sufficient to permit mining in spite of high dilution, sometimes greater than 20 percent, and poor recovery, rarely greater than 65 percent. Costs are high because 10 to 20 percent of the ore must be mined in development headings. Production and explosives costs are high, because ore must be blasted against caved waste to consolidate the waste and leave room for swelling. The potential for caving to surface may also limit application of the method.

7.2.9 Block caving

This method takes advantage of the caving that occurs naturally in fractured and weak ore and country rock following the mining of a sufficiently large undercut. An extraction level is developed beneath the panel of ore to be mined, and an undercut horizon is mined above this. Caving is triggered by the removal of pillar remnants from the undercut, and migrates upward to keep pace with extraction. Blasting is used only to start the process, after which gravity maintains the supply of ore to the drawpoints.

Block caving is an efficient method; not only are blasting costs minimal, but crushing and grinding requirements are reduced thanks to a natural grinding process as the ore moves toward the drawpoint. However, block caving is feasible only in large orebodies that have a vertical dimension greater than about 100 m. It is nonselective, so the orebody must be of uniform grade. Above all, it requires closely jointed ground with a minimum of three well-developed, closely spaced, and intersecting joint sets (Lorig et al., 1989).

Prediction of the *caving potential* (caveability) of an orebody is difficult: in some cases the ore block needs assistance to cave, and in others, it completely refuses to do so. A rock mass quality classification related to the caveability of an orebody has been proposed by Laubscher and Taylor (1976). An index in the range 1 to 100 is used to predict caveability, fragment size, the need for secondary blasting, and the undercut dimensions which are equivalent to the *hydraulic radius* of the flowing caving material.

7.2.10 In situ leaching

Leaching is removal of minerals by an aqueous solution of the leaching agent, from which the minerals can subsequently be recovered. The method can be applied either to broken ore in surface vats or to the rock in place. As conventional ore production becomes more difficult and expensive, in situ leaching becomes an increasingly attractive alternative. The method is often used to extract low-grade deposits and ore remnants from worked-out mines (Jackson, 1986).

A leaching solution is fed down drillholes into the orebody, which may be pretreated by blasting to increase the surface area exposed to the leaching medium. The metal-rich solution is then removed by pumping from a deep production well. Solutions include dilute sulphuric acid for removing oxidized copper ores, sodium hydroxide for removing alumina from bauxite, and cyanide for leaching gold. Some uranium ores are also mined in this manner.

Leaching is comparatively inexpensive and avoids the hazards associated with underground work in highly stressed, radioactive, or otherwise difficult ground. On the negative side, the method is limited to soluble minerals, and may require expensive treatments to avoid groundwater contamination.

7.3 Mine Planning and Design

7.3.1 Planning

Steps in bringing the mine into production include preliminary feasibility (conceptual) studies, detailed exploration and feasibility studies, engineering design, mine development, and mine and mill construction.

The conceptual study includes comparison of alternative stoping layouts, ore handling arrangements, equipment types and sizes, mine access locations, and the mine services required for each option. Factors to be considered include orebody configuration (size, shape, dip, and grade), geographical, environmental, and financial constraints, and characteristics of the ground. Highly fractured ground suggests caving techniques, whereas highly competent ore suggests open stoping (Folinsbee and Clarke, 1981). In the engineering study, concepts are modified and finalized; the detailed design stage provides drawings and specifications for mine construction.

Each mining method has its own set of operating characteristics including scale, production rate, selectivity, personnel and access requirements, and extraction flexibility, all of which must be considered along with safety and ground control requirements. Cash flow considerations encourage early and high rates of production. On the other hand, an excessively optimistic view of ground quality can lead to instability problems, dilution, unsafe working conditions, sterilization of reserves, and other serious consequences. Both long- and short-term objectives must be established; changes in technology and ore value may allow second-pass exploitation later in the mine life, and should be considered conceptually in planning.

The aim of the geomechanics contribution to mine design is to safeguard mine stability as a whole, including orebody and country rock; to protect the service openings; to provide secure access and safe working places in and around the centers of ore production; and to preserve the mineable condition of ore reserves.

Information on spatial variations in rock mass quality is applied to empirical and analytical design where appropriate. Three-dimensional computer graphics greatly assists in visualizing the complexities of stope geometry and geology (Fig. 7.11). The initial shafts and upper levels are monitored to check the assumptions and predictions of design, and mining methods are adjusted as the ground is opened up

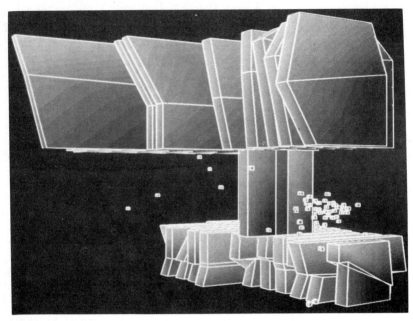

Figure 7.11 Three-dimensional computer graphics representation of Creighton Mine 400 orebody, Sudbury, Canada (Terry Wiles, Inco Ltd.). Superimposed on this geometry, in the rock around the stopes, are the microseismic events generated by the mining of Panel 3, and located by the mine's microseismic monitoring system. Only those events larger than one shot in a blast are displayed.

and experience gained. Interaction of analysis and field observations is an ongoing process in any mine, a learning process that must be recognized and carefully fostered because it leads directly to greater efficiency and safety.

7.3.2 Numerical modeling

Numerical models (*Rock Engineering,* Sec. 7.2) are useful particularly when evaluating alternative mining methods and strategies. They are valuable when used for *parametric analysis* rather than for precise predictions which can rarely be achieved because of data limitations. Mining sequences can be varied to study the consequences, and ground support, reinforcement, and backfilling alternatives can be examined (Starfield and Cundall, 1988).

In one study of a multiple-seam coal mine, *finite element modeling* showed that extraction of upper seams before lower seams could seriously affect the stability of barrier pillars, thereby leading to substantial loss of coal reserves (Su et al., 1984). Finite element modeling also helped to dimension crown pillars at the Rautuvaar iron ore mine, which uses open stoping methods (Sarkka, 1983).

In viscoplastic saltrocks, finite element models are available to simulate time-dependent behavior and to predict the creep rates around openings (Dusseault et al., 1987). High creep rates promote detachment of weakened roof slabs, a most serious design consideration. It is increasingly common to excavate a fully instrumented test chamber; large-scale and reliable rock mass properties can then be obtained by back-calculation and used for design elsewhere in the mine.

Finite element methods are ideal for analyzing a limited number of rooms and pillars or a few stopes. However, to investigate the behavior of an entire mine, *boundary integral* or *displacement discontinuity* methods are preferred. Ground behavior at the Mt. Isa lead-zinc-silver mine in Australia has been modeled using the program N-FOLD, a pseudo-three-dimensional program based on the displacement discontinuity method. Given the geometry, elastic properties of the ore and country rock, and stress field components, the model predicts behavior of the openings. It is used to determine regions of high stress and to study alternative stoping sequences, and can simulate a number of seams in the orebody (Bywater et al., 1983).

The two methods may be used interactively, the finite element method (FE) to give pillar stresses by analyzing mine cross sections, followed by the displacement discontinuity method (DD) using the FE results as input to study entire mine behavior. Entire salt and potash mines have been modeled in this manner using DD meshes of up to 85,000 elements. The mining horizon is divided into a regular mesh, and each DD element is accorded a certain behavior depending on whether it is in a pillar or a room and based on the FEM results. The method has been used for viscoplastic pillars interacting with viscoelastic roof and floor rocks.

Two- or three-dimensional analyses using *discrete element* numerical models may be used to study block interactions in complex fissured cases (Cundall, 1988; Hart et al., 1988).

The *limiting equilibrium method* is one of the best for studying the gravitational block movements that predominate close to surface. Hoek (1989) describes, complete with program listing, a simple yet effective limiting equilibrium analysis for examining the stability of surface crown pillars. Designed to permit a sensitivity study, the program gives a range of factors of safety and probabilities of failure corresponding to variations in material properties, in situ stresses, and groundwater conditions.

7.3.3 Physical modeling

Numerical methods are becoming more powerful, more useful, and more realistic day by day. Nevertheless, and in spite of their time-consuming construction requirements, physical models remain useful in their ability to explore mechanisms of rupture and instability. Large two-dimensional models with control of boundary loads are reasonable simulations of complex ground behavior (*Rock Engineering,* Sec. 7.3.2).

Physical models have been used effectively for design of West Germany's longwall coal mining operations. A 1:10 scale model of the longwall face gives information on the behavior of the strata and supports under simulated rock pressures. It aids mine planning by helping to determine favorable positions for cross-cuts, base roads, rise headings, and gateroads; the most suitable sequence of workings; the correct direction of advance; the dimension of solid coal between two panels; the behavior of protective seams; and, the optimum layout of protective workings (Grotowski and Irresberger, 1984).

7.3.4 Empirical methods

Empirical and observational methods are used almost exclusively for day-to-day design of stope geometries and ground control systems, because few mines have sufficient lead time to complete numerical analyses stope by stope. Opening geometries and excavation and support practices are adjusted day by day to accommodate the variations in rock quality and ground stress that occur at different levels and locations. Modifications to mining methods are thus introduced gradually under conditions of close monitoring and control.

As discussed in Chap. 5 and *Rock Engineering,* Sec. 7.1.5, judgment and experience are major ingredients in rock engineering design. Empirical design solves the problem of our own limited experience by making available the accumulated experience of others. It requires three steps beyond simple judgment: description of *ground quality* by a quantitative classification system, a description of *ground performance* by a formalized quantitative system which defines such parameters as unsupported stand time and maximum unsupported span, and a *correlation* of ground quality to performance by observations in a variety of mines over a full spectrum of ground conditions (case histories).

Many empirical design alternatives have been proposed, notably the *Q and RMR systems,* generally developed for tunneling and only then adapted to mining. These are reviewed in Chap. 5. In spite of their limitations, empirical predictions often give answers closer to the "truth" than the apparently more precise predictions of numerical modeling. The methods are in a state of evolution, and many mines employ their own variations developed to suit local ground conditions and specific mining methods. A particularly useful method-specific approach is the *caveability classification* of Laubscher and Taylor

(1976), discussed in Sec. 7.2.9. The authors have employed a simple *size-strength system* of rock mass quality classification, which can readily be adapted to suit local conditions (Franklin, 1986).

7.3.5 Room-and-pillar design

Room-and-pillar mines provide an example of the application of an empirical approach with simple rock mechanics calculations. When mining coal, potash, and metals, economic incentives exist to not leave behind substantial pillars of ore. Hence, every effort is made to increase the extraction ratio by reducing the size of pillars and increasing the spans of rooms.

Figure 7.12 gives an empirical prediction of the unsupported stand-up time and recommended roof span for coal mines as a function of Rock mass rating (RMR, Sec. 5.5.3.4). For example, with RMR = 40, the maximum span is 6 m, and if the span is reduced to 1.9 m, no support is required. U.S. federal law (CFR 30) stipulates that the span should not exceed 6 m with roofbolting or 9 m when using a combination of roofbolts and other supports such as timber posts. This law also specifies a minimum roofbolt length and maximum spacing of 1.05 m and 1.5 m, respectively.

The average pillar stress (MPa) (Bieniawski 1983) can be calculated from:

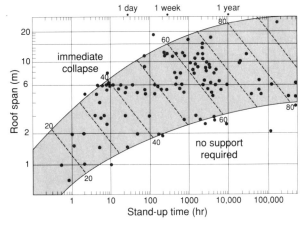

Figure 7.12 RMR classification and empirical design (Bieniawski, 1983). The shaded portion of the diagram represents 58 coal mining cases in the United States. Contour lines show Bieniawski's Rock Mass Rating RMR and the limits of applicability.

$$S = \frac{0.025\,H\,(w+B)(L+B)}{(wL)}$$

Where H = depth (m)
B = roof span (m)
w = pillar width (m)
L = pillar length (m)

This assumes that the vertical overburden pressure increases at a rate of 0.025 MPa/m. For square pillars with $w = L$, the equation reduces to

$$S = 0.025\,H \left(\frac{(w+B)}{w}\right)^2$$

The extraction ratio (%) for square pillars is given by

$$e = 100\left[1 - \left(\frac{w}{w+B}\right)^2\right]$$

Bieniawski recommends use of the following empirical equation for determining pillar strength in coal:

$$P = C\left(\frac{D}{h}\right)^{0.5}\left(0.64 + 0.36\frac{w}{h}\right)$$

Where P = pillar strength
C = uniaxial compressive strength of coal cubes, or cylinders of diameter 50 to 100 mm
D = specimen side length or diameter
h = pillar height, greater than 1 m in the calculations
w = pillar width

The term $(D/h)^{0.5}$ accounts for the size effect (Hustrulid, 1976), whereas the last term accounts for the shape effect, given as a function of pillar width-to-height ratio. Pillars become increasingly strong as their *width-to-height ratio* increases, and are thought to be almost indestructible for width-to-height ratios greater than 10 (Cook and Hood, 1978). This is because of the confining triaxial stress effect where the pillar meets the roof and floor, transmitted to the interior of the pillar throughout its height.

Bieniawski recommends designing to a *safety factor* of between 1.5 and 2.0, but the safety factor depends on the pillar dimensioning formula used (Mrugala and Belesky, 1989).

Fine tuning of design requires that the stress and strain levels in

Figure 7.13 Mine pillar in salt, reduced by stress-induced spalling. (*Photo courtesy U.S. National Committee on Tunneling Technology*)

pillars be monitored. Visual observation is not enough, because pillar or roof rock failures can be sudden, sometimes explosive. Bursting pillars shed their load to adjacent pillars, and in the worst case this can develop into a chain reaction leading to the collapse of a whole section of the mine. Brittle failures are more likely when pillars are slender (width-to-height ratio less than 0.6) because these tend to develop axial splitting and buckle, whereas squat pillars tend to yield and slab as large lateral strains develop (Fig. 7.13).

A comprehensive review of pillar design methods applied to longwall mining is provided by Mark (1990).

7.3.6 Predictions of water inflow

Although most mines are relatively dry, inflows can be a major concern in some underground operations. Predictions of water seepage are required as part of an initial mine plan, for the design of pumping systems or remedial grouting and drainage. Heavy inflows are often encountered during shaft sinking, in the upper more weathered rocks and at the boundary with the soil overburden (*Rock Engineering*, Chap. 18).

Winter et al. (1984) have reviewed over 300 computer models for their applicability to modeling underground mine water inflow. They chose UNSATZ, a finite element model, for its versatility, and used this technique to predict shaft and drift inflows at a uranium mine in New Mexico. Another method used at a lead mine in Missouri simulated nearly 16 years of water inflow quite accurately.

Clearly, inflows of water or unsaturated brine can be particularly hazardous in saltrock mines. Water-bearing strata overlying the mine must be identified during the initial exploration, and a mining strategy designed to minimize the differential deformations that could lead to inflows. Some potash mines exploit horizons where solution cavities or washouts developed in geological times; these often contain a core of granular rubble (pipe breccia) in hydraulic communication with overlying water-bearing strata (Barr, 1974). Careful exploration and planning will leave safety zones to prevent penetration of these features.

7.4 Mining Procedures and Equipment

7.4.1 Excavation and handling

Techniques of blasting employed in underground mining are discussed in *Rock Engineering,* Chap. 13, and boring machines are discussed in Chap. 15. Trends are toward increased mechanization, continuous mining or mucking, and remote control methods, which contribute toward both safety and productivity (Waddams, 1984).

Batch handling equipment is normally used to load and transport ore (Clarke, 1985). Loading equipment ranges from hand shovel to air and electric slushers, tracked and tired mucking machines, and load-haul-dump (LHD) vehicles. Trucks with a 36-t capacity are being introduced in some mines, and rail-mounted battery and trolley locomotives are also used. *Continuous handling systems* employ specialized loading equipment such as vibrating lip muckers, extensible conveyors, automatically adjusting belt benders, and low-profile belt storage units. Crushing and screening are often carried out underground to permit easier conveyance by belt, and waste rock can be rejected directly to abandoned stopes.

7.4.2 Shaft sinking

Shafts can be sunk either by blasting or by *blind boring*. Blasting is usually less expensive for shorter shafts, although raise boring can be more efficient when access is available from below.

Shaft boring, an extension of deep oil-well drilling technology (*Rock Engineering,* Chap. 15), makes use of a rotating cutter head to fragment the rock (Fig. 7.14). Reverse circulation is necessary to impart sufficient mud velocity to remove cuttings. Drilling mud supports the lower part or all of the shaft walls, cools the cutter head, and removes

Underground Mining 345

(a)

(b)

Figure 7.14 Shaft sinking. (a) Large cutter head armed with roller tools for blind boring; (b) hoisting a large steel liner to be lowered into the shaft. (*Photos courtesy Alberta Oil Sands Technology and Research Authority*)

the cuttings. At the ground surface, cuttings are separated from the drilling fluid, which is then recycled.

For *shaft blasting,* two techniques are available, benching and full face (*Rock Engineering,* Fig. 13.11). In the *benching method,* alternate sides at the base of the shaft are removed one after the other. The method works well in wet conditions because a sump is provided automatically. The *full-face method* makes use of an angle cut; deeper rounds can be pulled, but more overbreak is likely.

A compromise has to be struck between depth of round and overbreak. Increasing the depth of round gives a more rapid rate of advance, but also more overbreak, more mucking, more difficulty in maintaining accurate alignment, and a greater volume of concrete to be poured if the shaft is to be lined.

Conditions are usually wet in shaft sinking, so AN/FO explosives are almost never used. In this case, explosive slurries have for the most part replaced dynamites, and electrical firing is required for safety.

7.4.3 Raise blasting and boring

The three methods of raise blasting are *conventional raising,* in which short rounds are drilled upward by miners using ladders and working from a timber platform; a *mechanical climber method,* in which the platform and ladders are replaced by a hoist to transport miners and supplies; and *drop raising,* in which longholes are drilled the full length of the raise and then blasted either in their full length or in rounds advanced from the bottom upward.

The conventional raise can be fully timbered (cribbed) or lined for permanent access in poor ground or, in more stable ground, can be driven "bald" using no support. All blasting is electric, because a safety fuse may give insufficient time for miners to get out of the raise before the round goes off. Cartridge explosives are preferred to bulk-loaded ones such as AN/FO, because they are easier to transport and load; slurries are now used more often than dynamite.

The advantages of conventional raising are low capital cost, greater flexibility, and easy timbering for good ground control. However, the rate of advance is slow. A dead-end raise is difficult to ventilate, but this can be improved by drilling a pilot hole, which also helps keep the raise on track and reduces the need for surveying. Scaling in raises is hazardous and requires skilled miners. Costs are high if much timber is to be placed, and higher still if the timber must later be removed to install a more permanent liner.

Two types of mechanical climbers are in use in North America, the Alimak, which rides on rails attached to the hangingwall, and the

Joralift, which is hoisted on a wire rope that passes through a pilot hole. The Joralift is limited to short raises, whereas the Alimak can be used to drive blind raises long distances. Mechanical climbers mean less physical climbing and hauling, and greater protection for miners. Furthermore, rockbolting and timbering requirements are reduced because of the superior wall condition. The disadvantages include greater capital cost, long setting up and tearing down times, and the possibility of damage to the Alimak climbing rails during blasting.

The preferred drop raising method is to blast from the bottom up, in a manner similar to VCR mining (Sec. 7.2.6). Raise length is limited by the accuracy of drilling; special drills have been developed to allow a hole pattern to be drilled from one setup with increased precision. Large-diameter holes can now be driven 30 to 45 m. The method is safe, inexpensive, and rapid.

Raise boring is a safe procedure, and when ground stresses are low to moderate, the smooth walls of a bored raise provide great stability. They make superior chutes for ore and backfill, have good airflow properties for ventilation, and when extra stability is needed, they can be lined easily. However, blasting can give better results in highly stressed ground because it fractures and "softens" the rock around the shaft. The resulting more compliant rock near the opening deforms more readily under stress, moving the stress concentration outward into the more confined rock mass. Also, a prefractured rock is much less capable of storing the large amounts of strain energy, which can lead to bursting or high rates of spalling. Disadvantages of raise boring include high capital and operating costs and considerable preparatory work.

7.5 Rock Behavior and Ground Control

7.5.1 Mechanisms of instability

Five categories of underground rock behavior are:

Stable conditions in which the rock has reached a state of equilibrium. Rock displacements continue at decreasing rates and without damage to supporting systems.

Gravitational failures in which the roof and sometimes also the walls progressively collapse by raveling or caving (i.e., the loosening and falling of blocks). The blocks become detached by sliding along their bounding joints without much distortion or rupture of the rock.

Squeezing ground conditions, in which the crown, the sidewalls, and sometimes the invert of the excavation converge slowly and continuously by mechanisms of stress-induced viscoplastic flow. There is substantial distortion of intact rock material, often accompanied by creep along joints within the zone of overstressed rock mass.

Swelling ground conditions in which the rocks exposed near the excavation walls expand by physico-chemical mechanisms associated with the adsorption of water by clay minerals or anhydrite. Note that swelling is the result of mineralogical changes, whereas squeezing is caused by overstressing.

Bursting ground conditions, in which the rock fails explosively by propagation of fractures through intact rock. Stored energy is released suddenly and violently. Included in this category are gas outbursts, which occur as the violent ejection of gases and rock into an underground opening.

7.5.2 Gravitational raveling and caving

Except when mining by caving methods, initiation of raveling must be prevented; otherwise it develops upward and outward (Figs. 7.15 and 7.5*b*). Hangingwall or pillar collapse can lead to the open stope expanding to a width greater than the thickness of orebody, with con-

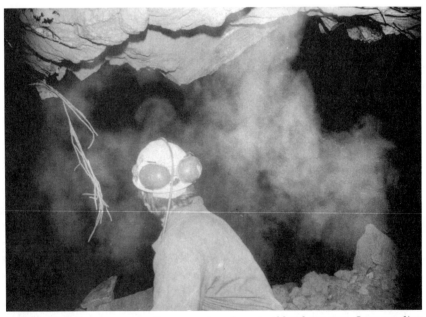

Figure 7.15 View across a raveling open stope from a sublevel ore pass. Once raveling starts, it may be difficult to observe directly, let alone to stabilize. It can usually be stopped only by natural bulking or arching or by rapid emplacement of backfill.

sequent dilution of the ore. This can be followed by upward chimneying that bypasses the reinforced crown of the stope, affecting and even immobilizing ore blocks at levels yet to be mined.

Gravitational raveling and chimneying are of greatest concern when they affect the *surface crown pillar* of shallow mine workings (the ledge of rock separating the crown of the stope from the overlying soil or ground surface). Caves to surface are not uncommon in shallow mining operations (Fig. 7.5). They present a serious hazard both at the surface and underground. They are most hazardous when mining below bodies of water or below water-laden soils that can liquify and flow into the mine (see also Sec. 7.6.2).

Betournay (1989) gives a comprehensive review and bibliography of surface crown pillar stability. Variables that govern arching and caving, discussed in the same conference proceedings by Franklin et al. (1989), include:

Properties of the rock: Raveling is most likely with small blocks and a large stope, with smooth, clay-coated joints, and with several joint sets.

Stress conditions: Raveling is most likely under conditions of low horizontal stress, and in deformable rocks that show little tendency to arch above the stope.

Groundwater conditions: Collapses are sometimes triggered by cracking of upper beds as a result of subsidence, throwing sudden water pressures on bedded roof strata (Maury, 1979). Erosion by downward percolating groundwater can also contribute to raveling.

Geometry of the stope: Initiation of raveling seems to depend on the narrowest dimension of the stope and not much on its length or height.

Stability is achieved in service and development openings by combinations of rockbolts, mesh, and if necessary, shotcrete. Rock reinforcement in the hangingwalls of stopes requires longer grouted cable anchors (Fig. 7.16), but because of access problems, they are not always possible or economical to install. Brady (1989) discusses optimum cable bolting patterns. In spite of improvements in cable bolting technology, basic gravitational stability is ensured primarily by an appropriate mining method and span width, by leaving pillars, or by using backfill support systems.

7.5.3 Squeezing conditions

7.5.3.1 Generalized squeezing.
In service and development openings that show a pattern of continuing convergence, long rock dowels are

(a)

(b)

Figure 7.16 Mount Isa Mine, Australia: graphite-coated bedding planes contributed to a 250-t wedge fall. The hangingwall was stabilized by rockbolts with steel straps, and the back by preplaced grouted cables, plus bolts and mesh. (a) View down the stope; the crown pillar is only 3 m thick here; (b) local roof support.

installed a few at a time until movements stop. Combinations of bolts, shotcrete, mesh and lightweight flexible steel ribs, as in the NATM method (Sec. 5.5.1.3), provide the best support. Convergence of as much as 10 percent (1 m in an opening 10 m in diameter) has been recorded in Alpine and Himalayan tunnels (Sec. 5.5.2.5). Here circumferential "windows" were left between panels of shotcrete to allow the rock to squeeze inward. Converging rock was trimmed from between the panels. Arching, assisted by bolting, resulted in eventual stabilization.

Squeezing ground conditions are normal for mine openings in deep underground potash and rocksalt deposits, and the rate of creep is controlled by optimizing long pillar and room widths (Mraz and Dusseault, 1986) and by use of yielding supports such as timber cribs. In the *stress control method,* the mining geometry is laid out and the sequence is adjusted so that active rooms are yielding slowly and safely whereas abandoned mined-out areas are progressing toward complete closure. Roofbolts are used to protect highly traveled drifts and to support roof slabs that have developed because of clay seams or rapid early closure.

The stress control method originated in coal mining in Germany and England with the use of "yield pillars" and "bearing pillars," and was adapted for potash mines in Saskatchewan, Canada (Barr, 1977). It is most commonly used to protect travelways, but can also be modified for development purposes. Outside rooms are cut first in a three- to five-room entry system. The central protected rooms are excavated 1 or several months later. Yielding pillars between the rooms deform in a controllable manner and are designed to carry the immediate roof, whereas the overburden loads are redistributed to the bearing pillars, which have excellent capacity because of the large width-to-height ratio (Mraz, 1984).

7.5.3.2 Floor heave. Many coal mines experience floor heave problems (Fig. 7.17). Slow heave requires only additional maintenance of roadway and coal handling equipment, but rapid heaving on a large scale risks the safety of mine personnel. Haramy and McDonnell (1986) describe a longwall coal mine 900 m deep with a floor of strong rock 1 to 3 m thick underlain by a coal bed. A major heave developed for a distance of 365 m along the tailgate and 91 m along the face, disrupting ventilation, stopping production, and damaging more than 40 longwall shield supports. It raised the floor to within 30 cm of the roof in some areas.

An early trial of a mechanized longwall in a synclinal coal seam in the Alberta Rocky Mountain foothills had to be abandoned because of a combination of floor heave and squeezing. The lateral tectonic

(a)

(b)

Figure 7.17 Floor heave at Bogdanka colliery, Poland. (a) Rupture of V-36 steel arches in the floor; (b) installation of supplementary lightweight OZW/V-29 steel invert arches to prevent floor heave in a roadway at depth 955 m. (Photos, B. Kozek, KPWWK Bogdanka and P. Gluch, Politechnika Slaska, Gliwice, courtesy M. Kwasniewski)

stresses were probably excessive for the design of the ground control system, and eventually the entire longwall with its equipment had to be abandoned.

Methods suggested to control heave include installing rockbolts in the floor, reorienting mine entries relative to the horizontal stress fields, and cutting slots in the floor to allow expansion. Heave has been controlled in the coal mines of France by installing resin-grouted wooden dowels that can be trimmed occasionally along with the floor rock itself to maintain the height of opening. In active longwall faces, use of full shields that almost entirely cover the floor and roof can eliminate heave, but floor heave in entries may continue to be a problem.

7.5.4 Rockbursting

7.5.4.1 Nature and history of bursting. A *rockburst* (or *burst*) is a sudden and violent rupture and expulsion of rock from the surface of an excavation (Figs. 7.10 and 7.18). Bursts occur in mine pillars or headings or in the roof or floor, occasionally even in the highly stressed floor of an open excavation. Rockbursts give little or no warning, and present a considerable threat to life and production. A single major

Figure 7.18 Rockburst in pillar wall, Heath Steele Mines Ltd., New Brunswick, Canada. (*Photo courtesy Noranda Minerals Inc.*)

rockburst in 1964 caused so much damage that a mine in Canada had to be closed (Blake, 1972).

Rockbursts are the extreme of mining-induced seismicity. Often their magnitudes approach those of moderate or even major earthquakes. Historical records show that a burst in the old tin workings at Altenberg, Saxony, registered as an earth tremor 45 km away in Dresden (Ortlepp and Steele, 1972). Rockbursts in the Sudbury and Kirkland Lake areas of Canada have reached Richter magnitudes of 5. In the North Staffordshire coalfield of England beneath densely populated Stoke-on-Trent, extraction has been accompanied by seismic activity which reached Richter magnitudes of 3.5 at the surface. In South African gold mines, Richter magnitude 4 events are common. Only a small fraction of the detectable tremors induced by mining are actually experienced as rockbursts in the underground workings.

In addition to earth tremors, severe airblasts can be generated by the sudden complete closure of an excavation when one or several pillars fail simultaneously, sometimes in a chain reaction triggered by the initial rockburst (Fig. 7.10).

Bursts became a problem in 1898 in the Kolar gold field of India, and in the gold mines of South Africa in the early 1900s. Problems developed from about 1930 in the hard rock mines of Canada. In Ontario mines between 1935 and 1978, bursts occurred at a rate of 20 to more than 300 per year, of which between 10 and 50 per year were heavy rockbursts (greater than 50 tons of rock displaced); 22 of the 208 fatalities caused by falls of ground between 1960 and 1979 were the result of bursts. During 1984 there were 105 rockbursts of magnitude 1.5 to 4.0 on the Mercalli scale; 82 percent could be classified as pillar bursts with the remaining 18 percent equally divided between strain and fault-slip bursts (Scoble, 1986).

Coal mine bumps associated with sudden pillar yield (different from gas outbursts, discussed later) resulted in at least 14 deaths between 1959 and 1984 in eastern U.S. coal mines (Iannacchione et al., 1987). The deep silver mines in Idaho are prone to bursting, and pillars in the Mt. Isa Mine in Australia have experienced intense damage from bursting triggered by a nearby production blast. One such pillar burst was considered as beneficial in the long run; the mine was empty, no resources were lost, and a large volume of potential ore was "conditioned" by the burst, leading to easier pillar mining. In general, however, rockbursts present a major danger to mine personnel and workings.

As mining penetrates deeper and extraction ratios increase, bursting tends to become more frequent and more severe. This tendency is fortunately being offset by improvements in ground control and safety practices.

7.5.4.2 South African mining experience. In South Africa between 1918 and 1931, fatalities from rockbursts ranged from about 20 to 60 per year. The average injury rate in mines has declined dramatically in recent years, but rockbursts and rockfalls continue to account for more than half of all mining fatalities (Chamber of Mines, 1977, 1988).

In the shallower mines, the regional mining span is limited to no more than one quarter of the mining depth, to reduce the risk of stope collapse. This is achieved by a system of appropriately spaced regional pillars, with intermediate yield ("crush") pillars to stabilize the stope hangingwall.

Gold from the slightly inclined orebody of the deep-level mines is extracted by longwall mining, which tends to eliminate stress concentrations associated with pillars. Highly stressed *remnants* must be avoided. As early as 1908, the South African Chamber of Mines recommended replacing solid pillar support with deformable waste packs. The types and arrangement of supports (rapidly yielding hydraulic props and timber or sandwich packs) are selected according to the predicted rates of energy release. The layout is planned so that stoping through faults and dikes releases energy as slowly as possible.

Energy release rates are reduced by regional support systems with stabilizing pillars or backfill or both. In a partial extraction method, the reef is mined in long parallel panels separated by thin strip pillars perpendicular to the face. Their width-to-height ratio is typically about 20, hence confinement is excellent. The stabilizing pillars become highly stressed and can never be mined safely. They occupy between 10 and 20 percent of the reef area, this percentage increasing with increasing depth below surface and decreasing rock strength. Such pillars can reduce the amount of energy released by between 65 and 75 percent (Salamon, 1974; Chamber of Mines, 1977, 1988).

Computer simulation has developed into a successful design tool, and linear elastic models are used routinely to calculate stresses and rates of energy release (Salamon 1974). Rockburst damage is closely correlated with the energy released per unit of mined rock. Longwall stopes are represented as narrow slots using the boundary element method, and alternative layouts and sequences are compared to select the least hazardous approach.

7.5.4.3 Causes and treatments. Rupture occurs when the stress level exceeds the strength of an element of rock. Whether this is violent or gradual (brittle rockbursting or ductile squeezing) depends on the strength and brittleness of the rock and on whether the element is stiffer or more flexible than the "loading system." The effect is similar to the bursting experienced by a rock specimen tested in the laboratory. Here, the loading system is a flexible testing machine which releases its stored energy to the specimen. In a mine, the loading system is the country rock which releases its stress to the near-field rock as the latter yields.

The rockburst hazard can be reduced by keeping the total energy change as small as possible; this may be achieved by backfilling, by partial extraction, or by adopting a mining method, such as the longwall method, that releases energy uniformly in space and time. Damage can be mitigated by employing yielding supports or backfill to limit load transfer to pillars, and the overstressed rock can be "conditioned" by preblasting to reduce its stiffness.

7.5.4.4 Preconditioning of the rock. Preconditioning entails blasting ahead of an advancing face. The aim is to fracture the rock mass, encouraging it to behave in a less brittle and more ductile manner. Blasting enhances the buffer zone of broken and yielding rock around an opening, which forms naturally under conditions of high ground stress. Stress concentrations are displaced deeper into the rock mass, where the rock is confined and less prone to catastrophic bursting. A buffer of broken rock separates the advancing zone of active fracturing from miners in the stopes.

Controversy still surrounds the effectiveness of this technique. Blake (1972) reported successful field trials in the United States; however, in 1977 the South African Chamber of Mines advised against preconditioning. Theory and seismic observations had indicated that the treatment produced little if any "softening," release of energy, or convergence. Brady and Brown (1985) remarked that the 3-m destressed zone did not provide a large enough buffer; to be effective, destressing would have to be implemented on a massive and costly scale. However, the principle remains attractive, particularly if the costs are partially recovered by reduced blasting and crushing during production.

7.5.4.5 Yielding support. The most effective way of limiting rockburst damage is to provide active support with rapidly yielding hydraulic props near the face, with crushable packs or with backfill in the void behind.

Hydraulic props are designed to yield faster than 1 m/s to accommodate rapid convergence between the hangingwall and footwall, while maintaining a steady support load. They can be installed as close as 1.5 m from the face without being damaged by the blast. *Pipe sticks* (timber props encased in a steel tube, with the wood protruding to give increased resilience) provide a less expensive but still effective alternative in some mines.

Backfilling is highly effective in reducing the incidence of dynamic rock instability (Jager et al., 1987). Sand or waste rock backfill can reduce convergence of a longwall stope by up to 50 percent. In some conditions, a fill of about 20 percent of the mined out area produces about 93 percent of the effect of a complete fill of the same density (Ortlepp and Steel 1972). The width-to-height ratio of the waste ribs must be sufficient to permit excellent confinement to develop; otherwise the backfill rib will simply be squeezed out into the open stope.

In South African practice, *rock-filled timber cribs* have now been replaced by *sandwich packs* made from alternating beams of timber and precast concrete. They are initially stiff and rapid-bearing but ultimately yielding, giving controlled yield up to 30 to 40 percent compression. They are also more resistant to blasting and fires than conventional timber packs.

In more permanent mine service and development openings, *yielding eyebolts* combined with *steel mesh* and *cable lacing* provide a highly resilient form of stabilization in conditions of dynamic loading (Fig. 7.19). Davidge et al. (1988) give specifications for typical lacing. Eyebolts are manufactured from 14-mm mild steel round stock, with the eye hot-formed to an internal diameter of 38 mm, and with two

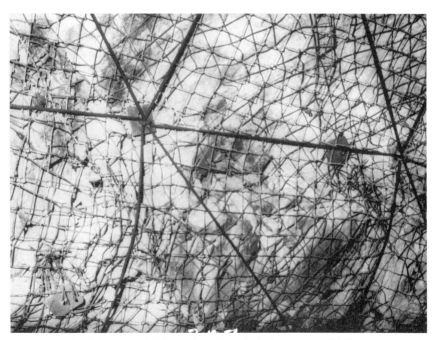

Figure 7.19 Cable lacing with bolts and mesh, typical of support used in burst-prone rock.

full legs 2 to 3 m long. The cable is a 1 by 19 classification flexible wire rope with a 13-mm nominal diameter. Galvanized woven wire mesh (chain link) is used in preference to welded mesh, because rockbursts tend to break welds. The bolts are installed in a 1.8-m staggered pattern, giving an effective spacing of 1.3 m and a coverage of 1.7 m^2 per bolt. The eyebolts are grouted into 32-mm drillholes using a water to cement ratio of 0.43 by weight. Three cables pass through each eye, so that no diagonals are omitted.

7.5.4.6 Microseismic monitoring. Microseismic monitoring is widely used to record the acoustic activity of the ground around mine openings. Data are recorded in terms of the frequency of occurrence of seismic and microseismic events or as complete wave trains. An array of several geophones can be used to locate the sources of seismic events (Leighton and Blake, 1970; Blake, 1975; Bourbonnais, 1984; Young and Franklin, 1990).

Most microseismic monitoring has until recently been aimed at predicting rockburst events. The philosophy behind this approach is that a buildup or change in the spatial or temporal pattern of microseismic activity can be used to give warning of a major burst. Documented successful examples can be found for coal outbursts in the United States and some deep mines in South Africa. On the whole, however, attempts at prediction have met with limited success, often resulting in unwarranted anxiety underground and in false alarms. A review by Blake (1982) showed that only in about 30 percent of cases did some form of precursor phenomenon precede a major rockburst, whereas 60 percent of recognizable precursor patterns were not followed by a rockburst.

An alternative approach nowadays is to use microseismic monitoring for the insight it provides into rock stress redistribution processes. It is used to help identify danger zones, without expecting it to give a reliable warning of individual events. The mine engineering department can then develop methods for prevention or at least minimization of the risk of rockbursts, such as by redesigning mine opening geometries, mining methods, and sequences or by softening or destressing the rock (Blake 1987).

Three levels, or scales, of monitoring system are commonly defined. The *seismographic system* is similar to those used in nationwide seismic networks, often with just one single-axis vertical sensor. The sensitivity threshold is such that earthquakes and bursts of Richter magnitude greater than about 0.0 are detected, but routine blasting is not. The *minewide microseismic system* has widely spaced sensors to cover an area of up to 5 km^2 around the mine. It will detect and locate, within about 30 m, events whose magnitudes are similar to those of routine blasting or larger. These systems are useful for determining the location of large events and pointing out where to look for damage or casualties. An auxiliary use is the identification of unstable areas that warrant further instrumentation and for whole waveform analysis of large magnitude events. A *local microseismic system* has a sensor spacing as close as 10 to 50 m, and is usually designed to monitor one or two stopes, up to a few hundred meters in extent.

7.5.5 Gas outbursts

7.5.5.1 Mechanisms and examples. The presence of gas in *coal deposits* clearly aggravates the problems of bursting under conditions of high ground stress. Coal and associated sandstones usually contain significant amounts of adsorbed methane. Khristianovich and Salganik (1983) describe an outburst in the Soviet Union that had 8400 m^3 of methane in only 40 t of coal. If a burst occurs, a large surface area is created suddenly because of fragmentation, and this increases the release of gas, aiding the gas-driven outbursting process. Gas is present not only in the coal seam being exploited, but also in roof and floor rocks.

Gas bursts in coal mines are becoming more severe because of increasing depths of mining, which lead to greater amounts of methane in solution. Investigations in Japan indicate that the gas content of coal seams that are prone to bursting amounts to 50 to 100 m^3/t.

In the Maritime provinces of Canada, coal mining operations have experienced outburst and bump activity at depths greater than 500 m. In October 1958, a major tragedy occurred at the Springhill mine, where a sequence of bumps culminated in the destruction of one entire production district and loss of 75 lives (Notley, 1984). A gas outburst in Japan in 1981 at the New Yubari Mine took 93 lives (Hiramatsu et al., 1983; Paterson, 1986). Hanes et al. (1983) report more than 450 coal and gas outbursts in Australia since 1895, the largest having displaced more than 1000 t of material. Most were associated with methane but some also with carbon dioxide.

In domal *salt mines* such as the Five Islands mines in Louisiana, petroleum and natural gas are found in traps on the flanks of the dome (Kelsall and Nelson, 1985). Small amounts of gas and light oil permeate into the salt over geological time, where they are stored in the intercrystalline moisture, or in fluid inclusions. Production blasts release the gas from solution, in the form of a *saltburst*.

7.5.5.2 Preventing gas outbursts. Outbursts can be alleviated by draining the gas, but only if the area can be drained for about 3 to 4 months and if the gas pressures are lowered to below 150 kPa in the danger zones. This requires either vertical wells from the surface or horizontal boreholes from the workings. *Hydraulic fracturing* can be used to enhance the drainage radius of wells. It is advantageous to inject a proppant, such as coarse-grained silica sand, to keep the fracture open.

The slow desorption of methane impedes rapid drainage, and pockets may be left behind. As gas bubbles evolve and grow in wet coal, they block flow channels, further reducing the effective permeability to both liquid and gaseous phases.

7.5.6 Mining with backfill

7.5.6.1 Functions of backfill.

Backfill is nearly always used where the mining method is other than by caving. Hydraulic filling was probably first employed in the coal mines of Pennsylvania in 1864 (Dickhout 1973). Stabilization of fill by the addition of portland cement was in common use by the mid 1960s. Since then, backfill technology has been steadily improving and has resulted in better ground control, decreased mining costs, and more efficient ventilation (Stephansson and Jones, 1981; Brawner, 1983).

Backfill serves to provide support, to control dilution, and to control subsidence on a larger scale. Although not strong in itself, fill maintains the integrity, and thus the load carrying capacity, of rock pillars and walls. Overall closure is reduced, joints are prevented from dilating and thus losing shearing resistance, time-dependent effects are reduced, and rockburst hazards are diminished (Stacey and Kirsten, 1987).

Secondary functions include provision of a working floor, fire control, improvement of mine ventilation, and disposal of tailings and broken rock wastes. Environmentally acceptable waste disposal is a matter of increasing public concern. Compared with timber and other forms of support, a filled stope is almost impervious to air flow, so more of the available air is directed to the working faces. Cemented tailings fill is superior in this respect to rockfill, which allows air to circulate easily within it, increasing overall ventilation requirements.

Pillar recovery from previously backfilled stopes can expose the backfill to substantial heights; therefore, the fill must be strong enough to stand unsupported; if fill collapses by slabbing, it dilutes the ore. In the Falconbridge nickel mines of Ontario, walls up to 76 m high of fill consisting of one part cement to 32 parts of tailings have been exposed and remained so for several months without collapse (Moruzi, 1978). Methods of analyzing the stability of freestanding pillars are reviewed by Brady (1989).

7.5.6.2 Categories of fill. Fills can be made of *mine waste, quarried and crushed rock,* or *natural sand,* but more often consist of *mill tailings* which are transported to the stope as a *hydraulic fill* slurry. A survey by the Canadian Advisory Committee on Rock Mechanics indicated that the 10.8 million tons of fill reported as placed in 1971 comprised 76 percent hydraulic mill tailings, 13 percent rock, 8 percent hydraulic alluvium, and 3 percent miscellaneous materials such as dry sand; 228,000 tons of cement were added to stabilize this fill.

Halite mill wastes are used as backfill in potash mines (Figs. 7.7 and 7.20). The granular halite compacts rapidly by creeping until the porosity drops to about 5 percent. It will not support significant loads until this point (Fordham et al., 1989). The design approach in potash mines is to carefully backfill rooms with a minimum of void space, and to accelerate closure to permit second pass mining. Acceleration of room closure may be achieved by pillar splitting or by pillar robbing in the most recently developed drift adjacent to a backfilled area. In Canadian mining operations, use of halite backfill is expected to increase the extraction ratio from 40 to more than 65 percent.

7.5.6.3 Fill placement. Rock fill requires a crushing plant, hauling equipment, and a shaft at least 2 m in diameter to avoid hangups when the fill is placed by gravity flow. Hydraulic fill is easier and cheaper to place, and will travel through 150-mm-diameter drillholes and 100-mm-diameter pipes.

Fill is most often used in steeply dipping stopes and not in flat-lying ones because of the difficulty in filling a horizontal stope right up to the back. However, if the fill is first dewatered, it can be placed pneumatically in a near-horizontal stope using a shotcreting type of placement procedure (Wayment, 1978) or a high-speed flinger which throws the granular material considerable distances at high velocity.

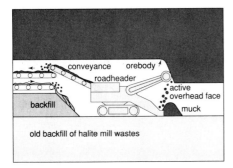
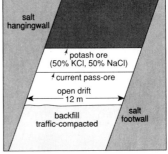

Figure 7.20 Cut and fill using halite wastes, PCA Mine, Sussex, New Brunswick, Canada.

The water needed to make tailings pumpable creates a problem underground. The fill is usually placed at about 40 percent water by weight and must dewater to around 20 percent before it can perform as ground support. Consequently, stopes must be provided with drainage systems and suitably designed bulkheads. Pressures on bulkheads may be reduced by geotextile filters, allowing the fill to drain freely, thereby consolidating and increasing its shear strength.

Usually the tailings are *classified* or *deslimed*, that is, some portion of the −200-mesh (silt and clay sized) material is removed to increase the permeability and rate of consolidation, making use of only the coarser-sized fraction.

7.5.6.4 Cemented tailings. Uncemented fill tends to drain more readily than cemented, but will not stand unsupported, and may also liquify as a result of blasting. The cost of a stronger cemented backfill is justified if it allows use of a more efficient mining method, and reduces dilution of the ore by backfill materials.

A fill consisting of cemented tailings provides an excellent working platform for personnel and machines; in overhand cut-and-fill mining (Sec. 7.2.4) a cement-enriched layer at the surface of each pour makes timber mucking floors unnecessary. Strength and stiffness increase with age as the fill consolidates under pressure, and as the cement cures. Miners can work on the fill after 8 h, and machines after 2 days (Nicholson and Wayment, 1964; Leahy and Cowling, 1978; Mitchell and Stone, 1987).

The *cement content* must be sufficient to prevent the fill layer from being scraped up from the floor during mucking operations, usually between 3 and 6 percent by weight. Little is to be gained by using a mixture either leaner than 1:32, which is essentially the same as straight tailings, or richer than 1:16, which appears not to warrant the additional expense. Costs can also be reduced by using cementing agents such as ground blast furnace slag or fly ash, which are often available at mine sites and have pozzolanic (cementitious) properties.

Fill properties vary according to fill type and cement content. In place mechanical characteristics can be very different from those measured in the laboratory, because the fill tends to form distinctly different layers in the stope. Cemented tailings are fragile and difficult to sample, and the properties of samples change if they are disturbed or allowed to dry out. Properties for modeling of backfilled mine openings are therefore best determined by in situ testing.

Porosities of in place cemented fill are typically 30 to 45 percent, and bulk densities range from 1.8 to 2.1 g/cm^3. Percolation rates range from 12 to 72 s/mm, and the addition of cement greatly reduces the permeability. The following is a rough guide to in situ mechanical characteristics of cemented tailings fill for a very lean mix and for a richer mix such as in the floor layer of a stope:

	Lean mix	Rich mix
Cement to tailings ratio	1:30	1:8
Compressive strength (MPa)	0.2	2.7
E (MPa)	40	250
Friction angle	30°	36°
Cohesion (MPa)	0.1	0.4

7.6 Subsidence

7.6.1 Subsidence processes

Underground excavation can place at risk not only the miners underground, but also the structures and services above. Potential hazards at surface include subsidence, drainage and lowering of the groundwater table, and earth tremors. Whether any or all of these will be experienced depends on the ground, the water and stress regimes, and the precautions taken (Kratzsch, 1983; Sendlein et al., 1983; Peng, 1986, 1987; Whittaker and Reddish, 1989).

Settlements above fully supported tunnels and caverns seldom exceed a few millimeters, and can be detected at the ground surface only by using sensitive instrumentation. Subsidence of greater magnitude results from absence or failure of support (*caving-induced subsidence*) or from drainage and consolidation of overlying soils (*drainage-induced subsidence*).

7.6.2 Caving-induced subsidence

7.6.2.1 Mechanisms and examples. Caving-induced subsidence is a settlement of the ground surface caused by closure or collapse of underground workings. It starts as small-scale roof raveling, followed by increasing quantities of rock fall. The cavity above the raveled debris migrates upward from the workings and becomes progressively smaller because of the "bulking" of broken rock. Ultimately, as in the case of most deep excavations, the cavity may disappear. Further long-term settlement may result from consolidation of the loosened rock fragments (Kendorski et al., 1983; see also Sec. 7.5.2).

The height to which caving will develop depends on the volume of the original opening (height and width), and the *bulking factor*. Field measurements give bulking factors of between 20 to 50 percent for coal measures rocks (Whittaker and Reddish, 1989), corresponding to upward migration of caving of between 2 and 6 times the thickness of the mined seam. This can increase to more than 10 times the seam thickness if the caving penetrates water-bearing strata.

We can distinguish two types of caving-induced subsidence. *Regional subsidence* (troughing) is characterized by an undulating topography with depressions above the caved openings, but with no surface scarp. Troughing is more common in soft, stratified rock because the strata deflect downward, rupturing at depths close to the excavation but bending at depths close to surface. *Chimney subsidence* (cave to surface) is characterized by the sudden appearance of a depression bounded by an abrupt scarp (Fig. 7.5), and is common when the rock cover is thin compared with the size of the mined opening. Chimney subsidence is also more likely where the rock contains near-vertical joints or faults.

Troughing can be detected in the early stages when the surface damage is still minimal. Even these small differential strains will affect rigid structures, and excessive longitudinal stretching on the flanks of the subsidence bowl may buckle or rupture pipes and services.

Chimney subsidence is particularly hazardous because there are few surface indications until the chimney forms. Where and when it will occur can seldom be predicted, especially if the geometry of underground workings is unknown. The risk is great when these caves occur above active mine workings and beneath sensitive overburden soils such as quick clays, unconsolidated mine tailings, or water-saturated silts and sands, because the soils can liquify and flow into the mine. Miners who escape injury during the caving itself may be injured by the airblast concussion or enveloped by the mud inflow. A disaster of this nature killed eight miners in Val D'Or, Quebec, in 1980.

Karfakis (1986) describes chimney subsidence at the Hanna coalfield in Wyoming, where mining began in 1868. The 89 chimneys surveyed were disk-shaped, probably because of erosion, typically 3 to 4 m deep, with a 20-m average diameter, similar to the room-and pillar-dimensions. The mined thickness was probably 3 to 6 m. Most of the subsidence had occurred where the thickness of overburden was less than 70 m.

Chimney subsidence continues to plague the City of Edmonton in Alberta, Canada, where turn-of-the-century coal mining from the river banks created a network of openings at depths usually less than 100 m. Problems arise as water or sewage pipes rupture by undermining, and roadway stability may be affected as well, particularly on hill slopes underlain by the old mine workings. Detailed site investigation is necessary because old records seldom locate openings precisely (Sladen and Joshi, 1988).

The problem of *abandoned mine hazards* is worldwide. Although some legislation aims at preventing future hazards, a legacy of problems remains from the past (Winters and Chen, 1986; Mackasey, 1989). Subsidence movements over room-and-pillar mines continue for hundreds of years after mining. Over longwall mines, however, subsidence typically lasts less than 10 years, with 90 percent of total subsidence occurring in the first few months as strata collapse behind the advancing longwall (Bhattacharya et al., 1986).

7.6.2.2 Prediction methods. Underground coal mining beneath populated areas has created a need for reliable surface deformation models and for prediction of damage to buildings of different flexibility. Foundation design in regions of active or abandoned workings and the subsidence tolerances of different types of structure are reviewed in Sec. 3.3.3. Subsidence control is now regulated by law in most countries. In the United States, for example, the Surface Mining Control and Reclamation Act of August 1977 requires that subsidence be prevented from causing material damage to the extent technologically and economically feasible, except in those instances where the mining method is one of planned subsidence in a predictable and controlled manner. The Act also gives authority to suspend underground mining if there is danger to inhabitants.

Subsidence predictions are carried out in two ways. The first is entirely empirical and based on the statistical best fit of years of detailed survey data without recourse to mechanistic ideas about ground movement. The second assumes that deformation of the rock mass can be predicted using continuum or discontinuum analyses (Saxena, 1978; Salamon, 1983).

The best known *empirical method* is that in the Subsidence Engineer's Handbook published by the National Coal Board of Great Britain (1975). It provides predictions of subsidence, tilt, and horizontal strain under the conditions prevailing in British coalfields. To be used elsewhere, such a procedure requires validation and recalibration under different sets of geological conditions. Fejes (1985), for example, conducted, for the U.S. Bureau of Mines, a 5-year study of subsidence above longwall panels in central Utah. The British NCB method generally gave a poor prediction of the subsidence profile. An empirical functional profile technique from the Donets coal region in Russia more closely approximated the Utah data.

The empirical approaches can be modified in some cases by using observations to define "influence coefficients" for the workings and overlying rock. These coefficients may then be used, often through a numerical integration approach, for predictions in other cases in the same or similar mining areas.

Karmis et al. (1984) describe the development of a model for the Appalachian region of the United States. A hyperbolic tangent function was chosen to mathematically describe the subsidence curves, such that the subsidence at any point, $S(x)$, is given by

$$S(x) = 0.5\ S\left(1 - \tan h\ \frac{cx}{B}\right)$$

Where x = distance from the inflection point to the point in question
S = maximum subsidence for the profile
B = distance from the inflection point to S
c = a constant, 1.4 for subcritical panels and 1.8 for critical and supercritical panels

Various predictive methods assume an *angle of draw* spreading outward from the underground excavations to reach the surface. The angle of draw defines the zone of influence. Bulking factors are taken into account in predicting the upward migration of strains and displacements through the rock mass.

Recently, modeling techniques have been developed based on *continuum mechanics* using more realistic materials properties. Mikula and Holt (1983) describe a finite element method of subsidence prediction that incorporates anisotropy and joint elements. Boundary element techniques may also be employed (Gambolati et al., 1987). However, without empirical "calibration," none of these methods can accurately simulate caving, bulking, and chimneying.

Discontinuum modeling (*Rock Engineering,* Sec. 7.2.5), either physical or numerical or a combination of the two, may be used to study caving mechanisms including chimney formation (Brady, 1989). It tends to overestimate bulking and exaggerate the beneficial effects of arching, because block rupture and crushing, which destroy arching and generate a denser mass, are particularly difficult to replicate. However, it can be coupled with other methods to give more faithful reproductions of actual subsidence behavior (Sutherland et al., 1984; Coulthard and Dutton, 1988).

For drainage and compaction problems, conventional finite element *consolidation analyses* may be used (Lewis and Schrefler, 1987). The time element becomes important in these cases because drainage in rocks and porous sediments is slow and consolidation may develop over years.

7.6.2.3 Control of caving-induced subsidence. There are two approaches to subsidence and surface works, to prevent subsidence or to design surface works assuming that subsidence will occur. Foundation designs used where surface subsidence cannot be avoided are described in Chap. 3 and Powell et al. (1988).

Normal practice when extracting coal or other flat-lying seams is to avoid mining at shallow depth beneath bodies of water, highways, railways, and major buildings, leaving pillars of unmined ore at these locations. This may be required by local legislation. Caving-induced subsidence can be controlled by choice of mining method, by careful design, and by use of backfill. To minimize subsidence, the fill is stowed tightly against the back of mine openings using hydraulic placement methods.

7.6.3 Drainage-induced subsidence

Drainage-induced subsidence results from the removal of pore fluids from soil, either directly or by draining the underlying rock. Soils that were once saturated shrink and consolidate when drained. The settlements are usually uniform over broad areas, although differential settlements and surface tilting can result from nonuniform soil thicknesses or types (Sec. 8.5 and *Rock Engineering,* Sec. 4.1.4.6).

The problem should be addressed as an aspect of mine planning. Once consolidation subsidence has occurred, it cannot be reversed by reinjecting water; the strains are permanent. The potential for drainage-induced subsidence damage exists whenever mining occurs at shallow or moderate depths beneath urban areas founded on soils.

To avoid consolidation, the pore pressures in the soils should be maintained at about the same value as before mining. The treatments used to limit infiltrations into tunnels and caverns (i.e., impermeable linings and extensive grouting of rock; Secs. 5.6.1.5 and 6.8.3.5) are impractical and prohibitively expensive for a mine stope. However, the underdrainage can be reduced by avoiding loosening of rock, which increases permeability of the zone between the opening and surface. Groundwater maintenance methods can be used, although this means continuing to recharge surface wells for as long as water is pumped from the mine.

7.6.4 Subsidence monitoring

Subsidence is monitored for safety, both of the mine and of the public. The greatest risks are when mining occurs beneath inhabited areas or bodies of water. Other uses of monitoring data are to assist in the assessment of subsidence damage claims, and to verify subsidence models. Observations often have the aim of determining whether the NCB method or a similar model is applicable to local rocks and mining methods, and to calibrate (i.e., adjust) the model to compensate for any differences (Bawden et al., 1990).

Instrumentation and techniques are described in *Rock Engineering,*

Chap. 12, and in Bawden et al. (1990). Current practice combines conventional geodetic surveying at the surface with rock mass deformation monitoring in drillholes or from underground, to provide a comprehensive picture of the overall ground behavior. In longwall workings, measurements of bed separation in the roof above the gob, except for the case of undersea mines, can be made in holes drilled from the surface. Techniques include settlement gauges based on the U-tube principle, multiple position extensometers, time domain reflectometry (TDR), and the use of radioactive markers in drillholes.

In 1978 the U.S. Bureau of Mines started a long-term mining subsidence research program to find practical methods and equipment for measurement. Conventional surveying techniques were chosen, supplemented by electronic theodolites. Measurements included traverses to obtain x, y, and z coordinates, geodetic measurement of angles and distances, and leveling combined with taping. In its subsidence research program in British Columbia, Canada, CANMET chose a tiltmeter consisting of two separate accelerometers with their sensing axes perpendicular to each other. The accelerometers were encased in a tube cemented into a drillhole in the bedrock.

Great benefits and economies have come from the use of electronic theodolites and computer processing of data. The need to gain subsidence data at remote locations and in severe environments and also to maintain a constant surveillance and to gain warning of movements has led to an increased use of remote data collection. Recently, the satellite Global Positioning System (GPS) has opened new possibilities for the monitoring of both horizontal and vertical ground movements.

Groundwater measurements are also required because groundwater lowering is the prime cause of drainage-induced subsidence, and water table fluctuations can be symptomatic of the onset of caving-induced subsidence.

References

Barr, C. A., 1974. "Geological Problems in Saskatchewan Potash Mining due to Peculiar Conditions during Deposition of Potash Beds." *4th Symposium on Salt,* North Ohio Geological Survey, vol. 1, pp. 101–118.

———, 1977. *Applied Salt-Rock Mechanics, 1*. Elsevier, Amsterdam, 294 pp.

Bawden, W., Chrzanowski, A., Barron, K., and Cain, P., 1990. "Mine Subsidence Monitoring Instrumentation. Chap. 3.4 In: *Mine Monitoring Manual,* J. A. Franklin, (ed.), Can. Inst. Min. Metall., Montreal.

Bétournay, M. C., 1989. "What Do We Really Know about Surface Crown Pillars?" *Proc. Int. Conf. on Surface Crown Pillar Evaluation for Active and Abandoned Metal Mines,* Timmins, Ont., pp. 17–31.

Bhattacharya, S., Singh, M. M., and Chen, C. Y., 1984. "Proposed Criteria for Subsidence Damage to Buildings." *Proc. 25th U.S. Symp. Rock Mech.,* Northwestern Univ., Evanston, Ill., pp. 747–755.

———, ———, and Huck, P. J., 1986. "The Time Relationship of Mine Subsidence and Its Implications on Subsidence Legislation." *Proc. 27th U.S. Symp. Rock Mech.,* Tuscaloosa, Ala., pp. 291–296.

Bieniawski, Z. T., 1983. "New Design Approach for Room and Pillar Coal Mines in the USA." *Proc. 5th Int. Cong. Rock Mech.*, Melbourne, Australia, sec. E., pp. 27–36.
———, 1984. *Rock Mechanics Design in Mining and Tunnelling.* Balkema, Rotterdam, 272 pp.
Blake, W., 1972. "Rock-Burst Mechanics." *Q. Colorado School of Mines,* vol. 67, no. 1, 64 pp.
———, 1975. "Design, Installation and Operation of Computer Controlled Rockburst Monitoring Systems." *Proc. 1st Conf. Acoustic Emission/Microseismic Activity in Geologic Strucs. & Mtls.,* Penn. State Univ., pp. 157–167.
———, 1982. *Microseismic Applications for Mining: A Practical Guide.* U.S. Bur. Mines Contract Rept. #J0215002.
———, 1987. "Microseismic Instrumentation." *Proc. Fred Leighton Memorial Workshop on Mining Induced Seismicity,* Montreal, Canada, pp. 3–4.
Bourbonnais, J., 1984. "A Research Application of Multi-Channel Microseismic Monitoring to Rockbursting at the East Malartic Mine in Northwestern Quebec." *Proc. Conf. Acoustic Emission/Microseismic Activity in Geologic Structs. and Materials.,* Penn. State Univ., pp. 252–268.
Brady, B. H., 1989. "Rock Mechanics and Ground Control for Underground Mining and Construction." Keynote address in: *Proc. 30th U.S. Symp. Rock Mech.,* Morgantown, W.Va., Balkema, Rotterdam, pp. 5–17.
———, and Brown, E. T., 1985. *Rock Mechanics for Underground Mining.* George Allen and Unwin, London. 527 pp.
Brawner, C. O., (ed.), 1983. "Stability in Underground Mining." *Proc. 1st Int. Conf. Amer. Inst. Min., Met. & Pet. Eng.,* New York, 1071 pp.
Budavari, S., (ed.), 1983. *Rock Mechanics in Mining Practice.* South African Inst. Min. Metall.
Bywater, S., Cowling, R., and Black, B. N., 1983. "Stress Measurement and Analysis for Mine Planning." *Proc. 5th Int. Cong. Rock Mech.,* Melbourne, Australia, vol. D, pp. 29–38.
Canadian Institute of Mining and Metallurgy, 1978. *Proc. 12th Can. Rock Mech. Symp., Mining with Backfill,* Sudbury, Ontario, 150 pp.
Chamber of Mines of South Africa, 1977. *An Industry Guide to the Amelioration of the Hazards of Rockburst and Rockfalls.* High-level Comm. on Rockbursts and Rockfalls. Chamber of Mines Pub. PRD216, 178 pp.
———, 1988. *An Industry Guide to Methods of Ameliorating the Hazards of Rock Falls and Rockbursts,* 114 pp.
Clarke, R. W. D., 1985. *Ore Handling Practices and Trends.* Dept. Energy, Mines & Resources, Ottawa, Canada, CANMET, Mining Res. Labs. Div. Rep., MRP/MRL 85-10, 13 pp.
Cook, N. G. W., and Hood, M., 1978. "The Stability of Underground Coal Mine Workings." *Proc. Int. Symposium on Stability in Coal Mining,* Vancouver, Canada, pp. 135–147.
Coulthard, M. A., and Dutton, A. J., 1988. "Numerical Modelling of Subsidence Induced by Underground Coal Mining." *Proc. 29th U.S. Symposium on Rock Mechanics,* Balkema, Rotterdam, pp. 529–536.
Cundall, P. A., 1988. "Formulation of a Three-Dimensional Distinct Element Model—Part I. A Scheme to Detect and Represent Contacts in a System Composed of Many Polyhedral Blocks." *Int. J. Rock Mech., Min. Sci., Geomech. Abstr.,* vol. 25, no. 3, pp. 107–116.
Davidge, G. R., Martin, T. A., and Steed, C. M., 1988. "Lacing Support Trial at Strathcona Mine." *Proc. 2d Int. Symp. on Rockbursts and Seismicity in Mines,* Minneapolis, Minn., 10 pp.
Dickhout, M. H., 1973. "The Role and Behaviour of Fill in Mining." *Proc. Jubilee Symp. on Mine Filling,* Mt. Isa, Australia. N.W. Queensland Branch, Aust. Inst. Min. Met.
Dusseault, M. B., Mraz, D., and Rothenburg, L., 1987. "The Design of Openings in Saltrock Using a Multiple Mechanism Viscoplastic Law." *Proc. 28th U.S. Rock Mech. Symp.,* Tucson, Ariz., Balkema, Rotterdam, pp. 633–642.
Fejes, A. J., 1985. "Surface Subsidence Resulting from Longwall Mining in Central Utah—A Case Study." *Proc. 26th U.S. Symp. Rock Mech.,* Rapid City, S.Dak., vol. 1, pp. 197–204.

Folinsbee, J. C., and Clarke, R. W. D., 1981. "Selecting a Mining Method." Chap. 5 in: *Design & Operation of Caving and Sublevel Stoping Mines,* D. Stewart, (ed.)., Soc. Min. Engrs., AIME, U.S.A. pp. 55–65.

Fordham, C. J., Dusseault, M. B., Mraz, D., and Rothenburg, L., 1989. "The Use of Relaxation Tests to Predict the Compaction Behavior of Halite Backfill." *Proc. 4th Int Symp. on Mining with Backfill,* Montreal, Balkema, Rotterdam, pp. 327–334.

Franklin, J. A., 1986. "Size-Strength System for Rock Characterization." *Proc. Int. Symp. Rock Characterization Techniques in Mine Design,* New Orleans, SME-AIME Ann. Mtg., pp. 11–16.

———, Butler, A. G., and Wang Baosheng, 1989. "Mechanisms of Caving and Crown Pillar Stability." *Proc. Int. Conf. on Surface Crown Pillar Evaluation for Active and Abandoned Metal Mines,* Timmins, Ont., pp. 79–88.

Gambolati, G., Sartoretto, F., Rinaldo, A., and Ricceri, G., 1987. "A Boundary Element Solution to Land Subsidence Above 3-D Gas/Oil Reservoirs." *J. Num. Analytic Meth. in Geomech.* vol. 11, no. 5, pp. 489–502.

Grotowsky, U., and Irresberger, H., 1984. "A Strata Control System and Its Application in West German Coal Mining." *Int. J. Mining. Eng.,* vol. 2.

Hamrin, H. O., 1982. "Choosing an Underground Mining Method." In: *Underground Mining Methods Handbook,* W. A. Hustrulid (ed.). Soc. Min. Engrs, Am. Inst. Min. Metall. & Petr. Engrs. New York, pp. 88–112.

Handewith, H. J., 1980. "Mine Applications of Tunnel Boring Machines." *Bull. Can. Inst. Min. Met.,* November, pp. 133–136.

Hanes, J., Lama, R. D., and Shepherd, J., 1983. "Research into the Phenomenon of Outbursts of Coal and Gas in Some Australian Collieries." *Proc. 5th Int. Cong. Rock Mech.,* Melbourne, Australia, vol. E., pp. 79–85.

Haramy, K., and McDonnell, J., 1986. "Floor Heave Analysis in a Deep Coal Mine." *Proc. 27th U.S. Symp. Rock Mech.,* Tuscaloosa, Ala., pp. 520–525.

Hart, R., Cundall, P. A., and Lemos, J., 1988. "Formulation of a Three-Dimensional Distinct Element Model—Part II. Mechanical Calculations for Motion and Interaction of a System Composed of Many Polyhedral Blocks." *Int. J. Rock Mech., Min. Sci, Geomech. Abst.,* vol. 25, no. 3, pp. 117–126.

Heuze, F. E., and Goodman, R. E., 1971. "'Room and Pillar' Structures in Competent Rock." *Proc. Symp. Underground Rock Chambers,* Phoenix, Ariz., Am. Society of Eng., New York, pp. 531–565.

Hiramatsu, Y., Saito, T., and Oda, N., 1983. "Studies on the Mechanism of Gas and Coal Bursts in Japanese Coal Mines." *Proc. 5th Int. Cong. Rock Mech.,* Melbourne, Australia, sec. E, pp. 7–10.

Hoek, E., 1989. "A Limit Equilibrium Analysis of Surface Crown Pillar Stability." Keynote address, in: *Proc. Int. Conf. on Surface Crown Pillar Evaluation for Active and Abandoned Metal Mines,* Timmins, Ont., pp.3–13.

Hustrulid, W. A., 1976. "A Review of Coal Pillar Strength Formulas." *Rock Mech.,* vol. 8, pp. 115–145.

Iannacchione, A. T., Campoli, A. A., and Oyler, D. C., 1987. "Fundamental Studies of Coal Mine Bumps in the Eastern United States." *Proc. 28th U.S. Rock Mech. Symp.,* Tucson, Ariz., Balkema, Rotterdam, pp. 1063–1072.

Jackson, E., 1986. *Hydrometallurgical Extraction and Reclamation.* New York.

Jager, A. J., Piper, P. S., and Gay, N. C., 1987. "Rock Mechanics Aspects of Backfill in Deep South African Gold Mines." *Proc. 6th Int. Cong. Rock Mech.,* Montreal, Canada, vol. 2, pp. 991–998.

Jeremic, M. L., 1985. *Strata Mechanics in Coal Mining.* Balkema, Rotterdam, 566 pp.

Karfakis, M. G., 1986. "Chimney Subsidence—A Case Study." *Proc. 27th U.S. Symp. Rock Mech.,* Tuscaloosa, Ala., pp. 275–282.

Karmis, M., Triplett, T., and Schilizzi, P., 1984. "Recent Developments in Subsidence Prediction and Control for the Eastern U.S. Coal Fields." *Proc. 25th U.S. Symp. Rock Mech.,* Northwestern Univ., Evanston, Ill., pp. 713–721.

Kelsall, P. C., and Nelson, J. W., 1985. "Geologic and Engineering Characteristics of Gulf Region Salt Domes Applied to Underground Storage and Mining." *Proc. 6th Int. Symposium on Salt,* Salt Institute and North Ohio Geological Survey, vol. 1, pp. 519–544.

Kendorski, F. S., Cummings, R. A., Bieniawski, Z. T., and Skinner, E. H., 1983. "Rock Mass Classification for Block Caving Mine Drift Support." *Proc. 5th Int. Cong. Rock Mech.*, Melbourne, Australia, vol. B., pp. 51–63.

Khristianovich, S. A., and Salganik, R. L., 1983. "Several Basic Aspects of the Forming of Sudden Outbursts of Coal (Rock) and Gas." *Proc. 5th Int. Cong. Rock Mech.*, Melbourne, Australia, vol. E, pp. 41–50.

Kratzsch, H., 1983. *Mining Subsidence Engineering* (English translation). Springer-Verlag, London, 543 pp.

Laubscher, D. H., and Taylor, H. W., 1976. "The Importance of Geomechanics Classification of Jointed Rock Masses in Mining Operations." In: *Exploration for Rock Engineering*, Z. T. Bieniawski (ed.), Proc. Symp. Johannesburg, 1975, Balkema, Rotterdam, vol. 1, pp. 119–128.

Leahy, F. J., and Cowling, R., 1978. "Stope Fill Development at Mount Isa." In: *Mining with Backfill, Proc. 12th Can. Rock Mech. Symp.*, Sudbury, Ontario, Special vol. 19, Can. Inst. Mining and Metallurgy, pp. 21–29.

Lee, M. F., and Bridges, M. C., 1980. "Rock Mechanics of Crown Pillars between Cut-and-Fill Stopes at the Mt. Isa Mine." *Proc. Conf. Application of Rock Mech. to Cut-and-Fill Mng*, Inst. Min. & Met., London, 376 pp.

Leighton, F., and Blake, W., 1970. *Rock Noise Source Location Techniques*. U.S. Bur. Mines Rept. Investigations 7432, 18 pp.

Lewis, R. A., and Schrefler, B. A., 1987. *Finite Element Method in the Deformation and Consolidation of Porous Media*. John Wiley, Toronto, 344 pp.

Lorig, L. J., Hart, R. D., Board, M. P., and Swan, G., 1989. "Influence of Discontinuity Orientations and Strength on Cavability in a Confined Environment." *Proc. 30th U.S. Symp on Rock Mech.*, Morgantown, W.Va., Balkema, Rotterdam, pp. 167–174.

Mackasey, W. O., 1989. "Concepts on Dealing with Abandoned Mine Hazards." *Proc. Int. Conf. on Surface Crown Pillar Evaluation for Active and Abandoned Metal Mines*, Timmins, Ont., pp.135–141.

Mark, C., 1990. *Pillar Design Methods for Longwall Mining.*" U.S. Bur. Mines Information Circular IC9247, 53 pp.

Maury, V., 1979. "Effondrements Spontanés. Synthèse d'Observations et Possibilité de Méchanism Initiateur par Mise en Charge Hydraulique." *Rev. Industr. Minerale*, October, pp. 1–12.

Mikula, P. A., and Holt, G. E., 1983. "Prediction of Mine Subsidence in Eastern Australia by Mathematical Modelling." *Proc. 5th Int. Cong. Rock Mech.*, Melbourne, Australia, vol. E., pp. 119–126.

Mitchell, R. J., and Stone, D. M., 1987. "Stability of Reinforced Cemented Backfills." *Can. Geotech. J.*, vol. 24, no. 1, pp. 189–197.

Moruzi, G. A., 1978. "Consolidated Fill Practice at Strathcona Mine." In: *Mining with Backfill, Proc. 12th Can. Symp. Rock Mech.*, Sudbury, Ont., Special vol. 19, Can. Inst. Min. & Met., pp. 10–15.

Mraz, D. Z., 1984. "Solutions to Pillar Designs in Plastically Behaving Rocks." *Can. Inst. Min. & Met. Bulletin*, vol. 77, no. 868, pp.55–62.

———, and Dusseault, M. B., 1986. "Effects of Geometry on the Bearing Capacity of Pillars in Saltrock." *CIM RMSCC Workshop Proc.*, Saskatoon, Sask. 23 pp.

Mrugala, M. J., and Belesky, R. M., 1989. "Pillar Sizing." *Proc. 30th U.S. Symp on Rock Mech.*, Morgantown, W.Va., Balkema, Rotterdam, pp. 395–402.

National Coal Board, Great Britain, 1975. *Subsidence Engineer's Handbook*, Production Dept., London.

Nicholson, D. E., and Wayment, W. R., 1964. *Properties of Hydraulic Backfills and Preliminary Vibratory Compaction Tests*. U.S. Bur. Mines, RI 6477, 21 pp.

Notley, K. R., 1984. "Rock Mechanics Analysis of the Springhill Mine Disaster (1958)." *J. Min. Sci. Technol.*, vol. 1, no. 2, pp. 149–163.

Ortlepp, W. D., and Steel, K. E., 1972. "Rockbursts: The Nature of the Problem and Management Countermeasures on East Rand Proprietary Mines Ltd." *Assoc. Mine Mgrs. S. Africa*, Papers and Discussions, pp. 225–278.

Paterson, L., 1986. "A Model for Outbursts in Coal." *Int. J. Rock Mech., Min. Sci., Geomech. Abstr.*, vol. 23, no. 4, pp. 327–332.

Peng, S. S., 1987. *Surface Subsidence Engineering*. West Virginia Univ., Dept. of Mining Eng., 485 pp.
———(ed.), 1986. *Proc. 2nd Conf. Subsidence due to Underground Mining*, W. Virginia Univ., 316 pp.
Potvin, Y., and Hudyma, M. R., 1989. "Open Stope Mining Practices in Canada." *Proc. 91st Ann. Mtg.*, Can Inst. Min. Met., Quebec City, 35 pp.
Powell, L. R., Triplett, T. L., and Yarbrough, R. E., 1988. *Foundation Response to High-Extraction Mining in Southern Illinois*. U.S. Bur. Mines Rept. Investigations RI9187, 54 pp.
Salamon, M. D. G., 1974. "Rock Mechanics of Underground Excavations." *Proc. 3d Int. Cong. Rock Mech.*, Denver, Colo., vol. 1B, pp. 951–1099.
———, 1983. "Linear Models for Predicting Surface Subsidence." *Proc. 5th Int. Cong. Rock Mech.*, Melbourne, Australia, vol. E, pp. 107–114.
Sarkka, P. S., 1983. "Interactive Dimensioning of a Crown Pillar in the Rautuvaara Mine." *Proc. 5th Int. Cong. Rock Mech.*, Melbourne, Australia, vol. D, pp. 101–108.
Saxena, S. K., 1978. "Subsidence—A Review." *Proc. Eng. Foundation Conf. on Evaluation and Prediction of Subsidence*, Pensacola Beach, Fla., pp. 1–25.
Scoble, M. J., 1986. "Strategic and Tactical Measures to Alleviate Rockbursting in Canadian Underground Mining." *U. Nottingham Min. Dept. Mag.*, vol. 38, pp. 47–53.
Sendlein, L. V. A., Yazicigil, H., and Carlson, C. L., 1983. "Subsidence," In: *Surface Mining, Environmental Monitoring and Reclamation Handbook*, Elsevier, Amsterdam, pp. 603–740.
Sladen, J. A., and Joshi, R. C., 1988. "Mining Subsidence in Lethbridge, Alberta." *Can. Geotech. J.*, vol. 25, no. 4, pp. 768–777.
Smart, B. G. D., and Redfern, A., 1986. "The Evaluation of Powered Support Specifications from Geological and Mining Practice Information." *Proc. 27th U.S. Symp. Rock Mech.*, Tuscaloosa, Ala., pp. 367–377.
Stacey, T. R., and Kirsten, H. A. D., 1987. "Mechanisms of Backfill Support in Deep Level Tabular Stopes." *Proc. 6th Int. Cong. Rock Mech.*, Montreal, Canada, vol. 2, pp. 1245–1249.
Starfield, A. M., and Cundall, P. A., 1988. "Towards a Methodology for Rock Mechanics Modeling." *Int. J. Rock Mech., Min. Sci, Geomech. Abstr.*, vol. 25, no. 3, pp. 99–106.
Stephansson, O., and Jones, M. J., 1981. "Application of Rock Mechanics to Cut-and-Fill Mining." *Conf. Proc., Inst. Min. & Met.*, London, 376 pp.
Stillborg, B. L., 1984. "Open Stope Design at the Research Mine in Kiruna, Sweden." In: *Design and Performance of Underground Excavations*, ISRM/BGS, Brit. Geot. Soc., London, pp. 273–283.
Su, W. H., Hsiung, S. M., and Peng, S. S., 1984. "Optimum Mining Plan for Multiple Seam Mining." *Proc. 25th U.S. Symp. Rock Mech.*, Northwestern Univ., Evanston, Ill., pp. 591–602.
Sutherland, H. J., Heckes, A. A., and Taylor, L. M., 1984. "Physical and Numerical Simulations of Subsidence above High Extraction Coal Mines." In: *Design and Performance of Underground Excavations*, ISRM/BGS, Brit. Geot. Soc., London, pp. 65–72.
Waddams, J. A., 1984. "Recent Developments in Hard Rock Mining in South African Gold Mines." In: *Design and Performance of Underground Excavations*, ISRM/BGS, Brit. Geot. Soc., London, pp. 285–293.
Wayment, W. R., 1978. "Backfilling with Tailings—A New Approach." In: *Mining with Backfill, Proc. 12th Can. Rock Mech. Symp.*, Sudbury, Ont., Special vol. 19, Canad. Inst. Min. Metal., pp. 111–116.
Whittaker, B. N., and Reddish, D. J., 1989. *Subsidence; Occurrence, Prediction and Control*. Elsevier, Amsterdam, 528 pp.
Winter, G. V., Bloomsburg, G. L., and Williams, R. E., 1984. "Underground Mine Water Inflow Prediction." *Proc. 13th Ann. Rocky Mtn. Groundwater Conf.*, Spec. Pubn., Montana Bur. Mining & Geol., vol. 91, Great Falls, Mont.
Winters, D., and Chen, C. Y., 1986. "Current Status of Federal Regulations and Rulemaking in Governing Subsidence due to Underground Mining." *Proc. 2d Workshop on Subsidence due to Underground Mining*, Morgantown, W.Va., W. Virginia Univ., pp. 1–5.

Wyllie, R. J. M., 1969. "Mechanized Cut and Fill Stopes—Big Machines in Narrow Places." *World Mining,* January and February.

Young, P., and Franklin, J. A., 1990. "Seismic Events," In: *Mine Monitoring Manual,* J. A. Franklin, (ed.), Can. Inst. Min. Met., Montreal.

Yuen, C. M. K., Boyd, J. M., and Aston, T. R. C., 1987. "Rock-Support Interaction Study of a TBM-Driven Tunnel at the Donkin Mine, Nova Scotia." *Proc. 6th Int. Cong. Rock Mech.,* Montreal, vol. 2, pp. 1339–1344.

Chapter

8

Oil, Gas, and Geothermal Energy

8.1 Oil and Gas Reservoirs

8.1.1 Where oil and gas are found

The drilling of wells specifically to find petroleum was probably first undertaken in China hundreds of years ago, using a cable tool as opposed to rotary methods. The modern oil industry is considered to have begun in North America with the sinking of the Drake well in Pennsylvania in 1857, although a well to find oil had been sunk 7 months before this, near the town of Oil Springs, Ontario.

An accumulation of oil, gas, water, or heat in the earth is called a *reservoir*. Whether it is worth exploiting depends on the quantity and value of the resource, its depth, and the costs of exploration, development, and production. Because of the variations in supply and demand, a reservoir that is uneconomical one year may become economical the next.

Oil and gas accumulate when a suitable *source rock* is in a favorable geological and hydrodynamic relationship to a permeable *reservoir rock*, with a *cap rock* to trap the hydrocarbon. The appropriate combination of conditions is found almost exclusively in sedimentary basins (Blatt et al., 1972). Traps are either closed structures (upside-down saucers), where a low-permeability cap rock overlies a porous formation, or lithostratigraphic changes, where one rock type grades into, or pinches out against, a different rock type. Folding and faulting create structural oil traps (Fig. 8.1).

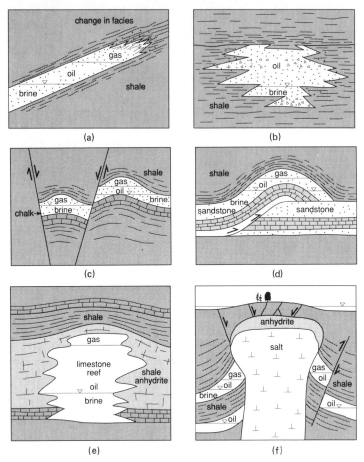

Figure 8.1 Types of oil and gas reservoir. (*a*) Lithology change; (*b*) channel; (*c*) normal fault block; (*d*) thrust fault and anticline; (*e*) carbonate reef; (*f*) salt dome.

Reef and tabular limestones and sandstones are the sources of most of the world's petroleum and natural gas production; however, huge but uneconomical deposits of oil shale contain about 90 percent of the organic content of sedimentary basins, largely in the form of kerogen and carbonaceous matter (Chapman, 1976; Tissot and Welte, 1978). Recent studies indicate that significant amounts of methane may be generated nonorganically by deep crustal processes (Gold and Soter, 1982). The potential of such sources remains unknown, and a deep hole drilled recently in Sweden was unsuccessful in its search for deep gas.

Conventional resources are oil and gas reservoirs in which the amount and viscosity of fluid, and the permeability of the rock, permit economic exploitation by primary and secondary extraction methods (defined later). More than 95 percent of the world's present consumption comes from the 3 to 4 percent of hydrocarbon energy reserves classified as conventional.

Unconventional resources include those exploited by enhanced oil recovery methods, such as tight gas sands (Haas et al., 1987), coal seam gas, oil sands, and oil shales. An uneconomical unconventional resource can become economical with advances in technology and changes in the price of oil. Oil sands and shales are the resources of the future; they will last for centuries, but will require large investments to develop the necessary technology.

8.1.2 Conventional reservoirs

The various ways in which sedimentary basins are formed determine the properties of the rocks within them (Fig. 8.2). *Subduction zones* adjacent to continental plates generate trenches and basins where large volumes of terrestrial detritus accumulate. Examples exist around the Pacific ocean, such as along the west coast of South America. Along continental margins such as the west coast of Africa, rivers dump their sediment to create thick and extensive *deltaic complexes.* Great thicknesses of *limestone reef* such as the Great Barrier Reef of northeastern Australia accumulate along the margins of stable tropical platforms. Shallow seas have in the geological past covered extensive areas, such as in North America from the Yukon to the Gulf Coast. These have generated sequences of shallow marine and terrestrial deposits including clastics, evaporites, and carbonate rocks.

Properties of reservoir rocks vary greatly. A shallow, porous, lithic sandstone from the Maracaibo heavy oil deposits in Venezuela can have a Young's modulus as low as 1 GPa, and behave essentially as a sand. An oolitic limestone of moderate porosity from Cretaceous beds in Texas typically has a Young's modulus of 10 to 30 GPa and a compressive strength of 20 MPa. A dense dolomite from a Devonian oil field in Alberta can have a Young's modulus as high as 70 GPa and a compressive strength of 150 MPa.

Because of mineral changes, compaction, and cementation, rocks have a tendency to become stronger, stiffer, and less permeable as they become older and more deeply buried (*Rock Engineering,* Sec. 2.2.3). The pore fluids within them become more saline and warmer and more highly pressured with increasing depth. Clay minerals, which control shale swelling and reservoir rock permeability, change from kaolinite-smectite-illite to illite-chlorite; smectite is rare below 4000 m and in rocks older than 300 million years.

Exceptions exist; reservoirs in which the pore pressure has been maintained above hydrostatic for millions of years may have proper-

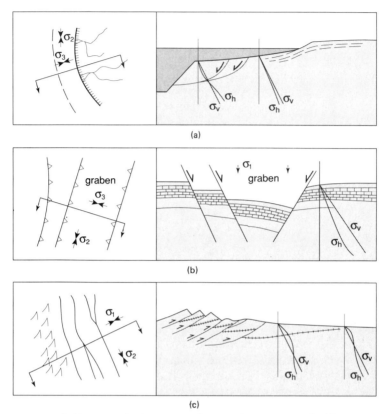

Figure 8.2 Sedimentary basin types and stress directions. (*a*) Stable continental margin (Gulf Coast); (*b*) extension basin (Rhine graben); (*c*) thrust basin (Alberta syncline).

ties similar to those of much shallower formations. The Ekofisk chalk reservoir in the North Sea retains a porosity as high as 48 percent in spite of its 2900-m burial depth, and it is highly compactible (Sec. 8.7.1.1). Recent uplift, such as in Southeast Asia, has brought stiff, strong rocks under unusual stress states close to the surface, and these present particular problems for reservoir development.

Sedimentary rocks are modified by processes of burial, diagenesis, dolomitization, uplift, and tectonics. Erosion exposes vast hydrocarbon deposits like the Athabasca oil sands and the Green River oil shales. Tectonism can fold and fracture a reservoir, creating a jointed and often strongly anisotropic fabric. Fractured reservoirs are sensitive to stress changes arising from production, and special techniques are required to analyze flow in these materials.

8.1.3 Oil sands

8.1.3.1 Occurrence. Certain sand deposits contain large amounts of extremely viscous oil, often called bitumen. Most are found within 1500 m of the surface.

The world's largest accumulations are the Faja Petrolifera del Orinoco in Venezuela (250 billion cubic meters), and the Athabasca Oil Sands in Canada (140 billion cubic meters; Fig. 8.3). Resources of the Athabasca deposit alone would be sufficient to supply the United States with all its oil for the next 70 years. These two deposits are both quartzose sands laid down in rivers and deltas flanking an igneous rock shield. The Athabasca deposit forms a single interconnected reservoir, whereas the Venezuelan one is a stacked sequence of

Figure 8.3 River outcrops of the 60-m-thick Athabasca oil sands, viscous oil in a porous uncemented sandstone overlain in places by smectitic clay shale. The oil sands are mined in open pits and by thermal recovery methods at depth. The Athabasca deposit may represent the single largest oil reservoir in the world, with more than 140×10^9 m^3 of oil.

hundreds of discrete reservoirs. Many smaller deposits have been discovered, and large ones remain to be explored, particularly in the Middle East, China, and the U.S.S.R.

8.1.3.2 Mechanical properties. Because they are often older than 10 million years, oil sands are rocks geologically, but they have physical and mechanical properties more like those of a dense sand (Barnes and Dusseault, 1982). Oil penetrating the sands early in their history prevented cementation between grains. However, subsequent pressure solution at grain contacts densified the material. With porosities of 27 to 33 percent, oil sands are much denser than ordinary sands. Their compressibility is low for a sand, usually in the range 0.2 to $2.0 \times 10^{-6}/\text{kPa}$. In situ sonic P-wave velocities of 2.5 to 3.0 km/s for oil sands are higher than those for sands, and within the range for sandstones (1.4 to 4.6 km/s). Stiffness and wave velocity increase with stress and therefore with depth of burial.

Oil sands dilate strongly when subjected to a shear stress, permitting arching and preventing collapse around boreholes. The peak shear strength envelope is curved like that of rock, with a high friction angle at low stresses (more than 45° from 0 to 3 MPa confining stress). Most deposits, with the exception of some in Utah, the U.S.S.R., and Madagascar, are unjointed and can be analyzed realistically by soil mechanics continuum methods.

8.1.4 Oil shales

8.1.4.1 Occurrence. Most fine-grained sedimentary rocks contain small percentages of organic matter, but not enough to yield a liquid product upon retorting. The term *oil shale* describes siltstones, shales, and mudstones with organic matter in excess of 10 percent by weight.

Oil shales originated in quiet, deep water, where organic-eating organisms and oxygen were absent. Clay minerals, silts, and organic matter were buried, heated, and compressed, generating a shale with a high organic carbon content in the form of *kerogen,* a carbon-rich organic semisolid that cannot be dissolved in ordinary solvents. Graphite-rich shales and slates are probably oil shales exposed to metamorphic conditions which drove off all the volatile oxygen and hydrogen, leaving behind graphitic carbon in a dense, fissile, fine-grained matrix.

Oil shales are widespread. The two best-known deposits, studied extensively as potential energy sources, are in North America. In the west, the lacustrine Green River oil shales of Wyoming, Colorado, and Utah form perhaps the largest single oil shale deposit in the world, and are found at depths from surface to 1 km. In the east, the

Devonian black shales in the basins west of the Appalachians in the United States and Canada underlie a vast area, at depths of up to 5 km. Oil shales are known to exist in many other countries including Australia, the U.S.S.R., China, and Brazil. Because they are not considered resources in the short-term, there has been little effort to find them and systematically assess their extent and properties.

8.1.4.2 Mechanical properties. The mechanical properties of oil shales depend on their mineralogy, organic content, and geological history. Some are true rocks, whereas others may be classified as very stiff soils. The Green River oil shales, for example, have kerogen contents locally exceeding 60 percent by volume and comprise mainly clay minerals, carbonates, and some unusual evaporitic minerals (nahcolite, dawsonite). The higher the kerogen content, the lower the stiffness and strength and the greater the tendency toward viscoplastic behavior. Shallow trial room and pillar mines, however, demonstrate great stability.

The eastern black shales in contrast have kerogen contents no greater than 25 percent, and contain quartz, illite clay, and pyrite. They are much stiffer and stronger than the Green River shales, and their mechanical properties show little effect of organic content (Dusseault et al., 1986).

Oil shales have an extremely low hydraulic conductivity, less than 10^{-12} m/s. Organic material fills pores and blocks the pore throats, and any water present is immobilized by the high surface area of the clay minerals. They are brittle enough to fracture when blasted. In stiff, strong shales, vertical jointing systems are well developed, and tend to govern excavation behavior, whereas in the younger and highly organic shales such as the Green River deposit, jointing is less intense and not as critical to rock mass behavior.

8.1.5 Other unconventional reservoir rocks

Diatomaceous earth is a siliceous material consisting of the shells of small organisms called *diatoms*. In California a large viscous oil deposit is found in this material (the Diatomite). Recovery of oil by surface retorting is being tested, following the same approach as for oil shales, and production at depth using steam injection has been partially successful. The material behaves as an incompetent rock.

In Alberta, Canada, *jointed limestones* of 5 to 15 percent porosity recently have been found to contain more than 50 billion cubic meters of bitumen, partly in the joints and partly in the pores of the rock (Harrison, 1984).

Deeply buried *coal seams* are often more economical to mine by gasification than by conventional mining methods. They contain substantial volumes of methane gas in the pore space. To avoid gas outbursts and explosion hazards in conventional coal mining, the coal seams must be degassed before mining by hydraulic fracturing and vacuum wells. The methane produced then becomes a gas resource, rather than a hazard.

In the United States, coal seams have been estimated by the U.S. Department of Energy to contain 8.4 trillion cubic meters of natural gas, mainly methane (Byrer et al., 1987). Seams deeper than 900 m contain more than half of this. In Canada, resources appear even greater, particularly in the deep seams of the Elmworth Gas Trend in Alberta and British Columbia (Wyman, 1984). Extensive methane-in-coal resources are likely to exist in many other sedimentary basins around the world.

Gas in seams deeper than 1.5 km presents special problems because of high stresses and the weakness of coal. Problems include extensive crushing, coal dust production, and plugging of equipment.

8.2 Geothermal Reservoirs

Geothermal reservoirs are sites of hot rock, hot water, or steam, which can be harnessed to generate electricity, or for heating (Armstead, 1978; Buntebarth, 1984). All geothermal resources are created by the earth's geothermal gradient, which averages about 1°C for each 40 m of depth (25°C/km). This gradient is maintained largely by conduction from the earth's interior, and locally high heat flows may be associated with vulcanism and hydrothermal convection.

Low-grade geothermal resources exist because of the normal geothermal gradient and are found at depth in all areas of the earth. They can be used for space heating, through a heat exchanger. The city of Reykjavik in Iceland is the only large-scale user of low-grade geothermal heat.

Anomalous temperature gradients are caused by geologically recent volcanic or igneous intrusive activity, by variations of thermal conductivity between formations, and by flowing groundwater. *High-grade* geothermal resources have a gradient of more than 30°C/km, and produce steam for electricity generation. They are often associated with hot springs and geysers, particularly within the worldwide seismic and volcanic belts that mark the boundaries between continental plates (Fig. 8.4).

High- and medium-grade geothermal resources occur in three forms (Fig. 8.5):

Hot dry rock, usually volcanic rock or a recent magmatic intrusion, from which heat can be extracted by circulating fluid through nat-

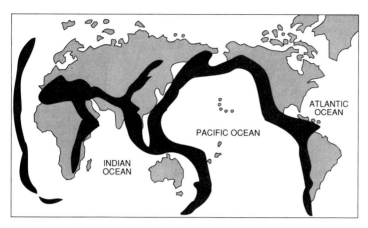

Figure 8.4 Geothermal areas of the world (Edwards et al., 1982).

ural or artificial fracture systems. Examples are the igneous plutons in northern New Mexico and sites of thick recent lava flows.

Wet permeable rock sited above a heat source such as a recent pluton. The water is usually in a state of active thermal convection, replenished by downward flow of cooler groundwater. These hydrothermal fields yield steam when penetrated by a borehole as a result of release of pressure. Most developed geothermal fields are of this type, such as the Cerro Prieto in Mexico, the Geysers in California, Monte Amiata in Italy, and Wairakei in New Zealand.

Hot water or brine in residence at depth in the pores of reservoir rock, usually because of a somewhat elevated geothermal gradient. An unusual example is the geopressured zone along the Gulf of Mexico, which contains brines at temperatures of 100 to 160°C at depths in excess of 4.5 km. The fluid pressures in the reservoir are high, often approaching 90 to 95 percent of the total pressure imposed by the overburden (Swanson et al., 1986).

The steam extracted from the reservoir may be dry (superheated) or wet. Deposits that yield significant amounts of dry steam (free vapor) are by far the rarest. Usually, the reservoir contains hot pressured water only, some proportion of which can be "flashed" to steam for turbine use (Armstead, 1978).

8.3 Exploration, Drilling, and Evaluation

8.3.1 Exploration methods

Reservoirs are explored mainly by drilling at locations where conditions favor economic accumulations of oil and gas (Figs. 8.6 and 8.7).

384 Rock Engineering

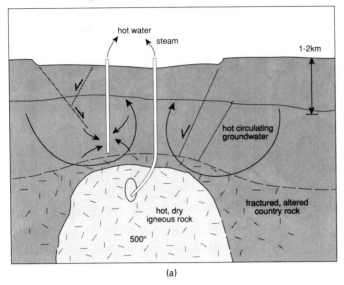

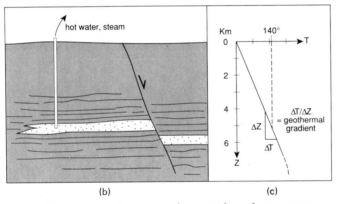

Figure 8.5 Geothermal energy—three modes of occurrence. (a) Igneous intrusion; (b) deep hot reservoirs; (c) geothermal gradient.

Suitable sites are suggested by *geophysical exploration* and by *regional modeling* of rock stratigraphy, structure, and pressure data from nearby wells. As noted earlier, the key requirements are a hydrocarbon source, a reservoir rock, an adequate cap rock, correct geometry to permit hydrodynamic emplacement, and appropriate juxtaposition of these elements in time and space.

Deeply buried rocks seldom outcrop; hence the main sources of data are surface seismic surveys and geophysical logs from previously drilled wells. Other techniques include gravity surveys (porous formations are less dense), electrical surface surveys (oil-bearing formations are more resistive), surface topographic surveys (salt domes and

Oil, Gas, and Geothermal Energy 385

(a)

(b)

Figure 8.6 Oil sand drilling. (a) Capping of triple tube plastic liner sections 1.5 m long and 135 mm in diameter, recovered from a depth of 350 m; (b) core extrusion from the liner ends is caused by gas exsolution.

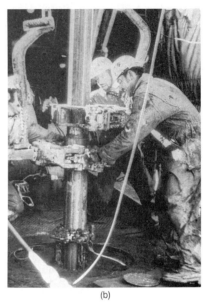

Figure 8.7 Drilling operations for Hot Dry Rock geothermal energy project, Fenton Hill, Valle Caldera, New Mexico. (a) and (b) Drilling operations; (c) control of drilling (weight on bit, torque, penetration rate). (*Photos courtesy J.-C. Roegiers* and *Los Alamos Scientific Laboratory*)

other structures may have surface expression), and detection of gas traces or organic traces in shallow wells and surficial sediments (light hydrocarbons migrate upward).

By far the greatest expenditure on geophysics worldwide is on seismic surveys for oil exploration. Methods have advanced dramatically along with use of the computer to analyze signals from digitally recorded data (*Rock Engineering,* Sec. 6.4; Geophysical Press, 1987).

A seismic source for geophysical exploration emits a sharp sonic signal that passes through the earth and is reflected and refracted back to detectors, usually a string of from 24 to several hundred receivers laid out in straight lines on a grid pattern. Sources may be dynamite explosions in shallow holes, special *sparkers* towed by vessels in *marine seismic surveys* or even up to six *thumpers,* large heavy trucks equipped to vibrate in unison. Single-axis electromagnetic geophones are the most common terrestrial receiver, but buried triple-axis geophones cemented into drillholes give superior data because surface effects are reduced, and because the devices are firmly attached to the rock.

Up to 20 seismic ray traces are "stacked" (digitally added, then averaged) to enhance the signal at the expense of noise. By mapping waveforms from each receiver against depth, good reflectors stand out clearly.

Borehole *tomographic methods* can be employed, in which both the source and the receiver are downhole, or *vertical seismic profiling* (VSP) surveys can be conducted in which either the source or receiver string is downhole. Geophysical borehole sonic logging equipment with up to 4 emitters and 12 receivers, axially spaced, permits detailed analysis of the sonic response of material in the vicinity of the wellbore, including velocities, attenuations, decay rates, amplitudes, and frequencies of compressional, shear, and Stoneley waves (Arditty et al., 1989). These techniques are used in development phases to refine a stratigraphic model based on larger-scale reflection seismic surveys. Logging methods are discussed further in Sec. 8.3.4.2.

The stratigraphic sequence is modeled to find the most probable spatial distribution of strata of different seismic properties that would lead to the measured seismic data, helping to identify promising exploration targets. Recent advances in three-dimensional seismic methods permit spatial resolution of velocities with remarkable accuracy.

8.3.2 Drilling

8.3.2.1 Drilling techniques. Techniques of drilling are reviewed in *Rock Engineering,* Sec. 14.1, including the rotary methods and equip-

Figure 8.8 Saddle-shaped disking in limestone recovered from a deep and highly stressed borehole. (*Photos courtesy V. Maury*)

ment employed in oilfield work. Some further considerations pertaining to oilfield drilling are discussed here, notably the problems and procedures of drilling at great depths, caused by the high temperatures and pressures encountered (Fig. 8.8).

Drilling companies continue to search for faster and cheaper ways to penetrate rock. Changing the bit is time consuming, requiring withdrawal of kilometers of pipe from the hole. The longer the bit life, the fewer these changes. Improvements have been introduced to the teeth and bearings of conventional tricone bits, and bits with no rotating parts which use tough artificial polycrystalline materials such as Stratapax (General Electric) or diamond impregnated ceramics and metals are revolutionizing drilling (Kerr, 1988). Penetration rates can be improved further by servo-control, using on-line computers to interpret and optimize drilling parameters such as thrust, torque, and speed of rotation (Fig. 8.7c).

More radical technologies include a continuous chain bit, fed link by link to the cutting end of the drill; carbide balls carried in a jet of fluid; vibrating drill bits; thermal (plasma) jets; and even small repeated blasts at the bottom of the hole. None of these is yet competitive with conventional rotary drilling. Bottom-hole turbines, rotated by drilling mud, were developed in the U.S.S.R. during the 1950s. New bearing materials and bits with longer lives make these increasingly attractive, and they are being used more and more, particularly where controlled hole deviation is required, such as in horizontal drilling.

The development of methods to deviate wells and drill at high angles with great accuracy was prompted by offshore drilling in the 1950s and 1960s. Because of the expense of moving a platform, it is more economical to drill many deviated wells from one platform, perhaps draining an area as large as 10 to 15 km². In the 1980s, these methods evolved into a strategy for deliberately developing reservoirs with horizontal drains because of their more efficient geometry and the benefits of lower flow gradients. The cost per meter of a horizontal well is now no more than 1.2 times the cost of vertical drilling, and is becoming the favored approach in many cases, particularly in reservoirs with heavy oil and in offshore development.

8.3.2.2 High pressures and weighted muds. Pressures in oil and gas reservoirs are often greater than hydrostatic pressure (the weight of a column of fresh water). In some regions such as the Texas Gulf Coast, southwestern France, the Persian Gulf, some North Sea fields, and the Niger Delta, these pressures approach the full vertical lithostatic stress (Fig. 8.9). This condition can result from a combination of faulting, incomplete consolidation, clay mineral conversion from smectite to illite, methane generation, thermal expansion of water, or accelerated diagenesis with reduction in porosity and generation of excess fluid (Rabia, 1985).

To prevent *collapse of the wellbore,* the mud density must be carefully controlled. Muds are designed to form an efficient *filter cake,* across which most of the pressure drop between the borehole fluid and the pore fluid takes place. Abnormal pressures are counterbalanced by muds containing ground barite ($BaSO_4$) to achieve densities as high as

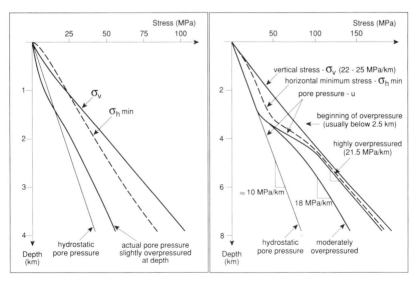

Figure 8.9 Stresses and pressures during deep drilling. (*a*) Gulf Coast, continental margin; (*b*) Alberta, continental thrust basin.

25.5 kN/m^3. In exceptional circumstances, ferric oxide additives (Fe_3O_4) give densities exceeding 28 kN/m^3, but these create severe abrasion problems in pump equipment.

If a long section of borehole remains uncased, the mud weight needed to counterbalance fluid pressure at depth can be sufficient to *hydraulically fracture* formations higher in the hole, where lateral stresses are lower. Therefore mud weight is maintained at the minimum required, with the option of increasing the weight rapidly if exceptional conditions are encountered. Casing strings (direct to surface) or liners (from bottom into the previous casing string) are installed and cemented above critical zones, the mud weight raised, and drilling continues.

In fully developed basins, the depths at which high pressures will be encountered are well known, and mud and casing strategies can be established in advance. In poorly explored regions, a conservative approach is taken to avoid blowouts.

The productive potential of a formation can be impaired if clay particles from drilling mud are washed into pores or if the liquid reacts with the porous formation. Drilling muds of the right properties are carefully selected to avoid damaging productive horizons, and brines, salt muds, or oil-based muds are often found to be useful for these purposes.

8.3.2.3 Lost circulation

Usually, nearly all of the drilling fluid pumped into a hole is recovered and recirculated after removal of rock chips. However, circulation is sometimes lost, for one of the following three reasons:

> High mud pressure can fracture the formation if the horizontal stress is much less than the vertical (this is typical of Gulf Coast conditions).
>
> Mud can escape through large vugs in a limestone or dolomite, or through buried karst features (e.g., Rospo Mare field in Italy, reefs in Alberta), or into large pores in a conglomerate or gravel.
>
> Leakage can occur through intense, open jointing if the interstitial fluid pressure is low.

Fracturing of the hole is avoided by reducing the mud weight. Losses through cavities and joints are avoided by drilling with air or foam (dangerous if the strata contain gas), or by adding to the mud platy, fibrous, or granular *bridging agents* (typically sawdust or ground walnut shells) up to a centimeter in size. Severe problems of lost circulation in dolomitized reefal carbonates have been solved by adding golf balls, baseballs, large foam chips, even gunny sacks directly into the open borehole, then flushing these into the formation until bridging occurs.

8.3.3 Borehole stability

The two major sources of borehole instability are swelling of sensitive clays because of invasion of the liquid phase of the drilling mud, and spalling because of stresses greater than the rock strength. The first is controlled by the use of special muds with inhibiting salts or chemicals, but the second is more common and more difficult to control.

8.3.3.1 Borehole overstressing. Borehole walls yield in all deep wells, and in weak shallow strata as well. Excessive yield can cause borehole sloughing (*wellbore breakout,* Fig. 8.10), wash-outs, bridging, stuck pipe, time-consuming cleanup during and after trips, contamination of bottom-hole cuttings, and borehole squeeze.

A damaged zone develops, within which the rock is cracked or plastically strained (Hayatdavoudi and Apande, 1986). The damage may occur during drilling, injection, or production. During drilling, the mud and the rotating drill pipe erode and remove crushed and sheared particles, eliminating any support provided by the yielded material. This erosion of material also happens during production as a result of

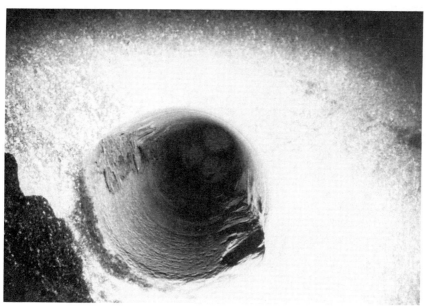

Figure 8.10 Example of wellbore breakout. (*Photo courtesy J. S. Bell, ISPG-EMR Calgary*)

inflow of high-velocity gas or viscous oil (Guenot and Santarelli, 1989).

Different yield mechanisms occur around boreholes depending on the stresses, and on whether the flow gradients are away from or toward the borehole. Instability problems are most severe where the difference between horizontal principal stress components is large. The stress distribution depends on virgin stresses, temperature gradients, and fluid pressures. It can be approximated by the Kirsch equations of elasticity (*Rock Engineering,* Sec. 7.1.7.2), but of course elastic solutions are of limited value when the rock yields or ruptures. Inelastic behavior, or a stress-dependent elasticity, $E = E_0 + f(\sigma_3')$, will cause a lowering of the stress concentration below that predicted by ideal elasticity (Fig 8.11).

8.3.3.2 Borehole swelling and solution.
Borehole instability can occur while drilling through poorly cemented swelling shales, which usually contain large quantities of smectite. The problem is worst when using fresh-water drilling muds, because salinity in the pores of the shale sets up an osmotic gradient that sucks in water, causing swelling of the shale and loss of strength. Swelling problems are avoided by careful design of the drilling mud, using saline fluids or ones with enough clay or polymeric agent to prevent flow across the borehole wall, or by using fluids with cations or organic molecules that reduce the swelling capacity of the formation clays.

Many borehole sloughing problems attributed to swelling are actually caused by a combination of high stresses, weakening, and yield.

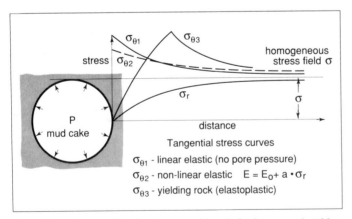

Figure 8.11 Stress redistribution around boreholes because of yield or nonlinear elasticity.

Borehole instability also can be caused by solution when drilling through soluble salt, potash, or gypsum. A huge washout can develop, even leading to the loss of chip flushing capability because of reduced velocity in the expanded section of hole. These and related problems are avoided by presaturating the mud with salt or $CaSO_4$ or by use of oil-based muds.

8.3.3.3 Stability of wellbores in coal seams.

Well completions in coal seams require special methods because of continual coal sloughing and deteriorating wellbore conditions (Logan et al., 1986). Sloughing occurs during drilling because coal is weak, and can worsen during production, when the well pressure is reduced. Continued production can be accompanied by collapse of the wellbore casing in the completion zone.

Casing and perforating are of little use because coal dust blocks the cleats (joints). Induced fractures also tend to become blocked by coal dust, and are difficult to keep open because the proppant sand embeds itself in the soft coal. To mitigate some of these problems, wellbores are usually completed without a casing where they pass through shallow coal seams, but this option is not available for deeper seams. New methods need to be developed to consolidate coal around the borehole so that it does not rupture, and to control drawdown so that effective stresses are increased only slightly, allowing the coal to consolidate without shearing. Another possibility is to perforate a cased hole below the coal seam, creating rising, propped hydraulic fractures to gain access to the gas resource.

8.3.3.4 Resolving problems of wellbore instability.

Most borehole overstressing problems are solved by increasing the weight (density) of drilling mud and by casing the hole before production. Design of a high-quality mud with good *filter-cake*-forming properties is essential. The filter cake forms soon after the drill bit has advanced past a permeable stratum because the mud is maintained at a pressure that is higher than the formation fluids; the solids are plastered on the borehole wall, and the fluid filters into the formation. If the filter cake stays in place, it supports the borehole wall in the same way a flexible liner supports a tunnel. The support pressure can be estimated as

$$p_{\text{support}} = \alpha \cdot (p_w - p_f)$$

where $p_w - p_f$ is the pressure difference, and α is the filter cake efficiency, 1.0 for a perfect cake of very low permeability against a high-permeability formation.

If hole stability problems in hot environments are severe, the mud can be chilled as much as possible at the surface so that it reaches the hole bottom in a cooler state. This causes thermal contraction of the rock, which reduces the borehole wall stress concentration (Guenot and Maury, 1988). Conversely, hot drilling mud rising from depth can heat up the borehole wall in an uncased hole, causing stress buildup and increased sloughing. With knowledge of the temperatures of the formation and of the drilling mud being injected at the hole bottom, the magnitude of the stress change can be estimated:

$$d\sigma_\Theta = \alpha_T \frac{dTE}{1 - \nu}$$

where $d\sigma_\Theta$ is the change in tangential stress (negative if the mud is cooler), α_T is the coefficient of thermal expansion of the rock, dT is the temperature difference, and E and ν are the elastic parameters of the rock.

Much insight into rock behavior around boreholes can be achieved through the use of linear elastic models and closed-form equations (Wang and Dusseault, 1991). However, because many of the processes are of a stress-dependent elastic and strain-weakening nature, modeling of transient and complex boundary value cases requires coupled nonlinear elastoplastic numerical analysis. These processes can be modeled as two-dimensional elastoplastic yield problems using Mohr-Coulomb or parabolic rock rupture criteria, with various degrees of direct coupling between the fluid behavior and the rock mass, including thermal aspects and dilatancy (Cheatham et al., 1986; Vaziri, 1988).

8.3.4 Formation evaluation

8.3.4.1 Evaluation during drilling. Lithological information traditionally has been obtained during drilling by examining drill cuttings under a microscope. More recently, data from rugged downhole sensing devices are conveyed to the surface by pressure pulses in the mud column or by electrical or magnetic telemetry. These data allow the driller to rapidly optimize bit weight and rotation speed, and provide some degree of online evaluation of the rock through which the bit is cutting.

8.3.4.2 Geophysical logging and other wireline methods. When drilling is complete, or just before setting an intermediate casing string, physical properties of the rock are measured by a combination of *geophysical logging* and *sidewall sampling*. Samples can be obtained by wireline sidewall methods if cores were not obtained during the drilling of the well. Small plugs are driven into the formation explosively or by microbit drilling, giving samples inferior to conventional core but useful nevertheless (Whittaker, 1985). Samples of the *formation fluid* can be retrieved by a wireline formation tester, which presses a pad against the drillhole wall and allows a fluid sample to enter into a low-pressure chamber.

Geophysical wireline logs (*Rock Engineering,* Sec. 6.6.2) provide most of the rock information, particularly at elevations of potential reservoir rocks, although the entire open hole is usually logged for purposes of regional geological correlation. In countries where mineral rights are government owned, logging of all holes is often mandatory. The logs and cores become government property and may be examined for a fee.

Hydrocarbon saturation can be estimated using multiple resistivity logs by combining information on porosity, salinity of the mud and connate water, and conductivity of the rock. The porosity of the reservoir is estimated from density (gamma-gamma) logs, neutron porosity logs, and sonic transit time logs, refined with knowledge of lithology, mineral and fluid specific gravity, and phase saturations. The depth of invasion of drilling fluid into a porous stratum can be used as an indicator of the flow potential of the rock. Multiple resistivity logs are used to evaluate this; different electrode spacings have different depths of penetration, and the conductivity of the drilling mud is different from that of the formation water. Service company literature is a valuable source of information on these topics.

No direct methods have been developed for measuring the strength and deformability of rock in holes 1 km or more deep. With limited reliability, mechanical properties can be inferred from the geophysical records, using correlations established at shallower depth and in the laboratory.

New geophysical logs are keeping pace with advances in microelectronics; individual radioactive species can be quantified, elemental analysis can be conducted downhole, the borehole wall may be examined through televiewers or ultrasonic images (Fig. 8.12), and moisture content of the drillhole wall can be measured using dielectric properties.

8.3.4.3 Drill stem testing. After target formations are identified, either during drilling or when the hole is complete, flow and fluid properties are measured by drill stem testing. A single bottomhole packer

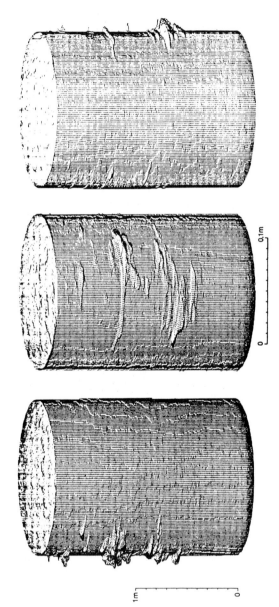

Figure 8.12 Images obtained by borehole televiewer (WBK Institut für Geophysik, Bochum, Germany).

or double (straddle) packers are lowered down the hole on the pipe (drill stem). They are inflated to isolate the desired interval, and the empty pipe is opened to allow fluid in the rock to enter. The pressure-time response and fluid chemistry are analyzed. Drill stem tests give measurements of the in situ pore pressure and rates of pressure buildup. Permeability tests of different types are used to assess the potential for long-term production. Later on, when the field has a number of wells, interference tests are carried out to evaluate large-scale permeability and interconnectivity between wells (Pollock and Bennett, 1986; *Rock Engineering,* Sec. 4.3.3.3).

8.4 Completion Technology

8.4.1 Casing, perforating, and acid treatment

Completion is the conversion of a borehole to a producing well (Fig. 8.13). After drilling and evaluation, casing is set to keep the drillhole from caving and to isolate the producing strata from other zones (Fig. 8.14). Only in extremely difficult rock conditions is a well cased to full depth before evaluation. Although at considerable cost, a well may be recompleted several times during its productive life, or even redrilled to gain access to new resources.

A *casing* or *liner* is a string of thin-walled tubes joined and cemented into the wellbore. To improve rock-cement bond, drilling mud *filter cake* is removed from the borehole wall by flushing the well, by mechanical swabbing, or by suddenly reducing the pressure and in-

Figure 8.13 A typical heavy oil well. About 60 barrels of oil per day are pumped into the tank at left.

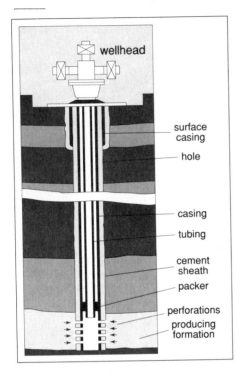

Figure 8.14 A completed well. Cemented casing isolates non-productive zones, and access to the producing horizon is obtained through peforations that penetrate the casing and cement.

creasing the flow gradient. The casing may be reciprocated and rotated with wire scratchers welded to its exterior. The *grout* is a cement-water mixture with additives to retard or accelerate set time, or to impart other properties. Casing is centralized in the hole and the grout is pumped in through the bottom, rising up along the outside. Hydraulic fracture and losses of grout can occur when the lateral rock stresses are low. To avoid this, the casing may be cemented in stages using special equipment or light-weight cements.

If the rock of the producing zone is competent and the permeability is unaffected by the drilling process, the casing is set just above the producing horizon for an *open-hole* completion. More commonly, the casing is set on the bottom, then *perforated* in the producing zone using a high-velocity shaped charge. This propels a projectile or a debris and gas jet through the casing and cement and into the rock, creating ports for production. Perforation patterns vary from a few closely spaced charges to tens of meters of widely spaced charges, in spiral, straight, staggered, or unidirectional arrays. These penetrate 200 to 500 mm beyond the cement into the formation, depending on rock type, strength, porosity, stress, wellbore pressure, and charge size.

The zone around the borehole wall is often a region of impaired permeability because of stress concentrations and clay particles washed into the pores during drilling, or because of fines migrating toward the well during production. Perforations can penetrate beyond this damaged zone, usually fracturing the rock and increasing the radius of drainage.

Perforation has little positive effect if the rock is pulverized; perforating in coal seams, for example, can impair productivity by blocking fissures with coal dust. Rock damage may lead to *solids production* problems (inflow of sand) later when the well is placed in production (Sec. 8.4.3). Design of perforation procedures remains empirical, based on large-scale experiments and previous successes.

Wellbore coatings can be penetrated by injecting acid mixtures through the perforations. *Acid treatment* is particularly effective in limestones, dolomites, and carbonate-cemented sandstones. Many acid types (HCl, CH_3COOH, HBF_4, formic acid) and chemical mixes are available, in aqueous or nonaqueous solutions, and careful consultation with a service company is required for particular applications. Sandstone acidization to remove clays and silica requires the use of hydrofluoric acid (HF), usually in combination with other acids and chemicals.

8.4.2 Fracturing

8.4.2.1 Why fracture formations? Hydraulic fracturing (Figs. 8.15 and 8.16) was first introduced around 1949 in the petroleum industry to enhance production rates and increase recoverable reserves. Fracturing alone is estimated to have added 15 percent to worldwide oil reserves (IOCC, 1983). The method is similar to that used for stress determination (*Rock Engineering,* Sec. 5.4.4.2) but the purpose is to stimulate production. The highly permeable conduits created by hydraulic fracturing serve to increase the volume of rock drained by the well, and in some cases to introduce large amounts of heat to reduce the viscosity of the reservoir fluids.

Benefits of enhanced production must be weighed against the cost of treatment and the risk of fracturing the cap rock. In appropriate conditions, hydraulic fracturing can reduce, by a factor of perhaps 4, the number of wells needed to develop a field. Fracturing is essential to develop low-permeability strata such as cemented oil sands, hot dry rock, tight gas sands, and certain chalks and limestones (Robinson et al., 1986).

8.4.2.2 Methods of fracturing. *Explosive fracturing* was once a common way of developing flow of oil into a well. Nitroglycerine was lowered to the desired location and detonated. The method was often successful, but occasionally holes were lost or the rock crushed, reducing production. The method became obsolete about 1960, although rapid impulse fracturing is of current interest (Christianson et al., 1988).

Figure 8.15 World's largest massive hydraulic fracture (MHF) stimulation treatment, Mobil Oil, Tex.: 70 million pounds of proppant, and 1.2 million gallons of fluid. (a) Proppant storage; (b) pump equipment. (*Photos courtesy J.-C. Roegiers*)

In *hydraulic fracturing*, water or a viscous gel is injected with a propping agent, usually coarse-grained rounded quartz sand, to keep the fracture open. Alternatively, the hole may be prefractured with liquid before a proppant is introduced. Gelled fracturing fluids can

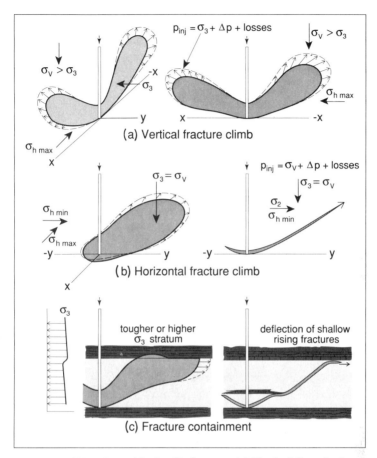

Figure 8.16 Directions of hydraulic fracture. (a) Vertical "angel wings"; (b) horizontal, curving upward; (c) contained beneath tough stratum.

carry sand far into the fractures, and they later degrade into a low-viscosity fluid which is cleaned up by back-flow to allow production. In water-sensitive formations, gas, foam, or petroleum-based liquids are used for fracturing. Recent trends have been toward larger and deeper fracturing, with injections of up to 30,000 m^3 of fluid and 200 kg/m^3 of proppant. This can provide drainage access to large volumes of reservoir rock. In *thermally enhanced recovery* projects, steam or hot water is often injected without proppant under conditions of continuous hydraulic fracturing.

Hydraulic or foam fracturing, in contrast to fracturing with explosives, avoids rock crushing but produces only a single fracture. Communication between natural joints in the formation may not increase significantly. *Tailored-pulse fracturing* techniques have been developed that avoid crushing yet produce up to eight major radiating fractures that increase the likelihood of intersecting natural joints. The method uses a charge of propellant (slow-burning explosive, usually a rocket fuel) tailored to produce rapid borehole pressurization to override the stress field without crushing the rock. The technique may be applied through cased, perforated boreholes (Cuderman, 1986).

Various methods have been tried in efforts to control fractures or to enhance recovery: cutting slots in the borehole wall using high-pressure *fluid jets* to control fracture direction; injecting *explosive slurries* by hydraulic fracturing, followed by detonation; high-energy *electrical discharge fracturing* at hole bottom; and even *nuclear devices*. These approaches have been rejected for reasons of poor control or high cost.

8.4.2.3 Behavior of hydraulic fractures. Hydraulic fractures on the large-scale propagate normally to the minor (least) principal stress (σ_3) direction, although locally they may follow open joints in a strong rock (Fig. 8.16). In regions where σ_3 is horizontal, a single vertical fracture plane is generated. If σ_3 is vertical, the fractures tend to propagate horizontally or at shallow upward angles.

Upward propagation occurs because the injected fluids have a much lower density than the surrounding rock. Typically, a fracturing fluid has a relative density of 1.0 to 1.3, whereas the lateral stress gradient in the rock is equivalent to a material of relative density 1.4 to 2.3. Thus, vertical fractures in an homogeneous reservoir climb upward and are often shaped like a pair of wings (Fig. 8.16a). For similar reasons, horizontal fractures tend to climb upward at angles as high as 30° if the principal stress difference (shear stress) is low (Fig. 8.16b).

A vertical induced fracture can be prevented from climbing only if the caprock is either stronger, or more highly stressed, than the rock in which the fracture is traveling (Fig. 8.16c). Fortunately, the caprock is usually a shale which is often both stiffer and more highly stressed than the reservoir rock.

In relatively shallow reservoirs where the minor principal stress is horizontal, rapid injection of viscous fluid creates a local increase in stress normal to the fracture. This leads to a new direction for σ_3 and to a change in the fracture plane orientation near the borehole. The direction of propagation can reverse again as the fluid drains, leading to cyclic pressure behavior (Dusseault and Simmons, 1982). Note that if the fracture is at first horizontal, this mechanism is unlikely because the maximum pressure across a horizontal fracture is governed by the overburden total stress and controlled globally by the free surface, which is free to move upward in response to injection.

Hydraulic fracture generates large increases in pore pressure, and weak rocks are induced to yield. Horizontal stresses are augmented, whereas the vertical stresses remain the same, because they are limited by the weight of overburden. Eventually, in shallow reservoirs, all induced fractures will become horizontal. This is the case for the in situ oil sand operations at Cold Lake and Athabasca, both in Alberta, Canada. Initially fractures were vertical at depths of 300 to 600 m, but after several months of thermal stimulation, they became horizontal climbing fractures.

8.4.2.4 Fracture modeling and simulation.
Various analytical and numerical models are used to estimate hydraulic fracture geometry, direction, and aperture from a knowledge of rock deformation properties, fracture toughnesses, and the stresses, temperatures, and fluid pressures around the well.

Analytical methods have been developed from elastic fracture mechanics; these require many assumptions as to homogeneity, constant pressure, two-dimensional plane strain, isotropy, and constant fracture toughness (Meyer, 1986). Finite element and boundary integral methods are used individually or in combination (Cheng and Liggett, 1984; Keat and Cleary, 1985; Detournay et al., 1986; Jones et al., 1986; Lam et al., 1986). The displacement discontinuity technique is a powerful means of investigating three-dimensional propagation from a well bore, particularly if coupled with finite element methods to account for material inhomogeneities (Vandamme et al., 1986a,b).

Fracture simulation is a challenging field of research because various processes interact and must be modeled in a coupled manner, and because an extensive fracture cuts through various rock types and many joints (Cleary, 1988). Fracture geometry predictions (height, extent, symmetry, and aperture) are difficult to verify because a fracture can rarely be observed; it can, with limited precision, be tracked using monitoring methods.

8.4.2.5 Fracture monitoring.
Near-surface strains caused by hydraulic fracturing can be detected and measured by extremely sensitive tiltmeters developed for monitoring of earthquakes. The method is applicable mostly to fractures at depths of less than 2 km, although some information on orientation and direction can be obtained for large fractures at greater depths. Tilt data permit only a rough estimate of fracture dimensions, but can discriminate between vertical and hori-

zontal fractures, and give the azimuths of vertical fractures to within a few degrees.

The propagation of a fracture from a wellbore can also be monitored by an array of geophones that locate acoustic emissions generated around the advancing fracture tip (Sorrells and Mulcahy, 1986; *Rock Engineering*, Sec. 12.2.6.4). Other methods include electrical field measurements, postfracture coring, observation wells, and geophysical temperature logging (Dusseault, 1986). As with all monitoring, the use of several independent techniques gives added confidence to the interpretation.

8.5 Production Technology

Methods of hydrocarbon production are by convention classified as primary, secondary, or tertiary.

8.5.1 Primary and secondary methods

Primary production makes use of the natural drive energy available in the reservoir. This pressure-maintaining energy can come from gas expansion (*gas cap expansion* or *solution gas drive*), from expansion of a liquid, from natural replenishment of underlying water (*natural water drive*), or from compaction of the reservoir rock (*compaction drive*).

Secondary production requires that pressure be maintained by injection of gas or water (Craig, 1971); water injection at some wells may be used to displace oil toward others (called *waterflooding*).

8.5.2 Tertiary methods

8.5.2.1 Definition. Tertiary or *enhanced oil recovery* methods, used mainly in oil sands, oil shales, and depleted reservoirs (IOCC, 1983), include all methods not defined as primary or secondary. They employ heat (Prats, 1982), miscible liquids or gases (Stalkup, 1983), carbon dioxide (Mungan, 1978), viscous polymers (Shah and Schechter, 1977), and other chemicals. Often, several methods are applied simultaneously (e.g., steam and CO_2 injection) or sequentially (e.g., waterflood followed by polymer flood and then fireflood). Tertiary methods usually work best in the absence of clays and gypsum, and when the reservoir is homogeneous and isotropically permeable with few joints and no channels.

8.5.2.2 Chemical methods. The effectiveness of the most common secondary method, waterflooding, is improved by adding chemical agents that alter the oil-water interfacial tension, or that change the condition of wetness of the reservoir mineral (oil-wet, water-wet, or mixed).

These agents are alkaline, and react with oil to create *surfactants* which improve waterflood performance (IOCC, 1983).

Polymer flooding (Chang, 1978) is the injection of a large volume of fresh water to flush out saline liquids, followed by a slug of water made extremely viscous by adding a polymer such as xanthate or guar gum, materials of high molecular weight. The viscous polymer displaces the oil toward production wells, increasing total production. Polymers are adversely affected by clays and by high-salinity fluids. Injection pressures are kept low to avoid hydraulic fracture, which would render the process ineffective. Polymer flooding is avoided under conditions of high temperature, shallow depth, thick gas zones, and high pore fluid salinity, where the polymer is rendered ineffective.

Micellar fluids (surfactant-stabilized emulsions of hydrocarbons in water) are used in a manner similar to polymers in polymer flooding. These materials are designed to give desired properties for individual cases, increasing sweep efficiency.

Miscible fluid flooding and an alternative, *miscible gas injection*, work mainly by reduction of capillary and interfacial forces, and reduction of the viscosity of the hydrocarbon phase. Fully mixed phases (self-dissolved) are created and displaced toward production wells.

Carbon dioxide flooding, either miscible or immiscible, is implemented in a variety of ways, depending on the recovery strategy. The major mechanisms contributing to increased recovery are reduced oil viscosity, increased relative permeability as the volume fraction of oil in the pores changes, volumetric expansion of oil as CO_2 is dissolved in it, lowering of interfacial tensions, and the driving pressure of the injected CO_2. Carbon dioxide flooding works best when the pay zones are thick and homogeneous without gas caps and when the reservoir is free from asphaltenes and similar materials that can cause precipitation of semisolid organic phases, blocking pores.

8.5.2.3 Thermal methods. *Steam flooding* (Fig. 8.17) takes advantage of the temperature sensitivity of heavy viscous oils. Several mechanisms aid in production, including viscosity reduction, thermal expansion, drive energy because of steam injection, steam distillation, and heat-induced molecular changes. Formations that require this treatment have low injectivity, and usually must be fractured to permit adequate steam injection. The steam is injected into individual wells in a pattern, flushing oil out radially or into a line of holes. Even a single well strategy is possible; steam is injected into a well for several weeks, the well is shut in for a while, then it is placed on production.

Fireflooding (combustion recovery) consists of injecting air or oxygen (with or without water) into the reservoir rock, igniting a fire downhole, and driving the flame front across the reservoir to another well. Viscosity

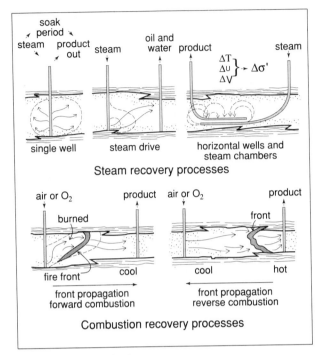

Figure 8.17 Thermal oil recovery.

of the oil is reduced because of heat, thermal cracking of hydrocarbons, and solution of gas in the oil. Drive energy is created by the combined pressure of injection and gas production. The large volumes of gas associated with this process can lead to difficulties because permeable channels may develop rapidly. Firefloods can be designed to move in the same direction as air flow (forward combustion) or in the reverse direction (reverse combustion). Often, a limited period fireflood is used to create a flushed area sufficient for steam to be injected.

The *underground coal conversion* (UCC) process is a promising method for developing deep coal deposits. A flow linkage is made between wells by reverse combustion, by hydraulic fracturing, or by directional drilling. Coal is ignited, and a gaseous mixture of oxygen and steam is injected continually. The coal partially burns, and produces a low-grade combustible gas which can be used to generate electricity, or can be cleaned for industrial use.

8.5.2.4 Monitoring of in situ processes.
Important considerations are the ability to predict, monitor, and control enhanced oil recovery pro-

cesses (Nyland and Dusseault, 1983; MacKinnon, 1986; Dusseault et al., 1988). Techniques to monitor in situ processes can be remote or proximal, active or passive, continuous or periodic. They make use of borehole logging and borehole transducers of various types, and seismic, electromagnetic, and direct strain measuring techniques, at the surface or in boreholes (Dusseault, 1986).

8.5.3 Solids production

8.5.3.1 The problem. Producing "solids" (sand) along with oil and gas costs the petroleum industry an estimated $250 million per year in Western Canada alone (Woodland, 1988). The problem is no less severe in oilfields in the North Sea, Gulf of Guinea, Indonesia, and other locations where poorly consolidated sandstones or weak rocks are exploited. The major consequences are abrasion of downhole and surface facilities, buckling of casing, plugging of casing by sand, and difficulties in disposing of an environmentally poor material after separation at surface.

Solids production is most severe in weak rocks subjected to large pressure drops near the wellbore during production, such as in gas production (high exit velocities) and heavy oil operations (large, near-wellbore pressure gradients) (Dusseault and Santarelli, 1989). Coal, chalk, oil sands, and high-porosity reservoirs are typical candidates for massive solids production, which can attain as much as 5 to 10 percent of the volume of fluid produced in exceptional cases. The problem develops in stronger rocks if a large damaged zone is created during drilling or perforation, and if high drawdowns are applied suddenly to the reservoir.

Most wells completed by perforation produce some minor amounts of solids during their initial stages and then stabilize. Solids production may be triggered later in the life of the well when free gas is produced along with the liquid or when the water content of the fluids rises significantly or as a result of attempts to maintain productivity by increasing the drawdown of the well.

Major problems develop when solids continue to be produced in large quantities. Continued erosion of the formation may lead to casing collapse when lateral loss of support is accompanied by downward drag as the overburden settles to fill the eroded cavity. Stress conditions in the reservoir are massively altered, approaching a condition where σ_v'/σ_h' is at the residual state, ≈ 3.0 to 3.5, over much of the reservoir. This has important consequences if the field is later subjected to hydraulic fracturing for tertiary recovery; lateral stresses are so low that wells "go on vacuum," and directionality of fractures controlled by virgin stress fields is destroyed.

8.5.3.2 Counteracting solids production.
Solids production often stops spontaneously because of arching or creation of a natural filter near the wellbore or from the reduction of pressure gradients that commonly accompanies production. The sensitive stratum may eventually be removed by erosion if it is thin, or, in thermal projects, it may become cemented by precipitated minerals or organic matter.

Suman et al. (1983) list the options for dealing with sand production:

Living with it and accepting the additional expense of frequent workovers, equipment replacement, and reduced safety around surface treatment facilities

Excluding the sand from the well bore by filters and screens placed around the producing perforations

Creating a granular filter (gravel-pack) around the bottom-hole production assembly, or forcing the gravel through the perforations into the formation

Using patented methods for epoxying or oxidizing the borehole before fracturing

Changing completion and production practice to prevent or stop solids production

Production practice, particularly the amount of drawdown and rate of production, can be adjusted to reduce destabilizing effects. However, this, like most other methods described above, often decreases oil production rates. Screens and gravel packs are the preferred methods, largely because operations personnel are familiar with them.

The problem of solids production is attracting more attention today than in the past, because heavy oil and natural gas are becoming more important, and solids production is usually most severe in these cases.

8.5.4 Production from oil sands

Oil sands (tar sands) represent a major fuel source for the coming centuries, and economical technology for developing them is a significant research goal (AOSTRA, 1969).

8.5.4.1 Surface mining.
Oil sands are extracted in large-scale open pit operations at the Suncor and the Syncrude mines in Canada (Fig. 8.18). Phases include overburden stripping, sand mining, transport to the extraction facility which uses a hot water separation method, and rejection of the waste sand to tailings ponds where it forms a sand beach, allowing thin sludge to run out into the pond (Fig. 8.19). The

Oil, Gas, and Geothermal Energy   409

(a)

(b)

Figure 8.18   Open-pit mining of oil sand. (a) Dragline creating a windrow of oil sand; (b) bucket wheel excavator transferring the ore to a conveyor.

Syncrude mine, which produces about 7 percent of Canada's oil needs, uses draglines and bucket wheel excavators operating from slope crest. Each day, Syncrude mines about 150,000 m^3 of ore, removes 30,000 m^3 (increasing each year) of overburden, and produces 22,000 m^3 of oil. The Suncor mine employs preblasting and bucket wheel

Figure 8.18 (*Continued*) Open-pit mining of oil sand.(*c*) conveyors meeting at the stockpile.

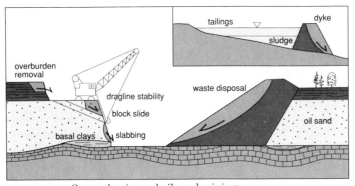

Figure 8.19 Geomechanics and oil sand mining.

mining on two levels. The purpose of blasting is not so much to fracture the oil sand as to loosen it to reduce frost penetration and reduce abrasion on the bucket teeth (Carrigy and Kramers, 1974; Spragins, 1978).

A 60-m-thick ore body consisting of uncemented sand (30 percent porosity) presents unique mining difficulties. Open-pit mining is feasible only within about 100 m of the surface, and conventional underground mining is uneconomical because 1 m^3 of sand must be extracted for every 0.18 m^3 of oil produced. Trials are in progress for a hybrid technology using mine access and horizontal steaming holes (Carrigy, 1986).

Problems of surface mining in tar sands include:

Removal of a clay shale overburden that has weak layers and swells when exposed to fresh water

Pit slope sloughing encouraged by high gas pressures in the pores, a problem that is reduced by preblasting and by keeping equipment away from the slope crest

High rates of wear on equipment caused by abrasiveness of the quartz grains, particularly in winter when the oil sand is frozen

Strength and toughness of the oil sands in cold weather which places great demands on mining equipment

The general difficulties of operating continuously in a harsh climate where winter temperatures regularly drop below $-35°C$ for weeks

Problems in disposing of large quantities of tailings produced as a by-product of the hot-water extraction process, including the slow consolidation rate of the sludge, and the need to construct kilometers of sand dikes on swelling clay shale foundations

Sampling and testing difficulties, caused by dissolved gases, make it difficult to accurately determine mechanical properties. Measurements of compressibility and shear strength are particularly sensitive to sample disturbance (Dusseault, 1980; Barnes and Dusseault, 1982; Dusseault and Scott, 1984).

Syncrude each day produces about 350,000 m^3 of *tailings;* the pond covers 11 km^3, retained behind sand dikes. Treatment of such a huge tailings stream, for example, by adding a chemical or employing filters, is usually uneconomical because of the low value of the ore. However, large-scale piping and blending of slurry wastes with sand can create a semi-solid material, allowing the open pit to be progressively filled, and eliminating the long-term need for tailings empoundments (Dusseault et al., 1989).

8.5.4.2 In situ extraction. The Canadian oil sands are being developed not only from open pits, but also in many in situ extraction "pilot projects," mostly in the Cold Lake and Peace River deposits (Fig. 8.20). Many of the problems discussed above in the context of open-pit mining also affect in situ extraction technology. The permeability of an oil sand deposit is higher than that of other oil reservoirs, but the viscosity of bitumen is also high, and attempts at conventional development are futile. The bitumen must be heated first, to reduce its viscosity.

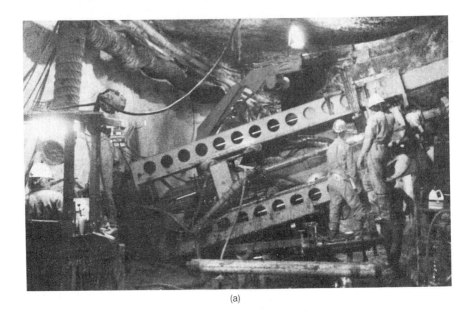

(a)

(b)

Figure 8.20 Underground recovery of oil from the Athabasca oil sand deposit, AOSTRA Underground Test Facility (UTF) at Fort McMurray, Alberta. (a) Drilling a 15° rising well; (b) completed well head. (*Photos, courtesy the Alberta Oil Sands Technology and Research Authority*)

Oil, Gas, and Geothermal Energy 413

(c)

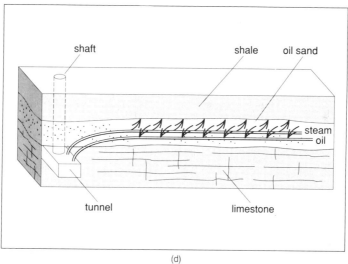

(d)

Figure 8.20 (*Continued*) Underground recovery of oil from the Athabasca oil sand deposit, AOSTRA Underground Test Facility (UTF) at Fort McMurray, Alberta. (c) 4-m-high drift with steam injection and extraction wells; (d) schematic of steam-assisted oil recovery. (*Photos, courtesy the Alberta Oil Sands Technology and Research Authority*)

Pilot projects in the United States, Venezuela, Europe, and Canada make use of steam injection, fireflood, electrical preheating, chemical emulsifiers, surface active agents, or solvent injection. Usually, combinations of several different processes are implemented sequentially, such as steaming followed by emulsion generation in situ (Carrigy, 1986). Major technical difficulties are associated with the weak and uncemented, but stiff, nature of the sands:

Hydraulic fracture propping agents are ineffective because of the granular nature of the reservoir rock.

Casing cement can easily fail because cement bonds poorly with bituminous sand.

Shear planes can develop, particularly in advance of horizontal rising fractures, and these can easily rupture the strongest casing.

Changes in the direction of hydraulic fracture propagation are induced by injection and thermal expansion, so that a technology developed for vertical fractures may be inapplicable later in the reservoir life when all fractures have become horizontal ($\sigma_3 = \sigma_v$).

Analyzing in situ extraction is challenging because of the complexity of the processes. Numerical analyses must rigorously couple fluid and heat flow aspects to elastoplastic behavior. The thermal model must simulate heat flow by convection, conduction, advection, thermal expansion, and contraction. The flow model must address changes in viscosity of oil and water, gases going into or out of solution, phase changes with pressure, generation of gases in situ, changing permeabilities because of dilation and shear distortion, and a heterogeneous and largely unknown flow system geometry in the reservoir. Geomechanical models must take into account shear distortion and volumetric dilatancy, stick-slip shearing of formations, thermal stresses, and effective stresses. Solutions call for simplifying assumptions as to reservoir geometry and properties (Dusseault et al., 1988).

As in all cases of complex geomechanics modeling, the predictions are inexact and must be checked by *in situ monitoring* during extraction. Relevant monitoring parameters include stress changes, strains and dilation, and pressure distributions. Monitoring is difficult because of depths (300 to 1500 m), and aggressive environments. Rock engineering can contribute to understanding of processes at depth in these difficult materials.

8.5.5 Production from oil shales

Oil shales, some of which were mined in the past, are not currently being exploited commercially because of the widespread availability of more readily extracted forms of oil and gas. However, salable by-products may affect the economics of mining. Oil shales can be used for making cement; the organic material is burned off in the clinkering process and reduces the cost of fuel. The Green River oil shales contain dawsonite ($NaAlHCO_3 \cdot H_2O$) and nahcolite ($NaHCO_3 \cdot 2H_2O$), industrial minerals of some value. Underground space generated by mining could be as valuable as the fuel in the ore, particularly near urban centers. Retorted shale can also be used as construction fill or in expanded form as a lightweight aggregate.

Oil shale cannot be burned efficiently as a heat source, and instead, interest is focused on extracting a liquid product. In *conventional mining* of oil shale, the extracted ore is retorted in a kiln at surface, and the waste material is returned to the cavities. A pilot underground mine has been operated successfully for a limited time at Rifle, Colorado.

Technical problems include the large amount of ore to be processed per barrel of liquid product, the viscous nature of the oil, and the low hydraulic conductivity of the shale. Disposal of spent shale is a major environmental concern because retorting renders heavy metals in the shale much more leachable, and the spent shale has a macroporosity in excess of 35 percent. The unit value of the ore is low, so environmentally acceptable disposal alternatives such as mine backfilling, which require further handling of materials, may prove uneconomical.

In situ extraction alternatives are variations of underground retorting or fireflooding. The impermeable shale must be fractured to create a material of sufficient mass permeability to allow injection of an oxidant, which sustains combustion and generates a gaseous or liquid product. Methods include hydraulic fracturing; injection and detonation of an explosive slurry; or mining, drilling, and blasting to create an in situ rubble retort which is then fireflooded. The shale must be uniformly fractured; otherwise channeling and bypassing of large unburnt areas renders the process inefficient. Only the richest oil shales, those of the Green River deposits with over 40 percent by volume organic matter, are likely to be amenable to economic development by in situ extraction in the foreseeable future.

8.6 Development of Geothermal Resources

8.6.1 Drilling and fracturing

Porous reservoirs with sufficient natural flow are developed in much the same way as oil and gas reservoirs. Elevated temperatures may lead to drilling problems, but these can usually be overcome with good drilling practice and careful control of drilling mud.

Flow through a natural jointing system can be enhanced by injec-

416 Rock Engineering

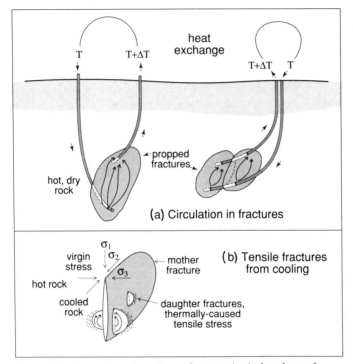

Figure 8.21 Fracturing of geothermal reservoirs in hot dry rock.

tion of a propping agent (sand or bauxite beads which are stronger and less soluble in hot liquids). In the absence of sufficient joints, the rock must be fractured to permit circulation of water (Fig. 8.21). Usually the stress conditions at depth dictate vertical fracture propagation (Sekine and Mira, 1980).

A volumetric shrinkage of several percent can result from years of injection of cool water into a hot hydraulically fractured granite mass, and this can lead to tensile fracturing of the reservoir rock. The thermal fracturing results in some loss of control over circulation of fluid, but the fractures also give access to additional volumes of hot rock, increasing the efficiency of the well (Fig. 8.21b).

8.6.2 Modeling of geothermal reservoirs

Numerical models permit evaluation of the economic potential of geothermal reservoirs, and design of systems for heat extraction. Modeling requires input values for thermal properties, coefficient of thermal expansion, specific heat, and thermal conductivity of the reservoir rocks (*Rock Engineering*, Sec. 10.3). At the depths of geothermal reservoirs, thermal properties are not affected much by stress variations, so nonlinear relationships with respect to stress are not required. However, the properties vary with temperature, and also depend on the intensity of jointing. Laboratory measurements on specimens therefore have limited relevance to the behavior of rock in situ. More information is now available on the thermomechanical characteristics of in situ rock as a result of studies for underground disposal of radioactive waste (Sec. 6.6).

8.7 Environmental Effects

8.7.1 Subsidence induced by extraction of fluids

8.7.1.1 Mechanisms and problems. Pressures in the pore fluids of a reservoir are reduced when oil and gas are extracted. This causes an increase in effective stress, followed by consolidation if the rocks are compressible. Substantial volume changes can occur also in the surrounding strata as the fluids drain from them (Donaldson and van Domselaar, 1983). Because compaction is spread throughout a wide area, it is mostly transmitted to the surface as uniform subsidence.

Consolidation creates a problem if subsidence has any significant environmental consequences, or if the behavior of the reservoir is impaired. Subsidence of reservoirs beneath lakes or the sea may present few concerns if adequately predicted and monitored, and if appropriate precautions are taken. Consolidation can even be helpful in that it pushes water and flushes oil or gas toward producing wells.

The chalk of southeastern England and northwestern France is a prolific oil producer in reservoirs beneath the North Sea. Chalk has unique properties; the small shells from which it is formed are susceptible to collapse under pressure. This condition can occur around a borehole subjected to large drawdown, resulting in impaired flow, and also in large surface subsidences.

The Ekofisk oil field is in a jointed chalk reservoir at a depth of 3 km. After 15 years of production, its center had subsided more than 5 m and was continuing to subside at 300 mm per year (Barton et al., 1986; Sulak and Danielsen, 1988). Excellent productivity has been maintained, perhaps assisted by these large vertical strains which serve as a "compaction drive" mechanism, as well as by the permeable network of joints. However, the subsidence has required extension of production platform legs at a cost of $450 million, and in-reservoir deformation has ruptured or deformed most casings, necessitating extensive well redevelopment. Had the high compressibility been understood initially, alterations to the platform design would have cost only a fraction of this amount.

Along the shores of Lago de Maracaibo in Venezuela, withdrawal of oil since the late 1920s has resulted in widespread subsidence by as much as 8 to 10 m, requiring the construction of long dikes and the relocation of facilities. Along the Texas Gulf Coast near Houston, withdrawal of groundwater from reservoirs 100 to 800 m deep has been followed by inundation of many square kilometers of low-lying agricultural land.

Areas of Wilmington Beach and San Diego in California have experienced up to 13 m of vertical subsidence and 3 to 4 m of lateral movement, as a direct result of petroleum production at depth in several "stacked" reservoirs in poorly consolidated sandstones. Land and buildings along the shoreline have been inundated, and other properties have been protected by the construction of dikes. Water mains have ruptured, railway lines buckled, and cracks have opened in roads and basements. Also in California, the San Joachin valley has experienced such severe subsidence because of deep groundwater withdrawal that flow in some canals has reversed, and other canals have overflowed or buckled. In Italy, the cities of Venice and Ravenna have banned further extraction of groundwater and natural gas, again because of the subsidence problem.

8.7.1.2 Prediction. Four mechanisms combine to cause volume changes in granular media: elastic grain distortion, grain rotation and sliding, yield and crushing at contact points, and pressure solution. The dominant mechanism in sandstone reservoirs is grain rotation and sliding, whereas in weak rocks such as the Chalk at Ekofisk, crushing of the microfossils and pore collapse are more important. In a subsidence-prone reservoir, the rock usually has porosity greater than 25 percent, poor cementation, and low quartz content (Dusseault, 1982). Geologically recent off-shore continental margin basins, or basins in areas of crustal extension, are the most susceptible.

Modeling of reservoir compaction is carried out to quantify the driving forces associated with compaction, and to make predictions of deformations in the reservoir and at surface. The mechanisms may be approximated using variations of consolidation theory, preferably with full coupling of stresses and pressures to account for effective stress changes (Biot, 1956). Initial conditions of stress state, porosity, geometry, and pressures must be known, as well as the mechanical characteristics of the reservoir rocks.

Nucleus-of-strain methods (Geertsma, 1973a,b) or similar elastic superposition approaches can be used to predict surface subsidence. They require that the volume change be specified through known pressure drops in reservoir blocks, giving a field of volume changes for the entire reservoir. To simulate a nonlinear compressibility, analysis can be done incrementally using small pressure drops, even though each analytic step is a linear one.

Finite element formulations can also be used. They allow simulation of more complex and realistic production history involving nonlinear compressibility and distributions of pressure drop, including anisotropic and heterogeneous overburden effects (de Waal and Smits, 1988).

8.7.1.3 Monitoring. Monitoring data are combined with modeling in order to better understand and predict reservoir behavior. Techniques are similar to those used to observe subsidence in mining (Sec. 7.6.4 and *Rock Engineering,* Chap. 12). Geodetic surveying is the most common basic method, supplemented by downhole measurements. Special oilfield methods include shooting radioactive bullets into the formations and logging using a gamma ray geophysical tool, surveys of casing collar locations using inductance logs, "behind-the-casing" logs of several types which can pick up formation interfaces or characteristic strata, and inclinometric surveys of wells to identify changes in curvature or deviation angle.

The following example illustrates the importance of monitoring, and its effectiveness when properly designed. Oil production from a reservoir on the coast of Lake Maracaibo in Venezuela began in 1926. Subsidence affecting 600 km^2 of low-lying coastal land now averages 5 m, and is progressing at up to 200 mm/year. To protect against flooding, an elaborate system has been developed including 47 km of coastal dikes, 59 km of inland dikes, and 500 km of canals. The dikes are of national importance, protecting about 200,000 people and one of the world's most productive oilfields.

The Maraven oil company, responsible for construction and safety of the dikes, is very much interested in monitoring their behavior to detect any deviations from predicted deformations. Some inland structures, particularly oil tanks, as well as some sensitive offshore oil platforms, are affected by the ground subsidence and have to be monitored to ensure their continuing stability and safety.

Leveling surveys now cover more than 1000 km^2 with 1600 benchmarks. The entire network is remeasured every 2 years, with some parts measured every 6 months. Contour maps of the subsidence basins are plotted automatically from computer-recorded field surveys. Parts of the time-consuming leveling surveys are being replaced by a satellite Global Positioning System. A network of seismic sensors with telemetric data acquisition covers the complete area. Horizontal strains are monitored with a Wild D14S instrument, and angles are monitored with a Wild T2000 electronic theodolite with automatic data recording and preprogrammed field checks of the observations.

Supplementary measurements include geodetic surveys of oil tanks using angular intersections with electronic theodolites, and horizontal rod extensometer measurements (up to 50-m rod length) with automatic data recording in selected areas of ground cracking (Bawden et al., 1990).

8.7.2 Disposal of spent fluids

8.7.2.1 Methods of disposal. Aqueous fluids produced during geothermal development or oil and gas extraction often contain large quantities of dissolved salts, and cannot be discharged into streams or lakes. Process water is often reinjected directly into the formation to maintain pressures and displace oil to producing wells. Otherwise, the most common disposal method is injection into nonproductive permeable formations, typically at depths of 500 to 1500 m. Target formations are investigated to prevent contamination of potable groundwater reserves. Many legal constraints and study requirements must be satisfied before licensing is allowed (Galley, 1968; Walker and Cox, 1976).

Criteria for successful injection are:

The reservoir must be sufficiently porous, permeable, and extensive to accept the volumes of waste.

The injected fluid must not react detrimentally with the reservoir rock and fluids (i.e., it must not cause plugging and pressure buildup).

Injection must not interfere with potable waters, nor impair other potentially productive horizons.

The well must be designed with a sufficient diameter and developed length to receive the desired flow rate once steady state is reached after many years.

Clear water disposal is relatively easy, but highly mineralized water can clog the formation. To develop adequate capacity, an injection well can be hydraulically fractured and propped initially, but may not be used continually at pressures approaching the hydraulic fracture condition.

8.7.2.2 Induced seismicity. Pressure injection of waste liquids can generate earthquakes, as it did near Denver in the late 1950s and early 1960s. To reduce the risk of induced seismicity, bottom hole injection pressures should not normally exceed 60 to 70 percent of the overburden pressure, although this requirement can be relaxed if the reservoir is very permeable and extensive, and if horizontal stresses are known to be close to the vertical stresses (Flak and Brown, 1988).

Seismometers are installed in shallow surface wells (penetrating rock beneath the overburden soils) to detect induced seismic events. Pressures and volumes of injection are then limited accordingly, to keep events below some level of magnitude or frequency.

References

AOSTRA (Alberta Oil Sands Technology and Research Authority), 1969 to present. Various reports and publications on oil sands and heavy oil technology. Alberta Oil Sands Information Centre, 500 Highfield Place, Edmonton, Alberta.

Arditty, P. C., Mathieu, F., and Staron, P., 1989. "Characterization of Fractured Hydrocarbon Reservoirs Using the EVA Acoustic Logging Tool." *The Log Analyst,* Special edition on full wave-form acoustic logging.

Armstead, H. C. H., 1978. *Geothermal Energy.* E. and F. N. Spon, London, 357 pp.

Barnes, D. J., and Dusseault, M. B., 1982. "The Influence of Diagenetic Microfabric on Oil Sands Behavior." *Can. J. Earth Sci.,* vol. 19, pp. 804–818.

Barton, N., Harvik, L., Christianson, M., Bandis, S. C., Mokurat, A., Chryssanthakis, P., and Vik, G., 1986. "Rock Mechanics Modelling of the Ekofisk Reservoir Subsidence." *Proc. 27th U.S. Symp. Rock Mech.,* Tuscaloosa, Ala., pp. 267–274.

Bawden, W., Chrzanowski, A., Barron, K., and Cain, P., 1990. "Mine Subsidence Monitoring Instrumentation," in J. A. Franklin (ed.), *Mine Monitoring Manual,*Can. Inst. Min. Metall., Montreal, Chap 3.4.

Biot, M. A., 1956. "Theory of Elasticity and Consolidation for a Porous Anisotropic Solid." *J. Appl. Phys.,* vol. 26, pp. 182–185.

Blatt, H., Middleton, G., and Murray, R., 1972. *Origin of Sedimentary Rocks.* Prentice-Hall, Englewood Cliffs, N.J., 634 pp.

Bowen, R., 1979. *Geothermal Resources.* Applied Science Publishing, London, 243 pp.

Buntebarth, G., 1984. *Geothermics.* Springer-Verlag, New York, 144 pp.

Byrer, C. W., Mroz, T. H., and Covatch, G. L., 1987. "Coalbed Methane Production Potential in U.S. Basins." *J. Petrol. Tech.,* vol. 39, no. 7, pp 821–834.

Carrigy, M. A., 1986. "New Production Techniques for Alberta Oil Sands." *Science,* vol. 234, pp. 1515–1518.

———, and Kramers, J. W., 1974. "Geology of the Alberta Oil Sands." *Proc. 1st Conf. Western Region, Eng. Inst. Canada,* Edmonton, Alberta, pp. 13–24.

Chang, H. L., 1978. "Polymer Flooding Technology—Yesterday, Today, and Tomorrow." *J. Petrol. Tech.,* pp. 1113–1128.

Chapman, R. E., 1976. *Petroleum Geology, a Concise Study.* Elsevier, Amsterdam, 302 pp.

Cheatham, J. B., Jr., Lin, Y.-H., and Pattillo, P. D., 1986. "Analysis of Borehole Stability Using a Strain Softening Model." *Proc. 27th U.S. Symp. Rock Mech.,* Tuscaloosa, Ala., pp. 552–561.

Cheng, H.-D., and Liggett, J. A., 1984. "Boundary Integral Equation Method for Linear Porous Elasticity with Applications to Fracture Propagation." *Int. J. Numerical Methods in Eng.,* vol. 20, pp. 279–296.

Christianson, M. C., Hart, R. D., and Schatz, J. F., 1988. "Numerical Analysis of Multiple Radial Fracturing." *Proc. 29th U.S. Symp. Rock Mech.,* Minneapolis, Balkema, Rotterdam, pp 441–451.

Cleary, M. P., 1988. "The Engineering of Hydraulic Fractures—State of the Art and Technology of the Future." *J. Petrol. Tech.,* vol. 40, no. 1, pp. 13–21.

Craig, F. F., Jr., 1971. "The Reservoir Engineering Aspects of Waterflooding." *Soc. Petrol. Engrs.* Monograph, Series 3, Richardson, Tex.

Cuderman, J. F., 1986. "Tailored-Pulse Fracturing in Cased and Perforated Boreholes."

Proc. Amer. Soc. Petrol. Eng., Unconventional Gas Technol. Symp., Louisville, Ky., pp. 525–534.

———, 1986. "Effects of Well Bore Liquids in Propellant-Based Fracturing." *Proc. 27th U.S. Symp. Rock Mech.*, Tuscaloosa, Ala., pp. 562–569.

Dake, L. P., 1978. *Fundamentals of Reservoir Engineering.* Elsevier, New York, 317 pp.

de Waal, J. A., and Smits, R. M. M., 1988. "Prediction of Reservoir Compaction and Surface Subsidence: Field Application of a New Model." *Soc. Petrol. Engrs. Formation Evaluation J.*, vol. 3, no. 2, pp. 347–356.

Detournay, E., McLennan, J. D., and Roegiers, J.-C., 1986. "Poroelastic Concepts Explain Some of the Hydraulic Fracturing Mechanisms." *Proc. Amer. Soc. Petrol. Eng. Unconventional Gas Technol. Symp.*, Louisville, Ky., pp. 629–638.

Donaldson, E. C., and van Domselaar, H. (eds.), 1983. *Proc. Forum on Subsidence due to Fluid Withdrawals.* U.S. Dept. Energy Forum, Checotah, Okla., 1982, 141 pp.

Dusseault, M. B., 1980. "Sample Disturbance in Athabasca Oil Sand." *J. Can. Petrol. Tech.*, vol. 19, no. 2, pp. 85–92.

———, 1982. "Identification of Reservoirs Susceptible to Subsidence." *Proc. Forum on Subsidence due to Fluid Withdrawal*, U.S. Dept. of Energy, Checotah, Okla., pp. 6–14.

———, 1986. "Monitoring in Situ Processes." *Proc. 37th Ann. Tech. Mtg., Petrol. Soc.*, Canad. Inst. Min. Metal, Calgary, Alberta, Paper 86-37-63, pp. 351–365.

———, Loftsson, M., and Russell, D., 1986. "The Mechanical Behavior of the Kettle Point Oil Shale." *Canad. Geotech. J.*, vol. 23, no. 1, pp. 87–93.

———, and Santarelli, F., 1989. "A Conceptual Model for Massive Solids Production." *Proc., ISRM/SPE Int. Symp. on Rock Mechanics at Great Depth*, Pau, France, Balkema, Rotterdam, vol. 2, pp. 789–797.

———, Scafe, D. W., and Scott, J. D., 1989. "Oil Sands Mine Waste Management: Clay Mineralogy, Moisture Transfer and Disposal Technology." *AOSTRA J. Res.*, vol. 5, pp. 303–320.

———, and Scott, J. D., 1984. "Coring and Sampling in Heavy Oil Exploration: Difficulties and Proposed Cures." *Proc. Am. Assoc. Petrol. Geol. Conf. on Exploration for Heavy Crude Oil and Bitumen*, Santa Maria, Calif., 22 pp.

———, and Simmons, J. V., 1982. "Injection Induced Stress and Fracture Orientation Changes." *Canad. Geotech. J.*, vol. 19, no. 4, pp. 483–493.

———, Soderberg, H., and Sterne, K., 1984. "Preparation Techniques for Oil Sand Testing." *ASTM J. Geotech. Testing*, vol. 7, no. 1, pp. 3–9.

———, Wang, Y., and Simmons, J. V., 1988. "Induced Stresses Near a Fireflood Front." *AOSTRA J. Res.*, vol. 4, pp. 153–170.

Edwards, L. M., Chilingar, G. V., Rieke, H. H., and Fertl, W. H., 1982. *Handbook of Geothermal Energy.* Gulf Publishing, Texas, 613 pp.

Flak, L. H., and Brown, J., 1988. "Case History of Ultra Deep Disposal Well in Western Colorado." *Proc. Int. Ass. Drilling Contractors and Soc. Pet. Eng. Drilling Conf.*, Dallas, Tex., pp. 381–394.

Galley, J. E. (ed.), 1968. "Subsurface Disposal in Geologic Basins—A Study of Reservoir Strata." *Amer. Assoc. Petrol. Geol. Mem.* no. 10, 253 pp.

Geertsma, J., 1973a. "Land Subsidence above Compacting Oil and Gas Reservoirs." *J. Petrol. Tech.*, vol. 25, pp. 733–744.

———, 1973b. "A Basic Theory of Subsidence due to Reservoir Compaction: The Homogeneous Case." *Verh. Kon. Ned. Geo. Mijnbouwkungid Genootschap* (Netherlands Royal Geological Society Proceedings), vol. 28, pp. 43–62.

Geophysical Press, 1987. Series on Geophysical Exploration, 14 volumes by various authors. London.

Gold, T., and Soter, S., 1982. "Abiogenic Methane and the Origin of Petroleum." *Energy, Exploration, and Exploitation*, vol. 1, pp. 88–103.

Guenot, A., and Maury, V., 1988. "Stabilité des Forages Profonds." In: *Thermomécanique des Roches, Manuels et Méthodes No. 16*, Pierre Bérest (ed.), BRGM Editions, Orléans, pp. 292–303.

———, and Santarelli, F. J., 1989. "Influence of Mud Temperature on Deep Borehole Behavior." *Proc. ISRM/SPE Int. Symp., Rock Mech. at Great Depth*, Pau, France, Balkema, Rotterdam, vol. 2, pp. 809–817.

Haas, M. R., Brashear, J. P., and Morra, F., Jr., 1987. "Historical Trends and Current Production of Gas from Tight Formations." *J. Petrol. Tech.,* vol. 39, pp. 77–88.

Harrison, R. S., 1984. "The Bitumen-Bearing Paleozoic Carbonate Trend of Northern Alberta." *Amer. Assoc. Petrol. Geol. Conf. on Exploration for Heavy Crude Oil and Bitumen,* Santa Maria, Calif., vol. 7.

Hayatdavoudi, A., and Apande, E., 1986. "A Theoretical Analysis of Well Bore Failure and Stability in Shales." *Proc. 27th U.S. Symp. Rock Mech.,* Tuscaloosa, Ala., pp. 571–579.

IOCC (Interstate Oil Compact Commission), 1983. *Improved Oil Recovery.* Oklahoma City, 363 pp.

Jones, A. H., Bell, G. J., and Morales, R. H., 1986. "Coalbed Hydraulic Fracture Treatment, Empirical Relationships and Computer Simulation." *Proc. Amer. Soc. Petrol. Eng. Unconventional Gas Symposium,* Louisville, Ky., pp. 699–710.

Keat, W. D., and Cleary, M. P., 1985. "Surface Integral and Finite Element Hybrid Method for Two and Three Dimensional Fracture Mechanics Analysis." Massachusetts Institute of Technology internal report.

Kerr, C. J., 1988. "PDC Drill Bit Design and Field Application Evolution." *J. Petrol. Tech.,* vol. 40, pp. 327–332.

Lam, K. Y., Cleary, M. P., and Barr, D. T., 1986. "A Complete Three-Dimensional Simulator for Analysis and Design of Hydraulic Fracturing." *Proc. Amer. Soc. Petrol. Eng. Unconventional Gas Symposium,* Louisville, Ky., pp. 673–684.

Logan, T. L., Seccombe, J. C., and Jones, A. H., 1986. "Hydraulic Fracture Stimulation Results and Diagnostics in Deeply Buried Coal Seams, Piceance Basin, Colorado." *Proc. 27th U.S. Symp. Rock Mech.,* Tuscaloosa, Ala., pp. 677–681.

MacKinnon, R. J., 1986. "Moving Grid Finite Element Simulation of Cavity Growth and Rock Response Associated with Underground Coal Conversion." *Proc. 27th U.S. Symp. Rock Mech.,* Tuscaloosa, Ala., pp. 716–724.

Maury, V. M., and Sauzay, J.-M., 1987. "Borehole Instability: Case Histories, Rock Mechanics Approach, and Results." *Proc., SPE/IADC Drilling Conference,* New Orleans, pp. 11–24.

Meyer, B. R., 1986. "Design Formulae for 2-D and 3-D Vertical Hydraulic Fractures: Model Comparison and Parametric Studies." *Proc. Amer. Soc. Petrol. Eng. Unconventional Gas Symposium,* Louisville, pp. 391–408.

Mungan, N., 1978. "Carbon Dioxide Flooding—Fundamentals. *J. Can. Petrol. Tech.,* vol. 17, no.1, pp. 87–92.

Nelson, R. A., 1987. "Fractured Reservoirs: Turning Knowledge into Practice." *J. Petrol. Tech.,* vol. 39, no. 4, pp. 407–414.

Nyland, E., and Dusseault, M. B., 1983. "Fireflood Microseismic Monitoring: Results and Potential for Process Control." *J. Can. Petrol. Tech.,* vol. 22, no. 2, pp. 62–68.

OGJ (Oil and Gas Jour.) Drilling Technology Report, 1988. *Horizontal Well Operations* (Collection of articles by Elf Aquitaine engineers), Pennwell, 42 pp.

Patigny, J., Li, T.-K., Ledent, P., Chandelle, V., Depouhon, F., and Mostade, M., 1987. "Belgian-German Field Test Thulin—Results of Gasification." *Proc. 13th Underground Coal Gasification Symposium,* 12 pp.

Pollock, C. B., and Bennett, C.; 1986. "Eight-Well Interference Test in the Anschutz Ranch East Field." *Amer. Soc. Petrol. Eng. Formation Evaluation J.,* vol. 1, pp. 547–556.

Prats, M., 1982. *Thermal Recovery.* Soc. Petrol. Eng. Monograph Series 7, Richardson, Tex.

Rabia, H., 1985. *Oilwell Drilling Engineering.* Graham and Trotman, London, 327 pp.

Robinson, B. M., Holditch, S. A., and Lee, W. J., 1986. "A Case Study of the Wilcox (Lobo) Trend in Webb and Zapata Counties, Texas." *J. Petrol. Tech.,* vol. 39, pp. 1355–1364.

Sekine, H., and Mira, H., 1980. "Characterization of a Penny-Shaped Reservoir in a Hot Dry Rock." *J. Geophys. Res.,* vol. 85, no. B7, pp. 3811–3816.

Shah, D. O., and Schechter, R. S., (eds.), 1977. *Improved Oil Recovery by Surfactant and Polymer Flooding.* Academic Press, New York.

Shi, Y., and Wang, C. Y., 1986. "Pore Pressure Generation in Sedimentary Basins: Overloading versus Aquathermal." *J. Geophys. Res.,* vol. 91, no. B2, pp. 2153–2162.

Sorrells, G. G., and Mulcahy, C. C., 1986. "Advances in the Microseismic Method of Hydraulic Fracture Azimuth Estimation." *Proc. Amer. Soc. Petrol. Eng. Unconventional Gas Symposium,* Louisville, Ky., pp. 109–120.

SPE, 1987. *Petroleum Engineering Handbook,* vols I, II. Society of Petroleum Engineers, Also, various texts in SPE Monograph Series, Society of Petroleum Engineers, Richardson, Tex.

Spragins, F. K., 1978. "Athabasca Tar Sands: Occurrence and Commercial Projects." Chap. 5 in: *Bitumens, Asphalts and Tar Sands, Developments in Petroleum Science.* G. V. Chilingarian and T. F. Yen, (eds.), Elsevier, New York, pp. 93–121.

Stalkup, F. I., Jr., 1983. *Miscible Displacement.* Soc. Petrol. Eng. Monograph Series no. 8, Richardson, Tex.

Sulak, R. M., and Danielsen, J., 1988. "Reservoir Aspects of Ekofisk Subsidence." *Proc. 20th Offshore Technology Conf.,* Houston, Paper 5618. See also related papers, same proceedings, nos. 5619-5623, 5652-5656, 5678.

Suman, G. O., Ellis, R. C., and Snyder, R. E., 1983. *Sand Control Handbook,* 2d ed., Gulf Pub., Houston, 89 pp.

Swanson, R. K., Bernard, W. J., and Osoba, J. S., 1986. "A Summary of the Geothermal and Methane Production Potential of U.S. Gulf Coast Geopressured Zones from Test Well Data." *J. Petrol. Tech.,* vol. 39, pp. 1365–1370.

Tissot, B. P., and Welte, D. H., 1978. *Petroleum Formation and Occurrence, a New Approach to Oil and Gas Exploration.* Springer-Verlag, New York.

UNITAR 1979, 1982, 1985, 1988. *Proceedings of UNITAR* (United Nations Institute for Training and Research) 1st, 2d, 3d and 4th Int. Conf. on the Future of Heavy Crudes and Tar Sands. Edmonton, Alberta; Caracas, Venezuela; Santa Barbara, Calif. Available through UNITAR, New York.

Upton, J. W., Jr., 1979. "Diamond Drilling Applications for Geothermal Exploration." *Proc. 36th Ann. Mtg. & Convention, Canad. Diamond Drill. Assn.,* Victoria, B.C., paper no. 4, 27 pp.

Vandamme, L. M., Jeffrey, R. G., Jr., and Curran, J. H., 1986a. "Pressure Distribution in Three-Dimensional Hydraulic Fractures." *Proc. Amer. Soc. Petrol. Eng. Unconventional Gas Symposium,* Louisville, Ky., pp. 663–672.

———, ———, and ———, 1986b. "Effects of Three-Dimensionalization on a Hydraulic Fracture Pressure Profile." *Proc. 27th U.S. Symp. Rock Mech.,* Tuscaloosa, Ala., pp. 580–590.

Vaziri, H., 1988. "Coupled Fluid Flow and Stress Analysis of Oil Sands Subject to Heating." *J. Can. Petrol. Tech.,* vol. 27, no. 5, pp. 84–91.

Walker, W. R., and Cox, W. E., 1976. *Deep Well Injection of Industrial Wastes: Government Controls and Legal Constraints.* Virginia Water Resources Research Center, Blacksburg, Va. 163 pp.

Wang, Y., and Dusseault, M. B., 1991. "The Pore-Pressure Gradient Effect on the Basic Stress-Diffusion Solution in Saturated Compressible Porous Media." *Water Resources Research,* in press.

Whittaker, A., 1985. *Coring Operations: Procedures for Sampling and Analysis of Bottomhole and Sidewall Cores.* Int. Human Res. Dev. Corp., Boston, 174 pp.

Woodland, D., 1988. *Public Report on Sand Control Costs,* Geomechanics Group, Petrol. Soc., Can. Inst. Min. Met. Calgary, Alberta.

Wyman, R. E., 1984. *Gas Resources in Elmworth Coal Seams.* Deep Basin Gas, Amer. Assoc. Petrol. Geol. Mem. No. 38, pp. 173–187.

Index

Abandoned mines, storage in, 279
Abrasion value test, 27
Aggregates, 21–29
 absorption of, 23
 crushing of, 26
 deleterious materials in, 22
 durability of, 23
 quality and testing, 21–28
Airblasts, 354
Alkali-aggregate reactivity, 24
Analytical design, tunnels, 245–248
Anchoring:
 of dams, 198–199
 of foundations, 149
Aprons for dams, 192
Aquifers for storage, 279–282
Arching above tunnels, 243–245
Armor stone, 17–19
Ashlar, 43
Avalanches, 61–63
 diversion structures, 101
 mobility of, 62

Back analysis, of tunnels, 246–248
Backfill:
 placement of, 361–362
 in underground mines, 17, 325, 357, 360–363
Basement excavations, 128–133
Bearing pressures:
 allowable, 136–137
 of dams, 178
Benches and berms, 77–78, 94
Bernold tunnel lining, 234
Blasthole stoping, 319, 323–325
Blasting:
 for aggregates, 28
 of caverns, 298–301
 of dam foundations, 189

Blasting (*Cont.*):
 of mines, 344
 of shafts, 346
 of stone, 42
 of tunnels, 221–223
Block caving, 336
Block shape in rockfill, 20
Borehole televiewer, 396
 breakout, 391–392
 casing and perforating of, 397
 stability of, 391–394
Boring machines for caverns, 301–302
Bouncing and rolling rock, 66, 83–84
Breakout, 391–392
Breakwaters, 17
Bricks, 30
Building stone (*see* Stone)
Bulking of ore, 324, 363
Bursting (*see* Rockbursts)

Cable lacing, 357–358
Caissons, 113
Canals, 76
Casing, 397
Catch fences, 94–101
Caverns, 204, 263–311
 construction of, 304–305
 design of, 292–298
 experimental, 304
 natural, 263–264
 shape of, 294–295
 size of, 263–264
 (*See also* Underground space)
Caves, natural, 263–264
Caving, 317–318, 323, 333
 potential, 336
 sublevel, 322, 335–336
Caving-induced subsidence (*see* Subsidence)

Cement manufacture, 30
Ceramics, 30
Chemical methods in oil production, 404–405
Chimney subsidence (see Subsidence)
Churchill Falls underground complex, 274
CLAB project, Sweden, 290
Classifications:
  for caveability, 336
  comparisons of systems, 241–243
  expert systems for, 242
  Laubscher system, 336, 340–341
  Q-system, 241–242, 296–297, 340
  RMR (rock mass rating), 241, 340–341
  RSR (rock structure rating) system, 240–241
  size-strength, 241, 341
  Terzaghi, 243–244
Clay mylonites, 71
Clints, 145
CN Tower, Toronto, 143–145
Coal:
  cutters, 332
  drilling, 393
  mining, 326
  underground conversion to gas, 406
Color of stone, 39–40
Compaction of rockfill, 14
Competence factor in tunnels, 238–239
Compressed air:
  as cushions, 275–276
  storage in caverns, 274–276
  use in tunnels, 227
Construction:
  control in tunneling, 254–255
  near-surface, 4
Contaminant migration, 287–288
Convergence in tunnels, 229
Conveyor systems, 226, 331
Core disking, 388
  storage of, 212
Cost:
  of caverns, 268–269
  of tunnel construction, 218–219
  of underground oil storage, 277–288
Creep of rock slopes, 66–68
Crown plates, 233
Crushing:
  of aggregates, 26
  of ore, 344
Cryogenic storage of liquified gas, 283–284

Cut-and-cover tunnels, 214
Cut-and-fill mining, 320, 329–330
Cutoff walls, 188

Dams and reservoirs, 4, 155–202
  case histories of, 163–168
  design of, 177–189
  foundation construction of, 189–199
  landslide-created, 52–53
  types of, 158–161, 177–178
Debris flows, 61
Deleterious materials in aggregates, 22
Design of mine openings, 337–343
Disking of core, 388
Ditches, rock collection in, 94, 97
Drainage of slopes, 90
Dressing of stone, 43
Drill stem testing, 395–397
Drilling methods:
  for deviated holes, 388–389
  for muds, 389–390
  for oil and gas, 387–394
Durability:
  of aggregates, 23
  of stone, 40–42

Earthquake-resistant design of slopes, 80–81
Earthquakes triggering landslides, 74
Ekofisk oilfield, 417–418
Embankment dam construction, 192–193
Embankments, rockfill, 13
Empirical design:
  of caverns, 295–297
  of mine openings, 340–341
  of tunnels, 239–245
Enhanced oil recovery, 404–407
Environmental effects:
  of oil reservoirs, 417–421
  in quarrying, 9
Erosion control, 90–91
  of reservoir perimeter, 172
  of rock slopes, 68–70
  and solution beneath dams, 165–166
  of spillways, 189
Expert systems, tunneling, 242
Exploration for oil and gas, 383–397
Explosive fracturing of well bores, 399

Factors of safety of slopes, 81
Failure modes underground, 206–207
Fan drilling of blastholes, 324

Faults:
  beneath dams, 168
  in foundations, 124–125, 189–190
Fences, catch, 94–101
Filter cake, 393
Final perimeter of open-pit mines, 78
Finger shields, 228
Fireflooding of oil reservoirs, 405–406
Floor heave, 351–353
Footings (see Foundations)
Forepoling, 228
Formation evaluation, oil and gas, 394–397
Foundations, 3, 111–153
  construction of, 145–151, 189–199
    on cavernous ground, 122–124
    on faults, 124–125
  design of, 134–145
  excavation of, 146
  failure mechanisms, 113–128
  treatment of, dams, 189–192
  types of, 111–113
Freezing:
  for caverns, 298
  tunnels, 227, 229, 251–252
Frost heave, 125–127
Full face tunneling, 222

Gas:
  exploration for, 383–397
  hazard in tunnels, 250
  outbursts in mines, 359–360
  storage, 283–284
Geophysical logging, 394–395
Geothermal resources, 375, 415–417
  drilling of, 386
  reservoirs of, 382–383
Gloryhole mining, 78–79
Gob, 333
Graben slides, 58
Gradation:
  of aggregates, 26
  of rockfill, 19–20
Ground response curves, tunneling, 245
Groundwater:
  around caverns, 297, 305–306
  control of, in foundations, 146
  in quarries, 11
  in tunnels, 248–254
Grouting:
  of caverns, 298
  of dam foundations, 187–188, 190–192
  of foundations, 148–149
  of oil wells, 398
  in tunnels, 251–252

Haulage of rock, 12
Hazards of abandoned mines, 364–365
Heave:
  of floors in mines, 351–353
  in foundations, 132–133
Highways, 76, 135
Himalayan tunnels, 238
History of tunneling, 209–210
Hong Kong landslides, 51. 72
Horizontal drillholes, 213
Hot water storage, 276
Hydraulic fracturing, 399–404
  mine props, 356–357
  roof supports (shields), 331
Hydroelectric facilities, 271–276
Hydrofracturing (see Hydraulic fracturing)
Hydrothermal activity, 314

Ice and clay damming of rock slopes, 68
Impoundment of reservoirs, 195
In situ extraction of oil sands, 411–414
Inclines (see Shafts and inclines)
Industrial minerals, 28
Inspection:
  of dams, 195–197
  of foundations, 149–151
Instability mechanisms, 347–348
Instrumentation (see Monitoring)
Itaipu dam, 194–195

Jetting, 189–192, 298

Karsts, 118–124

Lacing, cable, 357–358
Lagging, 229
Lake Maracaibo oil reservoir, 418
Landslides, 49–110
  activity, 75–76
  classification, 55
  dams, 52–53
  evidence of, 75
  hazards, 49–51
  investigations, 74
  mechanisms, 55–74
  prediction, 103–104
  velocity, 63
Lattice girders, 231–232
Leaching of ores, 336–337
Leakage of dams and reservoirs, 169–171, 186–187
Lighting, underground, 270
Lime, 31
Limiting equillibrium analysis, slopes, 79–82

Liners:
    concrete, 234–235
        cast-in-place, 237
        segmental, 237
    steel, 234
    stiff and flexible, 236
    thickness of, 236–237
    for tunnels, 234–248
Longwall mining, 331–333, 351–353
Los Angeles abrasion test, 22–23
Lost circulation, 390

Machinability of stone, 39
Magmatic differentiation, 314
Maintenance:
    of dams, 197–198
    of tunnels, 255–256
Malpasset dam, 167–168
Mesh:
    curtains on slopes, 96–98
    support in mines, 357–358
Microseismic monitoring, 358–359
Mine openings:
    design of, 337–343
    physical modeling of, 339–340
Mineralization, 5–6, 313–314
Mining, open-pit (see Open-pit mines)
Mining, underground, 313–373
    and civil engineering, 204–205
    exploration, 314–315
    methods, 337
    phases, 314–316
    planning, 337–338
    terminology, 316–317
Mining-induced seismicity (see Rockbursts)
Modeling:
    of hydraulic fractures, 403
    of mine openings, 338–340
Monitoring:
    of caverns, 303–306
    of dams, 193–195
    of hydraulic fractures, 403–404
    microseismic, 358–359
    of oil reservoirs, 406–407
    of slope stability, 101–106
    subsidence, 367–368, 419–420
    of tunnels, 246–248, 255
Mount Isa mine, 350
Mucking in tunnels, 226–227

NATM (New Austrian tunneling method), 229–234

Nuclear power plant foundations, 141–143

Observational method of tunnel design, 211
Offshore exploration for tunnels, 213
Oil and gas, 375–424
    exploration for, 383–397
    reservoirs of, 377–378
        modeling of, 384
    underground storage of, 277–285
Oil sands, 379–380, 408–414
    drilling of, 385
    well production, 404–415
Oil shales, 380–381, 414–415
Open-pit mines, 53, 71
    final perimeter of, 78
    site investigation of, 77
    (See also Slopes)
Open stoping, 319, 323–325
Orebodies, 5–6, 313–314
Orvieto stabilization, 72–74
Outbursts, gas, 359–360

Panama Canal, 76–77
Perforation of wellbores, 398–399
Petrographic number, 22
Physical modeling of mine openings, 339–340
Piles and caissons, 113, 128, 139–141, 145–146
Pillars:
    bursts of, 354
    design of, 341–343
Pilot tunnels, 213
Pipeline routes, 76, 135
Piping beneath dams, 166–168
Placer deposits, 313
Plug-and-feather splitting of stone, 42–43
Polished stone value test, 27
Polishing of stone, 27, 45
Portals and shafts, 214–216
Post pillar mining, 320, 330,399
Potash:
    backfill with, 361
    mining of, 326–327
Power lines, 76
Precast liner segments, 234–235
Preconditioning of rock, 356
Primary and secondary support, 227
Probabilistic design, slopes, 81–82
Progressive failure of slopes, 58–61
Pumped storage, 271–274

Punching failure, foundations, 114–116

Q system (see Classifications)
Quarrying:
  of aggregates, 28
  of building stone, 42–43
  design and costs of, 11–12
  exploration and planning of, 5, 8
  groundwater control, 11

Radioactive waste:
  annual production of, 385
  disposal in rock, 286
  research of, 289–290
  types of, 285
  (See also Waste disposal, Repositories)
Railways, 76
Raise boring, 225–226, 347
Raises, 316, 346–347
Raveling:
  and caving, underground mines, 348–349
  of rock slopes, 65–66
Recharge wells, 297
Recontouring of slopes, 89
Repositories:
  in salt, 288–289
  sealing of, 287–288
  (See also Radioactive waste)
Reservoirs:
  surface, water, 169–173, 205
  underground, oil and gas, 375–382
  water, 271
Resources, evaluation, 5–7
Retaining walls, 93–94
Riprap, 15–17
RMR (rock mass rating) system, 240–241
Roadheaders, 225, 328
Rock, behavior of, 206–207
Rock, preconditioning of, 356
Rock load on tunnel liners, 243
Rock mass rating (RMR) system, 241, 340–341
Rock quality designation for tunnel support design, 240
Rock slopes, creep of, 66–68
Rock structure rating (RSR) system, 240–241
Rockbolting:
  of slopes, 91
  in tunnels, 229–234
Rockbursts:
  in mines, 335, 348, 353–359
  in storage caverns, 303

Rockbursts (Cont.):
  in tunnels, 207
Rockfalls, 65–66, 116
  protection from 87–88
  warnings of, 102
Rockfill, 12–20
  compaction of, 14
  testing of, 17–19
Room-and-pillar mining, 319, 325–327
  design of, 341–343
Roughness of aggregate particles, 27
RQD (rock quality designation) for tunnel support design, 240
RSR (rock structure rating), 240–241
Rubble-filled caverns, 279

Salt domes, 314
  mining of, 327
Sandwich packs, 333
Sawing of stone, 43–44
Scaling:
  of foundation walls and footing, 146–148
  of rock slopes, 84–86
Seismic testing of dam foundations, 176–177, 193
Seismicity:
  induced by oil pumping, 420–421
  induced by reservoir filling, 172–173
Settlement:
  of embankments, 15
  of foundations 137–139
Sewage disposal in caverns, 292
Shafts and inclines, 214–216, 225, 316, 344–347
Shape:
  of caverns, 294–295
  of tunnels, 216–218
Shields, tunneling, 227
Shotcreting, 229–234
  in caverns, 302
  of slopes, 91–93
  in tunnels, 233–234
Shrinkage stoping, 320, 330
Sinkholes beneath foundations, 118–124
Site investigation:
  for caverns, 293–294
  for dams and reservoirs, 173–177
  for foundations, 133–134
  for tunnels, 211–214
Size-strength, 241, 341
Slab slides, 58
Slate for roofing, 45
Slide mechanisms, 55–74
Slopes, 2, 49–110
  blasting and excavation, 84

429

Slopes (Cont.):
  design of, 77–84
  drainage of, 90
  earthquake-resistant design of, 80–81
  engineered, 53–55
  instability, around reservoirs, 171–172
  stabilization, 84–101
  (See also Landslides)
Sluicing of rockfill, 13
Solids, production in oil wells, 407–408
Solution mining of caverns, 278–279, 297–298, 303
Soundness of aggregates, 23–24
South African mining, 333, 355
Specifications:
  for embankments, 15
  for tunnels, 219–221
Spiling, 228
Spillways, 188–189
Spoon-shaped slides, 57
Squeezing rock:
  around caverns, 302–303
  in foundations, 128–131
  in mines, 347–351
  in tunnels, 206, 238–239
Stability analysis, dams, 180–186
Stabilization of slopes, 84–101
Stand time in tunnels, 227
Steam flooding of oil reservoirs, 405
Steel sets, 229
Stepped-path slides, 57–48
Stone, 1, 5–47
  ancient uses for, 33
  building, 33–45
    production statistics for, 36–37
  color of, 39–40
  dressing of, 43
  durability of, 40–42
  machinability of, 39
  polishing of, 27, 45
  quality of, 38–42
  sawing of, 43–44
Stopes, 316
Storage:
  aquifiers for, 279–282
  of drinking water, 284
  of hot water, 276
  (See also Underground space)
Stress analysis:
  control in mining, 351
  of slopes, 82–83
Sublevel caving, 322, 335–336
  stoping, 319, 323–325

Subsidence:
  caving-induced, 363–368
  chimney, 323, 364
  control, 366–367
  drainage-induced, 367
  monitoring, 367–368
  oil and gas reservoirs, 417–420
  prediction, 365–366, 418–419
  processes, 363–365
  regional, 364
Superconductive energy storage, 276
Support in tunnels, 227, 248
Sweden, waste disposal projects, 290
Swelling:
  of boreholes, 392–393
  of foundations, 127–128
  in mines, 348
  in tunnels, 206, 238–239
Swimming pools, underground, 268–269
Syncrude mine, 408–410

Tailings:
  as backfill, 361
  cemented, 362–363
Talc, 31
TARP project, Chicago, 292
Terzaghi rock load method, 295–296
  classification, 243–244
Tieback anchors, 113
Timber supports, 228–229
Toe mining, 89–90
Toppling failures, 64, 83
Toxic waste, 290–292
Transportation:
  of ore, 344
  routes, 76–77
Transverse blasthole mining, 324
Trimming, 84–86
Tunnels, 203, 209–261
  alignment and grade of, 216
  blasting of, 221–223
  boring of, 223–226
  construction of, 254–255
    costs of, 218–219
  convergence of, 229
  cross-section shape of, 216–218
  leakage in, 249
  pressure in, 253–254
  road, 237–238
  single or twin, 218
  specification for, 219–221
  stability of, 207

Tunnels (*Cont.*):
  support of, 227–248
  unlined, 237–238
  utilization of, 210
  water induced blowouts, 249–250
  waterproofing of, 252–254
Tunneling:
  full face, 222
  ground response curves, 245

Umbrella grouting, 224
Underdrainage by tunneling, 250
Underground space, 204
  benefits of, 265–267
  uses of, 264–265, 267–268
  (*See also* Caverns)
Undermining of foundations, 116

Vaiont landslide, 63, 171–172
VCR (vertical crater retreat) mining, 321, 330–331
Ventilation underground, 227

Vertical crater retreat (VCR) mining, 321, 330–331

Warehouses underground, 267–268
Warnings, of rockfalls, 102
Waste disposal, 10
  chemical, 290–292
  fluids from oil wells, 420
  radioactive, 285–290
  toxic, 290–292
Water:
  inflow prediction of tunnels, 250–251
  jetting to remove clay seams, 189–192, 298
  of mines, 343–344
  pressures triggering landslides, 71–73
  of tunnels, 250–251
Waterproofing of tunnels, 252–254
Wedge slides, 58
Wire sawing, 43

Yielding support in mines, 356–358
  props, 333
  pillars, 351

ABOUT THE AUTHORS

JOHN FRANKLIN is a consultant and professor of geological engineering at the University of Waterloo in Ontario, Canada. Dr. Franklin's 20-year career as a consulting geological engineer has involved work in weak soils and hard rocks, at depths from surface to deep underground. Well known through his publications, he has served as a member of international commissions on rock testing and classification and as president of the international Society for Rock Mechanics during 1987–1991.

MAURICE DUSSEAULT is a consulting engineer and professor of geological engineering in the Earth Sciences Department at the University of Waterloo. Dr. Dusseault and colleagues do research mainly in the area of resource development and waste management, specifically in the geomechanics of potash, salt, coal, clay shales, open pit and underground mining, and heavy oil recovery.

DUE DATE

APR 4 1995			
NOV 1 2 1995			
DEC 15 1999			
			Printed in USA